Time and Its Adversaries in the Seleucid Empire

TIME
AND ITS
ADVERSARIES
IN THE
SELEUCID
EMPIRE

PAUL J. KOSMIN

The Belknap Press of Harvard University Press
CAMBRIDGE, MASSACHUSETTS
LONDON, ENGLAND
2018

Library of Congress Cataloging-in-Publication Data
Names: Kosmin, Paul J., 1984- author.
Title: Time and its adversaries in the Seleucid empire / Paul J. Kosmin.
Description: Cambridge, Massachusetts : The Belknap Press of Harvard
University Press, 2018. | Includes bibliographical references and index.
Identifiers: LCCN 2018007054 | ISBN 9780674976931 (alk. paper)
Subjects: LCSH: Time perception—Middle East—History. | Seleucids. |
Calendar—Middle East—History. | Imperialism and science—Middle East—
History. | Ethnoscience—Middle East—History. | End of the world.
Classification: LCC BF468 .K675 2018 | DDC 529.0935—dc23
LC record available at https://lccn.loc.gov/2018007054

For Sudeep

∾

CONTENTS

PREFACE

This book is conceived as a companion and completion of my first, *The Land of the Elephant Kings: Space, Territory, and Ideology in the Seleucid Empire*. Together, they should be understood as a single project of categorical history, a study of the spatial and temporal structures and concepts by which the Seleucid empire and its subjects made sense of their world.

It will become apparent that this book has further motivations: a preoccupation with the nature of my discipline and the conditions that make it possible (to ask what we do when we reckon with historical time, why we care about the past, and how we find meaning in it); a sense that the Seleucid empire's historical significance has, mostly for reasons of academic geography, fallen between the cracks; and a mission to bring back into classical ancient history the eastern worlds and religious texts that have long been marginalized. And while the battles of the past don't wound their historians, I have also been exercised by the blasts of apocalyptic violence currently erupting from the lands once under the Seleucid scepter. The volcanoes have not gone out.

I am aware that this book falls somewhere between *hybris* and *chutzpah*: not only will I be arguing for the massive significance of an overlooked phenomenon, but in order to do so I will also be trespassing in a number of distinct and long-established disciplines. So throughout this project I have relied on the generosity of many scholars and friends, to whom I have turned for advice, answers, and criticism. Emma Dench has been a true friend and generous mentor. Nino Luraghi remains a source of wisdom, kindness, and guidance. I am lucky to know Duncan MacRae. Aneurin Ellis-Evans and Johannes Haubold crossed the Atlantic to discuss drafts with me, and I have also benefited enormously from the feedback or assistance of Supratik Baralay, Gojko Barjamovic, Giovanni Bazzana, Andrea Berlin, Ruth Bielfeldt, Samantha Blankenship, Kathleen Coleman, John Collins, Stephanie Dalley, Rowan Dorin, Avner Ecker, David Engels, Alexander Forte, Janling Fu, Yonder Gillihan, Sylvie Honigman, Christopher Jones, John Ma, Elizabeth Mitchell, Christopher Moore, Ian Moyer, Gregory Nagy, Judith Newman, Carol Newsom, Monica Park, Anathea Portier-Young, James Russell, Mark Schiefsky, Felipe Soza,

John Steele, Rosalind Thomas, Yvona Trnka-Amrhein, Bert van der Spek, Marijn Visscher, Leah Whittington, Rachel Wood, and Bahadır Yıldırım. Albert Henrichs is missed. The hard work and insights of Alexandra Schultz saved me from typological and more serious errors, and Nadav Asraf assisted with the Bibliography. Isabelle Lewis drew the maps and plans, and I am grateful to Tom Boiy, Lisa Brody, Stefano de Martino, Gerald Finkielsztejn, Dietrich Klose, Vito Messina, Wilhelm Müseler, and Debora Pollak for image permissions. G. Nathan Harrison, Ian Stern, Danny Syon, Mkrtich Zardaryan, and Bracha Zilberstein kindly guided me round ruins or gave access to museum materials. Sharmila Sen and Heather Hughes have ushered this book into print with much kindness and good sense.

My colleagues at Harvard have offered encouragement, support, and the time to undertake this research, and my students there continue to inspire and inform. I am grateful to the participants in the Futures Past seminar, Memories of Kingship workshop, and Starr Fellowship in Judaica program. The Loeb Foundation and the Weatherhead Center for International Relations have funded aspects of this research. Ivy Livingstone, Alyson Lynch, and Teresa Wu have helped in ways too countless to list. Balliol College, Oxford, the Center for Hellenic Studies, Washington, D.C., the Radcliffe Institute for Advanced Studies, Cambridge, Massachusetts, and the Sardis Expedition have all made this research possible, and I am grateful for the faculty, staff, and colleagues who have made me feel so welcome or who welcomed me back. Parts of this project, or things related to it, were presented at Boston College, Columbia, Dublin, Durham, Edinburgh, Freiburg, Harvard, Nipissing, NYU Abu Dhabi, Oxford, Pennsylvania State, Princeton, and Tel Aviv Universities, and the Universities of Cincinnati, Pennsylvania, and Toronto, and this book is much improved for the feedback and criticisms of these audiences.

In a work of this kind—generalizing and interdisciplinary—there will, inescapably, be a falling short with respect to the demands and claims of each field I intrude upon; needless to say, I am responsible for all faults. My final hope is only that old questions may be seen in a new light and connections recognized between supposedly atomized phenomena.

This book could not have been completed without the support of my family and friends, who make the world beautiful. I dedicate it, in deepest gratitude and love, to my husband, Sudeep.

Time and Its Adversaries in the Seleucid Empire

Introduction

On the eve of the classical age the polymath Hecataeus of Miletus visits Thebes, the great temple city of Upper Egypt. Hecataeus, who would systemize the genealogies of archaic Greece's great families, boasts to the Egyptian priests of Amun-Zeus that his own began with a god sixteen generations back. In answer, and perhaps as comeuppance, the temple guides walk Hecataeus along a line of wooden statues. These, they declare, are the succession of our high priests, in their generations, father to son, an unbroken sequence, three hundred and forty-one.[1]

Two histories make contact. Our traveling intellectual finds himself among larger stories and older worlds. As with the discovery of geological time toward the end of the last millennium, this was a new immensity of past, a revelation of comparative chronology that rescaled Greek history.[2] It is a disorienting, horizon-moving moment, certainly; but it is also a temporary meeting. Hecataeus will sail home, carrying back with him, like some exotic memento, this new and provocative knowledge.

So what happens three or four centuries later, when the Hellenes stay? When the naïve tourist has become a tax man, a soldier, or a king? When the paradigmatic site of encounter is the colonial settlement and not the beach? What do history, relative antiquity, and newness mean in a world pressed by a permanent and strained propinquity?

This book explores the nature and function of historical time in such a setting. It proposes that the Seleucid empire, the dominant power in the Near East during the Hellenistic period, was a site of intense creativity in understandings of temporal duration, that it was the location and cause of something approaching a revolution in chronological thought and historical experience. I argue that the formal time structures projected by the Seleucid empire (above all, era-counting) and the local temporalities that responded to, resisted, and ultimately undermined these (apocalyptic eschatology, in particular) were, in important respects, new to their world and foundational for ours.

I am writing this introduction in the year 2017. Last year was 2016. Next year, when I hope this book is published, will be 2018. I am confident that a century ago it was 1917, and that in one thousand years it will be 3017, if there is anyone left to name it. Most of the world would concur. Such an accounting of time's passage by an accumulating, irreversible, and predictable number was introduced and institutionalized for the first time by the Seleucid empire at the end of the fourth century BCE; it is known as the "Seleucid Era." The consequences of this new era-time, along with the wider temporal regime that sustained it, can be detected in political life, economic behavior, historical thought, anti-imperial resistance, and the inward realms of sensibility, fantasy, and terror. Across the Seleucid empire, wherever evidence permits, we see local populations reckoning with durational time in new ways, turning with an urgency to history, elevating it as a subject of its own self-knowledge, and, ultimately and in agony, yearning for its end.

Contexts

To treat the interaction of time systems, Seleucid and subjected, as a unitary historical phenomenon brings together three fairly discrete fields of inquiry, which I introduce in turn: Seleucid imperial history; Near Eastern and biblical studies; and the history and philosophy of time.

The Seleucid Empire

The Seleucid empire was an improbable emergence, forged in the anarchic aftermath of Alexander the Great's intestate death at Babylon in June 323 BCE. Ancient witnesses divined miracles in its creation, reasonably. Its eponymous founder, Seleucus I Nicator, who had fought across Asia with Alexander, remained of second-rank importance up until the division of responsibilities at the conference of Triparadeisos in 320 BCE, when he was appointed to govern the vastly populous and hugely wealthy satrapy of Babylonia. Seleucus ruled Babylon, cultivating local support and warding off the hostile army of Eumenes, Alexander's former secretary, until 316 BCE, when he fled an expected punishment by Antigonus Monophthalmus, then the most powerful of the Macedonian warlords. For three or so years, Seleucus found sanctuary in Egypt with Ptolemy (I), son of Lagus, commanding his fleet in the eastern Mediterranean and joining in the defeat of Antigonus's son, Demetrius Poliorcetes, at the battle of Gaza in autumn 312 BCE. Seizing this opening, Seleucus crossed the wilderness with small numbers and much courage to regain Babylon in spring 311 BCE. As we shall see, this return would become the inaugural moment of the Seleucid Era. Over the next three decades, Seleucus brought under his authority territories first to the east, almost to the borders of China, and then to the west, up to the Mediterranean and Aegean shores. In 281 BCE he was assassinated, soon after crossing to Europe in a vain and final bid for his Macedonian homeland. It was left to his son and coruler, Antiochus I Soter, to reconfirm his inheritance and secure the Seleucid line.[3]

Such origins proved foundational for all that would follow. For two characteristics can be identified as the principal determinants of the empire's modes of government, ideological constructions, administrative compromises, and internal tensions: first, an astonishing ethnic, religious, and cultural diversity, with the kingdom extending at its maximal moments to "the peoples, nations, and tongues" from Central Asia to European Thrace and from Armenia to the Persian Gulf archipelago; and second, the diasporic situation of its imperial elite, with the ruling dynasty's original Macedonian homeland lying outside the limits and claims of imperial territory.

This double situation—a confrontation with cultural multiplicity and the sense of arbitrariness that follows; and a disembeddedness from the traditional forms and claims of both the European landscape of origins and the Asian territories of rule—made the Seleucid empire a work of intellection. That is to say, the formation of the kingdom demanded, alongside military conquests and political solutions, an explicit, self-conscious, and top-down determination of its architecture. We see, in consequence, a tendency toward abstraction in state design that is difficult to match in preceding, contemporary, or succeeding regimes. In my previous book, I explored how this played out at the spatial level: the early Seleucid kings bypassed the Achaemenid and pre-Achaemenid imperial centers to fashion new core regions of government in northern Syria and the middle Tigris; their colonial foundations employed a replicable, rigidly uniform urbanism and naming; and the landscapes of empire were measured, codified, and rationalized.[4] Here, in this book, we shall see that the empire's time system and affiliated institutions constitute an even starker and more troubling abstraction.

Further, such disembeddedness characterized the Seleucid dynasty's interaction with its various subject communities. In meaningful contrast to their Ptolemaic peers, who endeavored from the first to adopt the modes,

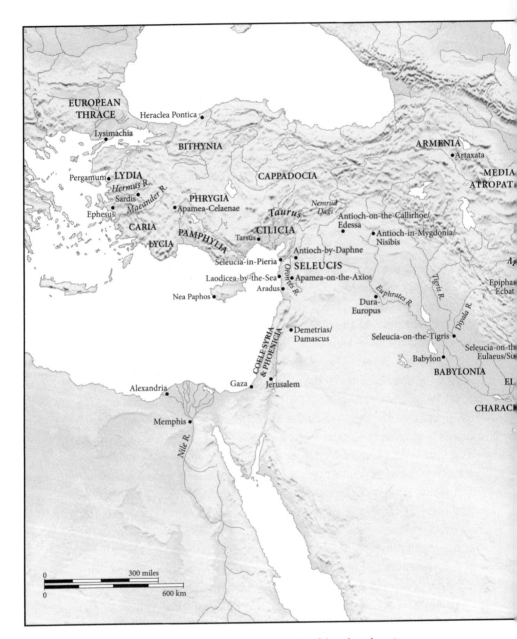

Map 1. The geography and topography of the Seleucid empire

4

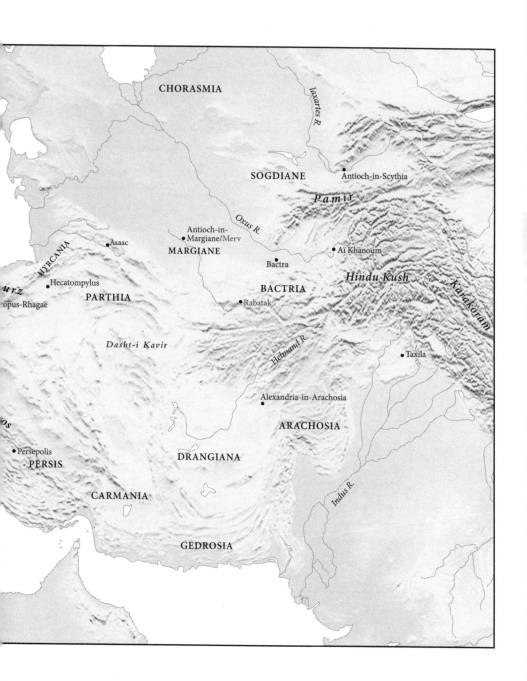

CHORASMIA

Iaxartes R.

SOGDIANE
• Antioch-in-Scythia

Pamir

Oxus R.

Antioch-in-
• Margiane/Merv
• Asaac
MARGIANE

• Aï Khanoum

HYRCANIA
Bactra •

urz
• Hecatompylus
BACTRIA
Hindu Kush

Karakoram

opus-Rhagae
PARTHIA
• Rabatak

Dasht-i Kavir

Helmand R.

• Taxila

Alexandria-in-Arachosia •

ARACHOSIA

os
• Persepolis
DRANGIANA

PERSIS

CARMANIA

Indus R.

GEDROSIA

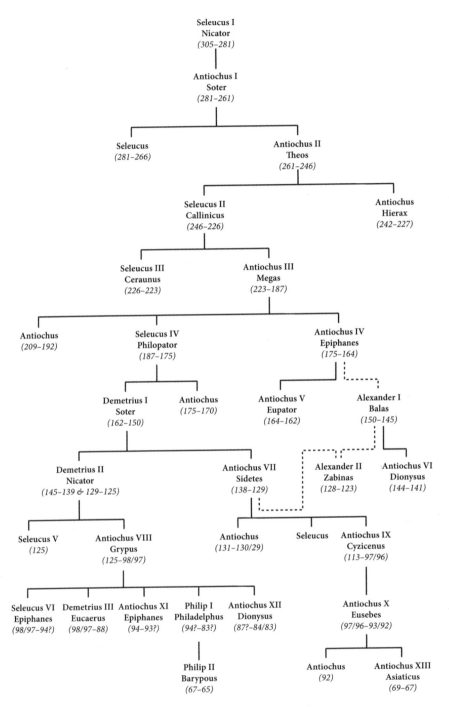

Figure 1. The house of Seleucus (simplified)

responsibilities, and legitimacy of pharaonic kingship, the Seleucids never attempted to occupy their subject communities' ancient monarchic positions: the Macedonian kings of Syria never became Babylonian or Persian or Jewish rulers. Instead, outside of the new colonial centers, the Seleucid imperial state, its monarchic interventions, and its more general cultural attitudes appear as if suspended above the indigenous political landscape, "balloonlike in mid-air," in Peter Thonemann's vivid phrase.[5] Seleucid participation in the traditions of long-defunct local monarchies was infrequent, temporary, unpredictable, visibly costumed, and fractured by misunderstanding, ritual failure, condescension, and resentment. These enactments were not the permanent inhabiting of a preexisting paradigm, as we find in Hellenistic Egypt, but rather temporary hypostatizations, avatar-like descents into local form, bungee jumps from the balloon.

Accordingly, a consensus view of the Seleucid empire has begun to emerge in recent scholarship: that, beyond its imperial heartlands in northern Syria and the middle Tigris, Seleucid rule was supple, nonstandardized, and attuned to regional particularities; that its ambitions were limited to fiscal and military extraction; and that this was achieved by bargaining with local elites, maintaining preexisting political and religious structures, and tolerating—even depending upon—regional self-assertions.[6] Conversely, violent opposition to the Seleucids in Judea (the Maccabees/Hasmoneans), Asia Minor (the early Attalids), Persia (the *fratarakā*), and elsewhere has been newly characterized as renegotiations of dependency within the mutually beneficial framework of acknowledged Seleucid suzerainty.[7]

I am much impressed by this recent approach. In comparison to older assumptions of unbridled Seleucid autonomy, exogenous political logics, and inescapable ethnic polarities, such treatments recognize a more complex landscape of distributed agency, constrained by inherited institutions and interwoven with multiple, nested sovereignties. Yet this "night-watchman" model[8] also flattens out the impact of the Seleucid state and underplays the opposition it could generate.

However limited, irregular, and regionally specific was the reach of the kingdom's fiscal, political, and military institutions, the ideas of empire could penetrate widely, deeply, and with consequence. The kingdom's new temporal technology, in particular, was unconstrained by the normal limits of centralized power and, as we shall see, capable of fixing horizons of potentiality and experience for many of its subjects. Indeed, this book will suggest that the clustering and interrelation of responses among the various communities of empire, across regional differences, and beneath apparently contradictory attitudes—a historical positioning, apocalyptic eschatology, and antiquarianism, among other things—can usefully be seen to constitute a Seleucid imperial culture.

Moreover, the emerging scholarly consensus tends to be overly eirenic, sanitizing the violence and terror of empire. We need not submit to anachronistic accounts of a nationalist sentiment, anti-Hellenism, or total implacability to recognize that the Seleucid kingdom and its worldview troubled and were opposed. The negotiated compromises and actualized outcomes of the slowly splintering kingdom must not be equated with its subjects' radical visions, deepest ambitions, and embittered hatreds: "God heard his prayer and answered half." Indeed, the apocalyptic works studied in this book, with their insistence on the irrational, the irreducible, and the unattainable, stand as a necessary corrective to a historical method that runs the risk of being teleological, narrow, and aligned with the logics of central power. The anguished end-of-kingdom fantasies of the book of Daniel, for instance, are at least as much a part of the Maccabean moment as the royal epistles and institutional delegations recorded in 1 Maccabees and so often preferred by historians. Just as we must not reduce Seleucid imperialism to a material domination alone, so we should conceive of anti-Seleucid resistance not only in terms of force and counterforce—of battles, bargains, coercion, and consent—but also as the contest and entanglement of different imaginaries.

So in this book we will see how some of the Seleucids' subjects met the time of empire with their own, deeper reserves of past and future. This was a specific response by certain groups to key elements of the Seleucid imperial system, and not the kind of generalized indigenous resistance explored by Samuel Eddy over half a century ago in his far-sighted, if often incorrect, *The King Is Dead: Studies in Near Eastern Resistance to Hellenism 334–31 B.C.*[9] Furthermore, as we shall repeatedly see, the counter-ideas and -practices under discussion arose in the social and institutional contact zones of imperial life; they did not exist before or beneath the empire.

The Near East

The core territories of the Seleucid kingdom, from the Levantine coast to western Iran, had already shouldered the yoke of empire for centuries when Macedonian invaders first darkened the horizon. To distinguish the particularities of the Seleucid situation from the more general horrors and glories of Near Eastern empire, the very briefest scene setting is necessary. Needless to say, all regions, each population, every city, sanctuary, village, and watering hole of this most ancient world had its own historical trajectory, interests, dynamics, and tensions; what follows is a severe, if unavoidable, simplification.[10]

Babylonia, the canal-gouged, loamy flats of central and southern Iraq, had formed the core of the Neo-Babylonian empire from its foundation by Nabopolassar in the 620s BCE through its expansion under Nebuchadnezzar II in the first decades of the sixth century down to the defeat of Nabonidus and his son Belshazzar by Cyrus the Great in 539 BCE. Massive regional revolts against Cyrus's successors, Darius I and Xerxes, were crushed, and Babylon served as a seasonal residence for the line of Persian kings until the battle of Gaugamela in 331 BCE. After Darius III's flight from the field, the city surrendered to Alexander the Great for benefactions, peaceful incorporation, and perhaps hope of renewed capital status. But the Wars of Succession followed hard on Alexander's premature death and were especially cruel to this land; a fecund security returned only under consolidated Seleucid rule at the very end of the fourth century BCE. The region's ancient cities—primarily Babylon and Uruk, but also Borsippa, Nippur, Larsa, Kutha, and Kish—continued to flourish, albeit in the new institutional, economic, and demographic shadow of Seleucia-on-the-Tigris, Seleucus I's vast imperial foundation. Such urban and temple vitality has laid down for us archive beds of clay cuneiform tablets, impressed with fiscal, administrative, cultic, literary, astronomical, and historical matters, from which we will pick; everything papyrus or parchment is gone.

Western Iran had the most to lose from Alexander's conquest. After exulting for two centuries in its succession of Persian kings, from Cyrus the Great to Darius III, the region suffered massacre, looting, palatial destruction, and provincial demotion. While certain diagnostic markers of the Achaemenid kingship survived the fall to resurface within and beyond the Seleucid empire's political landscape, no extensive corpus of evidence survives from Hellenistic Iran, a meaningful absence in itself: inscribed Old Persian fell with Persepolis; no cuneiform archives have been found and likely never existed; and numismatic, archaeological, and epigraphic data are scanty and disputed. With the very greatest caution, we can incorporate textual traditions from many, many centuries later (such as the post-Sasanian *Zand ī Wahman Yasn*) or from distant lands (primarily classical and Jewish historiography).

Judea's prominence in this book is a necessary anachronism, overenlarged with respect to its Hellenistic realities, yet fit to the proportion of our evidence and the future scale of the Judaism and Christianity that would preserve it. The region's successive subordination to the Neo-Babylonian, Achaemenid, Ptolemaic, Seleucid, and Roman empires makes it an especially helpful testing ground for establishing the nature and impact of particular imperial systems. Judea had become a small, upland temple state, centered on its one city, Jerusalem, deprived of its Davidic kingship since the destruction and exile under Nebuchadnezzar II, and so led by the Zadokite line of high priests. Cyrus the Great's conquests and the subsequent return from the Babylonian captivity had restored a territorial integrity and inaugurated the so-called Second Temple period, named after the Jerusalem sanctuary that was rebuilt with Persian support. The region was fairly inconsequential on the world stage: it was twice bypassed by Alexander, sustained without major change under the Ptolemies, and incorporated into the Seleucid empire only at the opening of the second century BCE, after Antiochus III's victory over the forces of Ptolemy V at the battle of Panium. Ptolemaic rule is little attested in textual sources and remains something of a historical dark age—in all likelihood, it was during this period that Chronicles, Qoheleth/Ecclesiastes, and the earliest parts of 1 Enoch, the so-called *Book of the Watchers* and the *Astronomical Book*, were composed. Seleucid rule, by contrast, brought swifter and more unsettling transformations—not only the desolating atrocities of Antiochus IV Epiphanes's persecution in the mid-160s BCE, but also the more sustained interventions of the empire's fiscal regime, colonization, and temporal system. The textualized responses encompass wisdom literature (Ben Sira/Ecclesiasticus), apocalyptic eschatology (the book of Daniel, the *Book of Dreams* and the *Epistle of Enoch* of 1 Enoch), more traditional historiography (1 and 2 Maccabees), the novel (the book of Judith), and many other miscellaneous texts, including the earlier Dead Sea Scrolls. This Seleucid-period library is joined by an ever-expanding data set of archaeology and epigraphy.

Conversely, the Phoenician and Greek city-states of the Levant and southeastern Anatolia were of greater historical importance than their extant traces imply. What survives, mostly epigraphic and numismatic, indicates the fierce retention of corporate identities, the progressive extinction of city monarchies, a certain degree of cultural convergence, and a willed incorporation into the international *polis* network. While these old communities, together with the newly founded Graeco-Macedonian settlements, show episodes of hostility to the Seleucid house, they seem not to have developed the kinds of history and eschatology we find in Babylonia, Iran, and Judea.[11] Their disputes with empire remained, for the most part, "between men, not gods,"[12] and so will occupy a less central place in this book than the other regions and peoples.

This core Seleucid territory, the approximate area of the earlier Neo-Assyrian empire, formed an interconnected, integrated unit. Throughout this book, I will use the term "indigenous" as a heuristic shorthand, rather than a normative or descriptive label, for the various peoples of this area who had predated (or claimed to predate) the Seleucid order and emphasized that temporal priority; indigeneity is a relational strategy, not an inherent property.[13] My comparative treatment of these populations, privileging common reactions and horizontal linkages over the successive analysis of bounded provincial experiences, is not only a metahistorical scale of analysis in a "universal history" mode; for as we shall see over and over again, this was also a substantive scale of historical process and an assumed scale of local thought.[14] Of the region's many significant developments in this period, three have attracted most academic interest (and dissent) and run as seams through this book. The first is the accumulation of changes in language, naming practices, and literary tastes, philosophical, religious, and technical concepts, political and social organization and aspiration, and domestic, public, and burial practices, spaces, and objects commonly known as Hellenization. The second is the simultaneous, parallel, polar phenomena of antiquarianism, nostalgia, and the self-ancestralization of local practices. And the third is the innovations of apocalyptic eschatology—that is, of the apprehension of history's totality and meaning, from beginning to end, divinely predetermined, revealed, periodized, and just. This book will suggest the significant role of the Seleucid temporal regime in prompting or organizing these dynamics.

Time

For well over a century, scholars have recognized that time—the very postulate of history and the prime vector of its ideas, actions, and processes—can also itself be an object of historical investigation. Important studies have explored the relationship between particular social orders and the temporal systems that absorb, coordinate, conventionalize, and restrain their rhythms, values, and tensions: politically, from ancient royal enthronement and the water clock of classical Athenian democracy to the decimilized calendar of Revolutionary France, the supposed timelessness of nationalism, and the evaluative modernity of European empires; and economically, from the task-based times of preindustrial agriculture and the seasonality of the medieval merchant's trade to the clock-time of the factory and the new now-time of our own information age.[15] The most productive of this crop have seen temporal schemata not as derivative reflexes or epiphenomena, but as generative of coherent worldviews, coordinated to other systems of practice and thought, sustaining of hegemonic regimes, yet vulnerable to internal contradiction and external critique.

This is a rich hermeneutic, and it has achieved splendidly inspiring and subtle scholarship. I take from it three lessons. First, all historical situations simultaneously interweave or overlay multiple historical modes and temporal attitudes—a plurality phenomenon termed *die Gleichzeitigkeit des Ungleichzeitigen*, "the contemporaneity of the non-contemporaneous," by the Marxist philosopher Ernst Bloch.[16] Yet these times are never equal: certain temporal regimes will be privileged and promoted by the economically dominant or politically powerful (capitalist time thrift, for instance) and so become targets of embittered criticism; other times will be relegated to peripheral groups, marginal functions, and uninstitutionalized spaces (peasant field rhythms, say) and so made available for insurgent revival.[17] It is the uneven weighting of these times, the friction of incompatible systems in contact, that generates the kind of heat worthy of historical analysis.

Second, time, unlike cartographic space, is not scalable. For both heuristic and substantive reasons, we must distinguish between the durational time that is our concern ("historicality") and the periodic time that is not ("calendricality").[18] Astronomical periods and subannual, calendrical expressions of time—hour, day, week, month, and their equivalents—find their ultimate referent in the cycles of nature: the journey of the planets, the moon's phases, seasonal change, and so forth. By contrast, historical years pass to a different logic: not the patterns of repetition and return that structure the cosmos, but the inescapably human order of societies. By accumulating, years divest themselves of nature. According to Reinhart Koselleck, years must always refer to "a historically and philosophically impregnated experience of time."[19] The accumulating year is the base unit of history. Such opposition of calendrical and historical is, in the taxonomy of Aristotle, that of *zoē* and *bios*,[20] in materialized form, that of the sundial and the chronicle.[21]

Third, triumphal narratives of temporal modernization and progress must be answered with a cautious suspicion. A tradition of antihistoricist thought—represented by, among others, Franz Rosenzweig, Walter Benjamin, and Mircea Eliade—has recognized in linearized, transcendent time an alienated individualism, brazen impiety, and debilitating meaninglessness.[22] These thinkers' openness to radical possibility and temporal rupture, their valorization of sacred timelessness and mythical images, and their profound aspiration to escape history alert us to the presence, within a supposedly homogeneous and rationalized "brownshirt Time," of a yearning for striation, heterogeneity, and enchantment.

If this book can do more than borrow from the broader field of temporal studies, I hope it is by exposing to discussion the buried roots of its own concepts: the double birth, within the Seleucid empire, of both state-institutionalized era-time and apocalyptic total history. Crucially, my argument reverses their standard ordering in accounts of modernity: not empty, transcendent time liberating itself from a premodern, apocalyptic, enclosed worldview; but rather apocalyptic time as a secondary, insurgent response to the Seleucid temporal regime, like Jacob grasping Esau's heel.

The Seleucid empire was neither long-lasting enough, like its Roman conqueror, nor "national" enough, like the Iranian succession, to be transposed into disciplinary form, with the result that the boundaries, demands, and evaluations of the modern academy occlude its unitary phenomena. So I would suggest that this book's combination of fields is more than the additive hybrid (worthy like Esperanto) or the unmotivated selection (picked like oranges) that it might seem to be. For this trinity of disciplines has an integral coherence that corresponds to the historical phenomena under investigation—a shared political space, a common imperial past, and similar cultural concerns. Its benefits extend beyond a widening of horizons and reorienting of perspective. The most obvious contribution is an expanded evidentiary base for Seleucid history. As we shall see, whole corpora of data, typically segregated in their woodland clearings, can be understood as relevant and related. Beneath the dapple of surfaces, we can identify for a range of biblical and Near Eastern works not only a common imperial context, but also an assumed and determining Seleucid horizon of reference. Conversely, the imposing of a Hellenistic, political periodization—the founding of the Seleucid empire—on fields typically bounded as Late Babylonian (from the end of indigenous kingship) or Second Temple (from the exile and return) makes visible a temporal distinction easily hidden by scholarly reflex, yet deeply felt by these communities. Finally, the incorporation of biblical texts can, like Hecataeus's Egyptian priests, open an alternative and far older intellectual genealogy for Seleucid studies. While Hellenistic history is seen as a discipline of recent birth, with the nineteenth-century Prussian historian Johann Gustav Droysen habitually invoked as its founding father,[23] the bizarre obscurities of the book of Daniel and its debated location in the Bible provoked an early tradition of historical exegesis in Latin, Greek, and Syriac that lasted until the Reformation and beyond.[24] This is a knot of religious reception and detailed analysis, focused on the period *post Alexandrum usque ad Caesarem Augustum,* that still requires careful unraveling.[25] But I would defy any Hellenistic historian to consider, say, the fragmentary twelfth book of Porphyry's *Against the Christians* or Calvin's commentary on Daniel 11 and not intuit, across the immense differences of orientation, some kinship of intellectual project. Needless to say, such a continuous line of theological-historical interest cannot hold true for archaic or classical Greek history, which hardly impinged on Judea's scribes and visionaries.

Argument

Time and Its Adversaries in the Seleucid Empire is organized into two parts of three chapters each.

Part I, "Imperial Present," explores the establishment and functioning of a distinctive Seleucid temporal regime. This Seleucid "present" is not the kind of instantaneous moment that has already disappeared by the time one observes it, as famously described in Augustine's *Confessions;*[26] rather, it is the period-time of the regime, a durational present of power, the dynastic here and now. Chapter 1, "The Seleucid Era and Its Epoch," investigates the texture of the Seleucid Era count and the intellectual and political contexts within which it arose. Combining classical and cuneiform accounts, it demonstrates that the Era's Year 1, its epochal opening, was retrospectively fixed on Seleucus I's heroic return to Babylon in spring 311 BCE. It suggests, further, that this "Big Bang" may have been sacralized by the founding or repurposing of a Babylonian sanctuary, the so-called Temple of Day One, and commemorated in various other ways by the Seleucid court.

The second chapter, "A Government of Dating," looks at how and with what consequence the empire's rulers and subjects inhabited public and private worlds that were comprehensively marked by this new era-time. Three case studies—of fiscality (taxes, archives, and their authorizing procedures), trade (coinage, weights and measures, and the agora), and imperial communications—demonstrate how the Seleucid temporal regime could model administrative procedures and penetrate deep into social life.

Chapter 3, "Dynastic Time," explores the Era's impact on the less obviously formalized expressions of temporal duration. It argues that the separation of imperial time from the person and life cycle of the king worked, in sometimes paradoxical ways, to downplay the individuality of Seleucid monarchs in favor of a dynastic, corporate identity. This, together with the absence of any opportunity for temporal restart, removed the ideological "rebooting" and social therapies that typically accompanied the start of new reigns elsewhere. We see how the Seleucid court expressed this unbroken time in its historiography, cult, and royal ambition. The chapter concludes by tracing the rejection or replication of the Era system within and beyond the Seleucid empire.

Part II, "Indigenous Past and Future," turns to the Seleucids' subject communities and their common reactions to this time regime. Chapter 4, "Total History 1: Rupture and Historiography," argues that an experience of fissure or break, promoted by the Era count and various other Seleucid practices, created the conditions for new attitudes to the pre-Macedonian past. Seleucid-period historiography from Babylonia and Judea, including the *Babyloniaca* of Berossus and various biblical and parabiblical works, testifies to a shared conception of a distanced, numericalized, and segmented past. It is as if the Seleucid empire functioned like a point of Archimedes in time, from which subject communities could objectify, narrate, and bring to a close their own histories.

The following, fifth, chapter, "Total History 2: Periodization and Apocalypse," extends this analysis to a kind of historical thinking that emerged for the first time within and in opposition to the Seleucid empire, apocalyptic eschatology. The biblical book of Daniel, the apocryphal 1 Enoch, the so-called Dynastic Prophecy from Babylonia, and the *Zand ī Wahman Yasn* from Iran share central concerns—the periodization of history and the positing of its end—that both correspond closely to and work to undo the Seleucid temporal

15

regime. We shall see that these texts staged a battle between Seleucid king and God over the control of time and the architecture of history, exposing the claims of the Seleucid empire as hollow and illusory. In place of the depersonalization, commodification, and vertiginous endlessness of the Seleucid Era, they offered divine determinism, historical meaning, and finitude.

Finally, Chapter 6, "*Altneuland:* Resistance and the Resurrected State," examines how movements of local self-emancipation in Babylonia, western Iran, Armenia, and Judea weaponized such historical and apocalyptic discourses to rise up, rooted, against the empire. It explores how the actions of these elites were pre-scripted, self-consciously engaging with antiquarian traditions and memories, reviving long-defunct forms of violence, reconstructing ancient monuments, and so claiming by action that their history either had not ended or could be reopened.

The book's big hypothesis lies at the hinge between Parts I and II, for I am proposing an important dialectical relationship between the empire's pervasive time regime and the temporal thinking and historical practice of its subjects. Only occasionally will an ancient source make the connection explicit, for the interaction of these temporal systems was not a thing that can be held as much as a mode of relations that must be inferred.[27] The inference, as in all historical arguments, is permissive, not compulsive, to employ R. G. Collingwood's distinction.[28]

We must be conscious throughout this book of two pitfalls. First, there is the siren call of unitary causation. So I want to emphasize as strongly as I can that I offer only one hermeneutic for making sense of the various political, social, and intellectual developments under review. Of course, certain of these dynamics can be interpreted in other ways and identified in other places or times. That being said, I maintain that this particular constellation of phenomena clusters with a heightened charge, clarity, and urgency around the Seleucid temporal regime, as if aligned to a gravitational center or organizing principle. The second temptation is to teleology. The book deals with new temporal understandings in their emergence, almost at the dawn of their first day. Immeasurably influential as they have been, we must not assume the identifying physiognomy or function that mark their more adult forms. In particular, we must be on guard against naturalizing or embracing the imperial time system, given its affinity (and, arguably, contribution) to contemporary historicist method.

PART ONE

Imperial Present

CHAPTER 1

The Seleucid Era and Its Epoch

For Eunapius of Sardis, the late antique historian, accurate dating should concern only bureaucrats and accountants. Specifics of chronology were as bothersome as uncalled witnesses at a trial, who show up of their own accord and add nothing to the case. "What does chronology contribute to the wisdom of Socrates or the brilliance of Themistocles?," Eunapius demands.[1] The Sardian's contempt is widely shared, even among historians: the precise fixing of dates has long been derided as an incurious virtue, a lingering for ideas, a myopia of the antiquarian, and an obliviousness to bigger stakes and larger stories.

Yet dates are the stuff that history is made on. In two respects, they turn the past historical: dates allow things to happen only once; and dates insist on the ordering and interrelation of all happenings. An event must be chained to its place in time before it becomes an available object for historical articulation. And the modes by which we apprehend such historical time frame how we experience our present, conceive a future, remember the past, reconcile with impermanence, and make sense of a world far wider, older, and more enduring than any of us.

Three types of public annual dating were employed in the pre-Seleucid Near East and elsewhere in Hellenistic west Asia and the Mediterranean. First, year names: in Mesopotamian kingdoms years could be named after an outstanding event, selected by royal authority and archived in lists of date formulae—for instance, "the year when the harp 'Dragon of the Land' was fashioned,"[2] or "the year when Zimri-Lim went to the help of Babylon, a second time, to the land of Larsa."[3] These are milestone, one-off events, supplied by historical circumstance and royal achievement. Second, eponyms: in both the Mediterranean and Near Eastern worlds years could be named after an individual or individuals who occupied an annual office of state (the *limmu*-ship in Neo-Assyria, the eponymous archonship in

Athens, the consulship in Rome, and so forth), again, archived or publicly displayed for reference. These eponymous dates were like a metronome, marking an even and repetitive temporal beat but lacking in direction or accumulation. Third, and most common in the ancient world's monarchic states, regnal years: the count of the years that the king had been on the throne, starting anew with each successive monarch. Of course, these annual reckoning systems could be combined, reshaped, and exchanged in various and complicated ways.

It takes an enormous effort of the imagination to unthink ourselves back into a world dated only by unique events, annual offices, or regnal rhythms. But two famed examples, one from the club of Greek city-states and the other from the succession of Mesopotamian kingdoms, can at least go some way to demonstrating the extraordinary difficulty of fixing temporal location or simultaneity. The Athenian historian Thucydides attempted to date the outbreak of the Peloponnesian War by assembling a series of coinciding temporal markers. The second book of his history opened:

> The Thirty Years' Peace, which was entered into after the conquest of Euboea, lasted fourteen years; in the fifteenth year, in the forty-eighth year of the priesthood of Chrysis at Argos, and when Aenesias was ephor at Sparta, and there still being two months left of the archonship of Pythodorus at Athens, six months after the battle of Potidaea, and at the beginning of spring, a Theban force a little over three hundred strong . . . at about the first watch of the night made an armed entry into Plataea, a Boeotian town in alliance with Athens.[4]

Thucydides has synchronized this first shot of war to diplomatic, religious, civic, military, seasonal, and hourly data points. Undoubtedly, his triangulation is semantically "thick," reaching the different audiences of his history and underscoring the conflict's international scale, major players, and eruption from a still unhealed dispute. But it is also a decentered, multilocal chronography that dates events to calendars, year counts, and eponyms embedded in highly particular and nonoverlapping social orders.

The second case is a mid-sixth-century BCE Akkadian cuneiform inscription from the northern Mesopotamian city of Harran, or Carrhae. This purports to record the autobiography of Adad-guppi, the long-lived mother of Nabonidus, the last king of Babylon.[5] The inscription traces Adad-guppi's full life, with attention to the sack of Harran at the fall of the Neo-Assyrian empire, the elevation of her son to the Babylonian throne, and his reconstruction of the city's moon-god sanctuary, the Eḫulḫul of Sîn:

> From the twentieth year of Aššurbanipal, king of Assyria, in which I was born, until the forty-second year of Aššurbanipal, the third year of Aššur-etillu-ili, his son, the twenty-first year of Nabopolassar, the forty-third year of Nebuchadrezzar, the second year of Amel-Marduk, the fourth year of Neriglissar—for ninety-five years I cared for Sîn, the king of the gods of heaven and earth, and for the sanctuaries of his great divinity.[6]

Adad-guppi's ninety-five years are calculated by totting-up the reigns of successive monarchs; seven separate time references are needed. This king-list chronology is neither accurate nor politically neutral: the year numbers do not accord with modern reconstruction; the dates cross over from one kingdom, the Neo-Assyrian, to another, the Neo-Babylonian that toppled it, perhaps implying imperial continuity or the transfer of Adad-guppi's allegiance; the last two Assyrian kings, Sîn-šarru-iškun and Aššur-uballiṭ II, are omitted; and no mention is made of Labaši-Marduk, the young Neo-Babylonian monarch murdered in the palace coup that brought Nabonidus to power.[7]

Thucydides's history and Adad-guppi's autobiography illustrate the complications of pinning down the flow of history in the absence of a transcendent and translocal time system. Their dates are intimately tied to central state institutions, dependent on a bureaucratic list making and scribal literacy, applicable only within a self-limiting geography, and highly sensitive to political change. Indeed, they are not so much dates as synchronisms between multiple events, coordinating a relational network of better- and lesser-known occurrences: what is being dated and what dates it belong to the same order of things.

A new kind of dating system was introduced and propagated by Seleucus I, the founder of the Seleucid empire. It is known to scholars as the Seleucid Era, but it has carried almost as many names as the populations who employed it: "the Years of the Kingdom of the Greeks," "the Era of the Hellenes," "the Era of Contracts," "the Syro-Macedonian Era," "the Era of the Macedonians," "the Era of the Two-Horned One," and "the Era of the Astronomers of Babylon." This Seleucid Era was a continuous, unbroken count of years. Its epoch, or Year 1, was placed in 312/311 BCE, six years before Seleucus I took the diadem and proclaimed himself king, as we shall see below. At Seleucus I's death in 281 BCE, his son and successor, Antiochus I, did not restart the count, and the subsequent kings let it continue unbroken. Accordingly, the Seleucid Era's time reckoning was uninterrupted, irreversible, paratactic, cumulative, endless, and directional. It did not represent a king's reign or office of state. It did not depend on or respond to actions in the real world. It was decoupled from things, objects, and events. It belonged to a separate order than the phenomena it dated. It was a regular, numericalized measure of ever-deepening duration. Put simply, it rolled (rolls) on and on.

This Seleucid Era was the world's first continuous tally of counted years and the unheralded model for all subsequent era systems, including the Common Era, by which this book is dated. For while there had been intimations of such a chronographic system in early second-millennium BCE Mesopotamia and the Phoenician city-states of the late-classical Mediterranean, these were shallow, highly localized, and easily displaced, with no observable or likely impact on the Seleucid invention.[8] It is in the Seleucid empire that era-counting was first fully realized.

The novel texture of the Seleucid Era can be illustrated by a small cuneiform tablet, inscribed on both sides, now housed in the British Museum (Figure 2, Table 1). The Saros Tablet lists the astronomical Saros cycle—that is, the eighteen-year intervals in which a lunar eclipse can take place on a particular day.[9] The tablet opens in the reign of Nebuchadnezzar II in the sixth century BCE and continues through the Neo-Babylonian dynasty, the Persian conquest of Babylonia, the Macedonian takeover, and on into the early first century BCE.

As we can see, the text is arranged into three columns: the right-hand column repeats in every row the number eighteen—that is, the lunar eclipse's eighteen-year periodicity; the middle column begins by identifying the king in whose reign the eclipse fell; and the left-hand column gives the ruler's regnal year in which the eclipse occurred. So as we read down the tablet, we see that an eclipse took place in the thirty-eighth regnal year of Nebuchadnezzar II (567/566 BCE); then, eighteen years later, in the seventh regnal year of Nabonidus (549/548 BCE); then, after another eighteen years, in the eighth year of Cyrus's kingship in Babylon (531/530 BCE); and so on, right up to the third independent year of the Macedonian general Antigonus Monophthalmus (315/314 BCE).

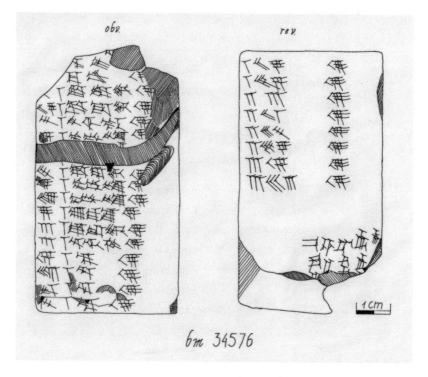

Figure 2. Saros Tablet (BM 34576)

At this point, the dating shifts from regnal counting to the new Seleucid Era system. The right-hand column continues, as before, to repeat the eighteen-year periodicity with each entry. The middle column no longer gives the name of a king, but the syllabic abbreviation (preceded by the male determinative marker) ᴵSe. This syllable is repeated without change despite the murder of Seleucus I, the accession of his son Antiochus I, the death of Antiochus I, and the accession of Antiochus II. In the left-hand column the Seleucid Era year numbers accumulate at eighteen-year intervals, without ever restarting or reversing: 15, 33, 51, and 69.

When the scribe turned over the tablet to write on its reverse side, the numerical calculations continued but the middle, political column, was simply omitted. Thus, in the right-hand column we have the repetition of the number eighteen and in the left-hand column the continuing Seleucid Era count: 87, 105, 123, 141, 159, 177, 195, 213 . . .

TABLE 1
Saros Tablet

Obverse	[38]	Nebuchadnezzar	[18]
	[7]	Nabonidus	[18]
	[8]	Cyrus	1⌜8⌝
	⌜9⌝	Darius I	1⌜8⌝
	27	Darius I	18
	9	Xerxes	18
	⌜6⌝	Artaxerxes I	1⌜8⌝
	[24]	Artaxerxes I	[18]
	[1]	Darius II	[18]
	19	Darius II	1⌜8⌝
	⌜18⌝	Artaxerxes II	1⌜8⌝
	36	Artaxerxes II	18
	8	Artaxerxes III	18
	3	Darius III	18
	3	Antigonus	18
	15	ᴵSe	18
	33	ᴵSe	18
	⌜5⌝1	ᴵSe	18
	69	ᴵSe	18
Reverse	87		18
	105		18
	123		18
	141		18
	159		18
	177		18
	195		18
	213		18

This little clay tablet offers a fairly stark visualization of the distinctiveness of the Seleucid Era vis-à-vis previous temporalities.[10] For the period from Nebuchadnezzar II to Antigonus Monophthalmus the Saros Tablet gives the ruler's full name along with a regnal year number that rises and falls according to life cycle or military conquest. But with the Seleucid Era count we witness a homogeneous time system unmooring itself from the world: first, on the final lines of the tablet's obverse side, with the use of the unchanging syllabic abbreviation ᴵ*Se*, an altogether unprecedented kind of political abstraction for monarchic states (see Chapter 3);[11] then, on the tablet's reverse side, with the total disappearance of the political referent. There is an almost Jacobin logic here for a temporal technology that can do away with the king.

Moreover, we do not know when, after the introduction of the Seleucid Era count, this Saros Tablet was composed. For the Seleucid Era, as a universal, freestanding, regularly increasing number, permitted an entirely new kind of temporal predictability.[12] It had been impossible for a subject of, say, the elderly Nebuchadnezzar II, in the thirty-eighth year of his reign, to confidently and accurately conceive, name, and hold in the imagination a date several years, decades, or centuries into the future.[13] Now, this sort of calculation was easy, unproblematic, and uniform for every subject of the Seleucid kings. The regularized directionality of the Seleucid Era gave future time the same texture as the present and the recent past, accessing a previously unreached chronographic extension. Time no longer had any obstacle in front. The development may be captured in a felicitous image from Karl Ove Knausgaard's recent novel: "It was as if a wall had been removed in the room they inhabited. The world no longer enveloped them completely. There was suddenly an opening. . . . Their glance no longer met any resistance, but swept on through more of the same."[14]

As the Saros Tablet dramatizes, the Seleucid Era proposed fundamentally new possibilities and problems of politics, history, and religion, with which this book will wrestle. This was not the "discovery" of linear time, for understandings of ordered and irreversible temporal succession had long existed. Rather, we are dealing with: a temporal transcendence; the automation of a generalized, "strong" time over governed, particular, "weak" times; the opening of a thinkable future; and the fixing of an absolute beginning.

In the Beginning . . .

The irreversible linearity of the Seleucid Era generated, for the first time in the chronographic imagination of states, a true beginning. Whereas the Year 1 of regnal dates was a replicable, not a singular, event and eponym or year-name formulae consisted only of directionless segments of identical time, the Era's year count identified an "epoch," a first year, as the originary moment from which all time flowed forward. This Seleucid Era epoch would remain the unique and foundational point of orientation for every subsequent year, implicit in all materialized, verbalized, and even interior, mental uses of the count. The invention—a numericalized Big Bang—was one of the most significant contributions of an era-based temporality. Today, with our experience of the Anno Domini, the Jewish Era of Creation, the Islamic Hegira, the calendar of the French Revolution, and so on, we are at ease with the structuring function of an era's epoch. But in a world in which a temporal origin point of this kind had never previously existed, how could Year 1 be conceived?

This Seleucid Era epoch was artificial. I mean this not only in the sense that the Seleucid court deliberately selected and privileged a particular year, that the Era's epoch was political and historical, not natural or inevitable. More particularly, the epoch never took place in fact. For the Seleucid Era had two beginnings—a real-world, messy, and graduated emergence amid Seleucus I's battle with Antigonus Monophthalmus for control of Babylonia; and a retrojected, utopian, and ritually meaningful origin point. The disjunction between these two births, actual and imagined, allows us to recognize the Seleucid Era's function, from the start, as a technology of historical idealization.

The Real Beginning

The political chaos that followed the death of Alexander the Great generated a chronographic confusion in date-bearing records from across the Near East. Village scribes, temple scholars, merchants, accountants, and whoever else needed to provide a year date for a document were confronted by multiple, mercurial sovereignties: the formal reigns of Alexander's two incapacitated heirs—his handicapped half-brother, Philip III Arrhidaeus, and his posthu-

26

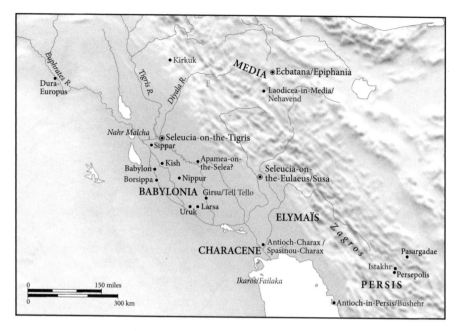

Map 2. Babylonia and western Iran

mous son, Alexander IV—as well as the informal but army-backed dominance of competing Macedonian warlords. Document year dates from this period are confusing and contradictory, reflecting not only the victories or failures of Alexander's would-be successors but also information time lags, regional loyalties, personal preferences, and, on occasion, a scribal disorientation in the face of political uncertainty.

Fortuitously, cuneiform tablets from Babylonia—the origin point of the Seleucid empire and its Era—allow a fine-grained reconstruction of year-date formulae in this post-Alexander period (Map 2). The administrative and astronomical practices of Babylonian scribes combined a scholarly culture of precise and ubiquitous dating with the imperishable medium of incised clay tablets. While the details of dating formulae may not seem the most exciting of ancient data, they provide explicit testimony to the creation of the Era count at the very place of its birth. These clay tablets—economic and bureaucratic records, astronomical reports, and historiographic narratives—demonstrate that four systems of year dating were used from the death of

Alexander the Great to the consolidation of Seleucus's rule.[15] They take us, in less than two decades, from the once standard regnal system to the new Era count:

Stage 1:   From the death of Alexander the Great in 323 BCE to Antigonus Monophthalmus's expulsion of Seleucus Nicator from Babylon in 316 BCE, records were dated by the years of king Philip III Arrhidaeus and then, after his posthumous eighth year, by the years of king Alexander IV.[16] This was a continuation of long-established regnal dating practices.

Stage 2:   From late 316 to 311 BCE, the period of Antigonus Monophthalmus's undisputed dominance in Babylonia and Seleucus's exile in Ptolemy's Egypt, records were no longer dated by the reigning monarch, Alexander IV, who was absent in Macedonia. Instead, scribes introduced a year count of Antigonus in his official capacity as ${}^{lú}rab$ uqu, the Akkadian term for the Greek title *stratēgos* (general) of Asia. This new formula acknowledged both Antigonus's on-the-ground authority and, by its avoidance of the royal title, his formal subordination to the distant king. The beginning of this count, in an important precedent for the Seleucid Era, was retrojected to 317/316 BCE.

Stage 3:   Following Seleucus Nicator's return to Babylon in 311 BCE, the interrupted regnal count of Alexander IV was restored. Thus, year 7 of general Antigonus was renamed year 6 of king Alexander IV.[17] The count was continued even after the young king's murder in 310 BCE. Seleucus's name and the title ${}^{lú}rab$ uqu, previously used for Antigonus, were added to the regnal year date of Alexander IV, thereby combining features of the stage 1 and stage 2 formulae.

Stage 4:   Finally, in 305/304 BCE, the year of Seleucus's formal self-coronation as an independent and legitimate king, the Seleucid Era system was introduced. The beginning of this year count was retrojected to Seleucus's return to Babylon in 311 BCE—that is, to the opening of stage 3.

This is presented schematically in Table 2.

TABLE 2

Year-date formulae at Babylon, from Alexander's death
to the Seleucid Era

Stage	Period of Use (BCE)	Formula	Year Count
1	323–316	Philip III	1–8
		Alexander IV	1–2
2	316–311	Antigonus the general	3–7
3	311–305	Alexander IV or Alexander IV and Seleucus the general	6–11
4	305–	Seleucus	7–

These frequent changes and experimentation indicate that the Macedonian warlords fully grasped the significance of dating formulae as political expressions. The attested choices of year date represent distinct strategies of legitimization in the post-Alexander context of inadequate and absent heirs, whether Antigonus's inauguration of his own, nonroyal count (stage 2) or Seleucus's initial attachment of himself, with Antigonus's former title, to the restored regnal years of Alexander IV (stage 3).[18] A fragmentary cuneiform tablet, composed by and from the perspective of the priests of Babylon's Esagil temple, a bureaucratic structure necessarily sensitive to chronographic change, confirms that we are dealing with directed manipulation. This historiographical text, known as the Successor Chronicle, records the major political events in Babylonia in the closing decades of the fourth century. The lacunose fourth column on the tablet's reverse recounts Seleucus's battle with Antigonus for control of Babylonia immediately following his return in 311 BCE. According to lines 3–4, "[Seleu]cus spoke as follows: 'You will count "Year 7 of Antig[onus the general" as "Year 6 of Alexander, son of] ditto and Seleucus the general'" ([ˡSi-lu-uk]-ku iq-bi um-ma mu 7.kám ˡAn-ti-g[u-nu-su ˡú gal.erínᵐᵉˢ mu 6.kám ˡA-lik-sa-an-dar a-šú šá] ki.min u ˡSi-lu-uk-ku ˡúgal ˡúerínᵐᵉˢ šidᵐᵉˢ)."[19] The entry records Seleucus's instruction, narrated as a speech-act, that Antigonus's dating system (stage 2) be replaced by that of king Alexander IV and Seleucus the general (stage 3).[20] Presumably, a similar directive was issued in 305/304 BCE, when the retrojected Seleucid Era was introduced.

The tablet characterizes Seleucus Nicator as an architect of time, an identity found elsewhere (see below) and, it seems, part of his own self-fashioning.

As we have seen, the retrojected Seleucid Era was first introduced in the year of Seleucus's coronation.[21] The *Babylon King List* and a couple of other sources connect these political and chronographic developments through a form of double dating: "Year 7 that is Year 1 of Seleucus the king (mu.7.ká[m] šá ši-i mu.1.kám ¹si-lu-ku lugal)."[22] The new Era system (stage 4) required both the phasing out of Alexander IV's regnal count and the retrojection of Seleucus's by six years. There was precedent for both steps in Antigonus Monophthalmus's short-lived Babylonian dating system (stage 2).[23] More generally, political chaos and the attendant chronographic muddle had already begun to unmoor temporal markers from their direct political referents. A year count of absent kings had culminated, following the assassinations of Philip III and Alexander IV, in the virtual years of dead kings. The temporal retrojection employed modestly by Antigonus (stage 2) and more dramatically by Seleucus (stage 4) was an equivalent fiction, just operating at the other end.

The beginning of the Seleucid Era presents a paradox. It was introduced by Seleucus Nicator only when he took the diadem of kingship: its creation can rightly be considered a marker of his legitimate and autonomous rule. This was the Era's real-world beginning in 305 BCE. Yet the significance of this coronation moment, of Seleucus's self-transformation from warlord into *basileus*, was downplayed by beginning the count in year 7. The Era was not a regnal dating system. Its beginning was artificially placed in 311 BCE. Why?

The Idealized Beginning

1 SE (Seleucid Era) functioned as a chronographically formalized origin myth for Seleucus and his successors. This epoch identified 311 BCE, the year in which Seleucus returned from exile to Babylon—not the death of Alexander the Great in 323 BCE, nor Seleucus's appointment as satrap of Babylonia in 320 BCE, nor even his coronation in 305 BCE—as the inaugural moment of imperial time. It has long been recognized that Babylon functioned as the archetypal city of kingship in Greek and Mesopotamian thought and so as a legitimizing location for Seleucus's new empire. Indeed, a number of folktale-like episodes from Seleucus's youth, taken as predictions of his

future kingship, were set in the city.[24] And there is little doubt that Diodorus Siculus's description of Seleucus's race through the Arabian desert, accompanied only by a small band of loyal followers, to indigenous acclaim at Babylon derives from Seleucid court traditions and demonstrated all the charismatic warrior virtues expected of a successor to Alexander.[25]

A recently published, postage-stamp-sized fragment of a cuneiform tablet, perhaps part of an Astronomical Diary, for the first time gives an indigenous account of Seleucus's momentous return:

> iti bar iti bi ^{I}s[i-lu-uk-ku . . .]
> ša ina mu 1.kám ^{I}an-t[i-gu-nu-su ^{lú}gal.erín^{meš} . . .]
> ta e.ki záh gur ud.1.k[ám]

> Month Nisannu. In that month, S[eleucus . . .]
> who in year 1 Ant[igonus, the general . . .]
> from Babylon had fled, returned. Day 1.[26]

The clay fragment narrates Seleucus's flight from Babylon in 316 BCE, nested, as a subordinate clause (ša . . . záh), within the framing main verbal sentence that recounts his return to the city in 311 BCE. It indicates that Seleucus arrived back at Babylon at the beginning of Nisannu, the first month in the annual cycle.[27] In other words, the event from which the imperial Era would count—Seleucus's foundational act of political independence—was precisely coordinated with the very start of the Babylonian calendar year. The fortuitous convergence of apparently unrelated circumstances, known by the Greek term *kairos*, was a concern of Hellenistic religious, political, and historical thought.[28] Such opportune combinations, particularly of military or political events with religious festivals, were considered well-omened and typologically meaningful. To cite only the closest parallel, the 296/295 BCE *adventus* at Athens of the "tragic king" Demetrius Poliorcetes coincided with the reception and installation of the god Dionysus for the City Dionysia festival.[29] In the case of Seleucus I at Babylon, the stage-managed chronography goes beyond mere calendrical convenience, for it allowed Seleucus, in his capacity as a usurping warlord, and his imperial Era count, as a temporal technology, to model themselves on and thus gain a cultural legibility from the key religious paradigm of first millennium Babylonia.[30]

Babylon celebrated its New Year festival, called the *akītu*, during the first two weeks of Nisannu. The festival confirmed the linked sovereignties of Babylon's chief deity, Marduk, and the earthly ruler of the city. Its complex program of cultic and political activities is known from normative descriptions in ritual scripts and historiographic accounts in royal inscriptions, Babylonian Chronicles, and Astronomical Diaries. It is clear from this evidence that the most publicly visible and cultically central aspect of the festival was a procession out of and then back into Babylon. In its ideal form, on 8th Nisannu, the city's chief god, Marduk, accompanied by the reigning king and the other major deities of central and southern Babylonia, paraded from the Esagil temple at the city's heart along the Ay-ibūr-šabû (the great processional way), through the Ištar Gate, and beyond Babylon's walls to the *bīt akīti* temple, at a still unlocated site in the steppe; three days later, they reversed the journey and reentered Babylon.[31] This departure and return has been understood, especially by scholars working in history-of-religions' tradition, as expressing a politico-religious ideology of the center. The pendulum-like movement from the city (*ālu*—the paradigmatic site of the correctly ordered, divinely ordained civilized life) to the *bīt akīti* in the steppe (*ṣēru*—the wilderness world of chaos, beasts, barbarism, and death)[32] and then back into the city ritually unmade the world in order to put it back in place.[33] At the New Year, the gods left Babylon so as to be able to return to the city in renewed triumph.

This physical movement has been mapped onto the festival's underlying mythic framework, the *Enūma eliš*. This cosmogonic hymn, Babylon's central statement of theology in this period,[34] was programmatically recited before Marduk by the *šešgallu* priest at the opening of the New Year festivities. It described in solemn, archaizing poetry the making and shaping of the universe: first, a *Chaoskampf*, in which Marduk, installed as king of the gods, went out to war against Tiamat, the ocean embodiment of primeval chaos;[35] then, following Marduk's victory, his creation of the properly ordered world. There is good reason to identify the parade out of Babylon to the steppe with the first theme of divine battle—a version of which was even depicted on the gates of the *bīt akīti* outside the ancient Assyrian capital Aššur[36]—and the triumphal return to the city with the second theme of demiurgic creation, the shaping of the physically and socially structured universe.[37] So at two levels (the kinetic out-in procession and the recited myth), the *akītu* festival celebrated and ritually enacted the dominant Babylonian theology of the world's beginning.

1 SE, the imagined origin point of imperial time, seems to have been assimilated to this archetype. For the vertical profile of Seleucus's actions—his rule in Babylon; his departure from the city in 316 BCE, escaping Antigonus Monophthalmus; and his triumphant return in Nisannu 311 BCE—closely

echoed the *akītu*'s thematizations of outward and inward ritual movement and the reestablishment of social and political order. Indeed, the new cuneiform fragment, quoted above, does not merely narrate Seleucus's arrival at the New Year of 311 BCE, the only contemporary event it needed to record; rather, and unusually for Babylonian historiography, the entry's subordinate clause glances back a few years to incorporate mention of Seleucus's flight. In other words, the fact of Seleucus's arrival was to be understood only in terms of his departure: this bonded doublet—sharply expressed by the juxtaposed Sumerograms záh gur (he fled, he returned)—functions as an integrated, singular action. While this narrative choice grossly simplified the complex warfare and politicking of the previous years, it reduced Seleucus's actions to the format of the *akītu*'s central ritual process. Indeed, the Sumerian verb gur (Akkadian *târu*) used in this fragment for Seleucus's return to Babylon is employed in some of the *akītu* ritual texts for the divine reentry to the city.[38]

According to this cultic paradigm, Seleucus's departure from Babylon could be mapped onto the ritual dislocation to the steppe and Marduk's mythic battle against oceanic Tiamat. The *Babylon King List* refers to the period of Seleucus's absence in terms that can signify only an unthinkable disorder for the Babylonian cosmology: "For [.] years there was no king in the country ([.] mu lugal *ina* kur nu tuk)."[39] It is worth raising the possibility (although we have no explicit testimony) that this mythic framework may have identified Seleucus's enemy Antigonus with Tiamat—much as, say, the Attalid kings of Pergamum would frame their defeat of the Galatians as the primordial Gigantomachy, or as Darius I modeled his victories over the Liar Kings on the Iranian hero Thraētaona's defeat of the dragon Dahaka.[40] Certainly, the Neo-Assyrian kings had made extensive allusions to the *Enūma eliš*' *Chaoskampf* in their annalistic campaign accounts,[41] and the famous Cyrus Cylinder modeled the actions of the new Persian conqueror on those of the myth's Marduk.[42] Seleucus in exile had served as Ptolemy's naval commander against the Antigonid fleet, and his dynastic blazon of the anchor represented, at its most obvious, the taming of the bitter waters;[43] it appears for the first time on coins minted at Babylon immediately after Seleucus's return to the city.[44] The ceremonies of the *akītu* festival closely identified the reigning king with Marduk, pairing divine and mortal rulers in various ritual gestures and transferring the god's epithets to the king.[45] This would only have been reinforced by the religious environment of the early Hellenistic period, with widespread ruler-worship and Seleucus's particular identification with Zeus.[46]

In any case, according to the *Enūma eliš* account, Marduk's first demiurgic act following his victory over Tiamat was the creation of structured, historical time. The beginning of the fifth tablet of the *Enūma eliš*—recited, as we have seen, at the opening of the New Year festival—described the shaping of Babylon's temporal system: "He defined the year, its bounds he fixed (*ú-ad-di* mu.an.na m[*i-iṣ-ra-ta*] *ú-ma-aṣ-ṣir*)."[47] In this programmatic chiasmus and the verses that follow Marduk establishes the ideal year, with months of thirty days, lunar opposition on the fifteenth day, and so forth. As David Brown has demonstrated, this is the paradigm underlying the Babylonian mathematical and nonmathematical astronomy that would flourish under the Seleucid monarchs.[48]

Thus, it seems that Seleucus selected for the beginning of his empire and Era an event—the spectacular *kairos* of his return to Babylon in Nisannu 311 BCE—that could be understood within the symbolic logic of the Babylonian New Year festival. If the actual introduction of the Seleucid year count in 305 BCE signified the consolidation of Seleucus Nicator's rule in Babylonia, the imagined epoch of 311 BCE retrojected this stability to the very moment of his heroic reappearance before the city's walls. In other words, it reformatted recent political history to the *akītu* mythic-ritual paradigm, in which the return to Babylon was correlated with the first establishment of legitimate sovereignty. Such clarity of symbolism allowed the co-opting of the *akītu*'s cosmological function: the beginning of the Seleucid empire could be modeled on the creation of the world and the origin of its Era count on the invention of structured time. It is precisely the epoch's status as a retrospective fiction—the deliberate selection, from all possible dates, of Seleucus's return in Nisannu 311 BCE—that should encourage us to understand it as motivated, at least in part, by the religious archetype in which it was anchored. We are dealing with what Marshall Sahlins calls "a structure of conjuncture," elegantly outlined in his exposition of the arrival of Captain James Cook at Hawaii during the Makihiki, the Polynesian New Year festival, and his assimilation, with unfortunate consequences, to the dying god Lono: "a set of historical relationships that at once reproduce the traditional cultural categories and give them new values out of the pragmatic context."[49]

As above, so below: the foundation myth that necessarily inhered in an irreversible and numericalized count of years transposed Babylonian religious notions of first beginnings to the new imperial formation. Put another way, Seleucus Nicator achieved in the temporal dimension, as indeed he would in the spatial, a periodizing distinction. I cannot decide whether it is the logic or the irony of history that places at the beginnings of linear era-counting—a way of thinking about durational time so often characterized as "empty" or disenchanted—an archaic myth of Marduk's victory over the Leviathan-like chaos-demon.

The Macedonian Beginning

There is one final, complicating "beginning" to discuss: the Macedonian epoch. We have seen that Babylon provided for the Seleucid empire a legitimizing location and for the Seleucid Era a conceptual paradigm. But the three decades of territorial incorporation that followed Seleucus's return to the city—in order, Iran and Central Asia, Syria and Cilicia, Asia Minor and European Thrace—ultimately pulled the imperial gaze toward the Mediterranean. Thus, at some still undetermined point in his reign, but certainly as a secondary phenomenon, the Seleucid Era was given a Macedonian epoch.

The Macedonian calendar had already been synchronized to the more sophisticated Babylonian one by the end of Alexander's reign. In a simple act of translation, Babylonian months had been assigned equivalent Macedonian names, reconfirmed by Seleucus at the foundation of Antioch-by-Daphne[50] (see below). Despite this month matching, however, the Macedonian New Year continued in its traditional position at 1st Dios, around the autumnal, not vernal, equinox.[51] In other words, the Macedonian calendar year began six months earlier than the Babylonian calendar year, so when the Seleucid Era count was expressed according to the Macedonian epoch, it counted from the autumn. Accordingly, we have evidence of two calendars in operation within the Seleucid empire: the original Seleucid Era (Babylonian), or SE(B), with the vernal epoch of 1st Nisannu 311 BCE, and the Seleucid Era (Macedonian), or SE(M), with the autumnal epoch of 1st Dios 312 BCE.

In the winter months, between the autumnal and vernal equinoxes, the SE(M) was ahead of the SE(B) by one year number; in the summer months, from spring to autumn, the count was identical. A single historical example may suffice to illustrate this: Antiochus IV died a miserable death somewhere in Persia in the winter of 164 BCE. This is reported in the *Babylon King List* as occurring in the month Kislimu, 148 SE(B).[52] But according to 1 Maccabees, the Hasmonean account of the Judean uprising against the Seleucids, the persecutor king perished in 149 SE(M).[53] As Table 3 illustrates, the king died toward the end of the Babylonian calendar year but after the beginning of the Macedonian one.

TABLE 3

Comparison of the Seleucid Era (Babylonian) and Seleucid Era
(Macedonian) dating of the death of Antiochus IV Epiphanes

	SE(B)				SE(M)	
Nisannu (Spring)	148	149 . . .	Artemisius		148	149 . . .
Ayyaru	148		Daisius		148	
Simānu	148		Panemus		148	
Du'ūzu	148		Loios		148	
Abu	148		Gorpiaeus		148	
Ulūlu	148		Hyperberetaeus		148	
Tašrītu	148		Dios (Autumn)		149	
Arahsamnu	148		Apellaeus		149	
Kislīmu (death)	_148_		*Audnaeus* (death)		_149_	
Tebētu	148		Peritius		149	
Šabātu	148		Dystrus		149	
Addaru	. . . 147	148	Xandicus	. . . 148	149	

The significance of this calendrical offset is unclear. Was the SE(M) merely
a technical development of the SE(B)? Or was its epoch also constructed as a
kairos, perhaps with the victory at Gaza over Antigonus's son Demetrius Poliorcetes?[54] Were the two calendars spatially distributed across the landscape
of empire, either between eastern and western hemispheres or between Near
Eastern and colonial populations? Did the respective calendars carry a cultural or ethnic identity? Would dates in the neighboring cities of, say, Babylon
and Seleucia-on-the-Tigris have used SE(B) and SE(M) years, respectively? If
so, would someone traveling the short distance from the ancient Mesopotamian capital to the new Seleucid one between autumn and spring advance a
year on entering the colony?

These two "year zones" established a binary and perhaps patchwork landscape of imperial time that no doubt detracts from our impression of the
Seleucid Era's overall chronographic simplification. Presumably, the beginning of the year, whether in spring or in autumn, was a basic temporal assumption too deeply inscribed in the empire's constituent populations to be
displaced. But there are two major qualifications. On the one hand, ancient

populations were long used to juggling multiple calendars (religious, civic, and fiscal), and we should take care not to overplay the dislocating effect of this temporal disjunction. On the other hand, the vast majority of the empire's populations was stationary, and so their experience of the Seleucid Era stable. Even for the more mobile of imperial subjects, major interregional movement occurred in the spring and summer months, when the year number was identical across the empire's full territorial extension.

Commemoration and Competition

Returning to the inaugural Babylonian epoch, a necessary consequence of the idealized chronographic beginning of 1 Nisannu 311 BCE would have been the resignifying, at least from the imperial perspective, of the city's New Year festival. Henceforth, every *akītu* would become, over and above the age-old celebration of Marduk's triumph and the world's creation, an anniversary commemoration of Seleucus Nicator, his miraculously timed return to Babylon, and the grand conquests that followed. Each *akītu* would mark the occasion when the empire aged by a year and the Era increased by one unit—a festival of political duration. It is within this context that we should interpret the participation of Seleucus's successors in later iterations of the festival— firmly attested for Antiochus III in person and Seleucus III at a distance, and with the participation of Antiochus I likely as well.[55] This is of broader importance, for these episodes have been taken as foundational proofs for the paradigm of role-playing or "chameleon" kingship, the overly eirenic model, inherited from studies of the Achaemenid empire, according to which the Seleucid rulers temporarily playacted the responsibilities of their subjects' defunct local dynasties.[56] While we must in no way doubt Seleucid royal sponsorship of indigenous cults in money or kind, the dynasty's unusually intense involvement in Babylon's *akītu* should rather be explained, at least in part, by the festival's new imperial-chronographic identity.

Cuneiform sources indicate that cultic actions directed toward the imperial family were incorporated into the festival program. A Babylonian Chronicle for the brief reign of Seleucus III reports:

Year 88 (SE), king Seleucus, month Nisannu. That month, the 8th day, a certain Babylonian, the *šatammu* of Esagil, established, according to the command of the king, precisely in accordance with the parchment letter which the king had sent before, as [the offer]ing of Esagil [.] silver from the royal treasury (é lugal),[57] from his own house, eleven fat oxen, one hundred fat ewes, eleven fat ducks for the offering within Esagil to Bel, Beltiya, and the great gods and for the ritual of king Seleucus and his sons (a-na dul-lu šá ¹si-[lu]-ku lugal u aᵐᵉˢ-šú).[58]

Paralleling the cult offerings for Bel-Marduk and his fellow Babylonian deities is a dullu—an unspecified religious ritual—for a king Seleucus and his family. Since this cannot be the reigning Seleucus III, who sired no children,[59] we should take the Sumerian aᵐᵉˢ to mean "descendants," not "sons,"[60] and so interpret the cultic recipients as Seleucus I and his successors. Note that the phrase "of their descendants" (šá li-pi-šú-nu) appears in a similar context from the akītu of 205 BCE (the year of Antiochus III's participation; see below). According to Seleucus III's own epistolary instruction, the royal treasury was to fund sacrificial victims for both the local divine pantheon and the imperial family. If the "dullu of king Seleucus" were indeed for Seleucus I Nicator, then it imbeds the imperial founder within the cultic framework with which the Era was coordinated. This is supported by a fragmentary Astronomical Diary entry for 11th Nisannu 246 BCE, which mentions Antiochus II's children Seleucus (II), Antiochus (Hierax), and Apame (['si]-lu-ku ¹an-ti-'u-ku-su u ˹a-pa-am-mu dumuᵐᵉˢ-šú ina é.sag.íl) in the context of the New Year cult at the Esagil temple; in all probability, the reigning Antiochus II, and perhaps also his predecessors, were included at the beginning of the broken list.[61] 11th Nisannu, the day when the imperial family appear in the context of the Esagil temple, was also the day on which the New Year procession returned to the city from the bīt akīti for the annual "decreeing of the destinies."[62] With all the caution that the scanty and partial state of our evidence demands, it is suggestive that our only extant accounts of Seleucid-period akītu celebrations identify the imperial dynasty with the precise mythic-religious paradigm on which the Era epoch was modeled: the grand parade to the steppe on 8th Nisannu and the return to the city and its main temple on 11th Nisannu.

Much more intriguing is the Astronomical Diary from Nisannu 205 BCE (107 SE), which records the personal participation of Antiochus III in Babylon's akītu festival. It dates to this great king's westward return from the seven-year anabasis that had reasserted Seleucid rule in the Upper Satrapies. The relevant parts of the diary read:

That [month (Nisannu)], on the 8th day, king Antiochus (III) and the [.] went out [from] the palace [.] In front of them, at the Pure Gate [of] the Esagil he performed the *ḫarû* of the year (*ḫa-ru-ú šá mu*) [.] the Esagil. Offerings were made [.] of their descendants (*šá li-pi-šú-nu*). They entered the Temple of Day One (*ana é ud.1.kám ku₄*) and made [offerings] to Ištar of Babylon and the life of king Antiochus (([d]15 tin.tir^ki *u bul-ṭu šá* ^l*an lugal*).[63]

Two aspects of this Diary entry suggest the Seleucidization of the Babylonian New Year. First, this *akītu* marked the precise centennial of Seleucus I's self-coronation as *basileus* and the real-world inauguration of the Seleucid Era count. That is, Antiochus III, the most historically conscious of all the dynasty's kings, incorporated himself into the Babylonian New Year ritual paradigm at the precise moment when the empire and Era had been born one hundred years earlier (see Chapter 3). Indeed, it is noteworthy that the "*ḫarû* of the year (*ḫa-ru-ú šá mu*)," performed by Antiochus III at the gate of the Esagil temple, is previously unattested in the millennia-long cuneiform record.[64] While the ritual's precise significance remains unknowable without extant commentary or parallel, the formula is suggestive—is it some kind of celebration of the Seleucid Era number or of the dynastic centennial?

Second, perhaps also for the first time in the extant cuneiform record the Astronomical Diary mentions a sanctuary in Babylon named the "Temple of Day One" (é ud.1.kám in Sumerian), within which the Seleucid entourage made sacrifice for the king's life. In the published version of this Diary, this shrine's name was interpreted as an alternative for the *bīt akīti*, Babylon's ancient New Year sanctuary, and translated accordingly; and it would be hard to doubt that in this particular context, on 8th Nisannu, the é ud.1.kám was being employed in the traditional manner of the *akītu* temple.[65] Possibly it is the same complex or a part of it or an addition to it. But Parthian-period administrative documents demonstrate that the Temple of Day One and the *bīt akīti* could be considered separate entities.[66] Moreover, the numericalized, calendrical name of this sanctuary is without close parallel in late Babylonia.[67] So, what is this Temple of Day One (é ud.1.kám)?

Wherever the Temple of Day One appears in the Astronomical Diaries it functioned as the privileged cultic location within Babylon for Seleucid actors and their imperial religion. Thus, after the above-quoted Nisannu 205 BCE episode, in Šabāṭu 188 BCE, in the course of his third, final, and fatal *anabasis*, the same king Antiochus III again entered the é ud.1.kám sanctuary, offering there oxen and sheep.[68] Then midway through 172 BCE, on 13th Abu, the commander of the Seleucid army (^{lú}gal erín^{meš}), arrived in Babylon, entered the Temple of Day One, sacrificed there to Bel, Beltiya, Ištar of Babylon, the great gods, and for the life of the kings, prostrated himself, and then continued to Seleucia-on-the-Tigris.[69] The following year, the Temple of Day One appears in a fragmentary Astronomical Diary in the context of "merry-making in the land (ni-[gu]-⌐tú⌐ ina kur)," phrasing elsewhere associated with the arrival of a king or royal representative.[70] There is another fragmentary reference to a sacrifice of bulls there in Ululu 159 BCE.[71]

At the Arsacid conquest of Babylonia in 141 BCE the Temple of Day One disappears from record as a site of sacrifice and dedication. Instead, in what may represent a significant reshaping of Babylon's landscape of imperial worship, parallel agents from the Parthian court in similar circumstances perform sacrifices at the Esagil temple, the city's central temple complex, most frequently at the ká.dumu.nun.na, the "Gate of the Son of the Prince."[72] In the decades following the Parthian takeover the Temple of Day One appears in only two reconstructable episodes. The first, recorded in the Astronomical Diary of 127 BCE, makes reference to "the sacrifices which had been stopped in the Temple of Day One (sískur šá ina é ud.1.kám ba-aṭ-⌐lu-ú⌐)."[73] The tablet, too broken for anything more to be recovered, testifies to a formal pause in the sanctuary's sacrificial cult. The second case requires a little context. After the Parthians had conquered Babylon in 141 BCE, the copycat Arsacid Era (AE), with its epoch of 1 Nisannu 247 BCE, had been joined to the Seleucid Era count as a new double dating formula for cuneiform tablets—for example, "Year 115 (AE), which is year 179 (SE), king Arsaces"[74] (this new Parthian chronography is discussed in Chapter 3). But in 127 BCE, in the political and military disorder that had followed Antiochus VII Sidetes's *anabasis*, defeat, and death, Babylon was briefly occupied by Hyspaosines, the former Seleucid satrap and now independent king of Characene, the sealand region of southern Mesopotamia. Hyspaosines strongly identified with and continued to use the forms, images, and institutions of his erstwhile Seleucid overlords. Notably, he struck the Seleucid Era date onto the precious metal coins minted in his name and, during his kingship in Babylon, he abandoned Arsacid Era dating to restore the chronographic monopoly of the Seleucid Era.[75] An Astronomical Diary reports that, during his short-lived takeover of Babylon, Hyspaosines took the throne of the king—explicitly identified as a Greek *thronos*—from the royal palace and dedicated it as a votive offering in the

Temple of Day One; later, in Ṭebētu 125 BCE, following the firm restoration of Parthian control,[76] the *politai* of Babylon opened the sanctuary's doors and took away this dedication.[77] Hyspaosines's gesture, as far as it can be reconstructed, joined together in a single religio-political complex his kingship in Babylon, the royal throne, the restored Seleucid Era count, and the Temple of Day One.[78]

Evidence for the Temple of Day One is infrequent, lacunose, and without clear precedent or parallel in the Babylonian tradition, so any interpretation or identification must be tentative. Yet the sanctuary's very name points to its celebration of and identification with an inaugural moment, a first beginning of time. Furthermore, none of the attested activity at the sanctuary took place on the first day of a month or a year, suggesting that its "Day One" was a singular, historical event and not a calendrical recurrence. The temple was privileged by Seleucid officials as the site of sacrifice for the royal family. Such offerings were discontinued and relocated under the Parthians. The sanctuary was markedly and singularly reused by the Seleucidizing, Era-restoring king Hyspaosines. Accordingly, I propose that the é ud.1.kám, the Temple of Day One, may have been founded or repurposed to commemorate in cult the first beginning of the Seleucid empire—that is, the idealized Seleucid Era epoch of 1st Nisannu 311 BCE. Such a sanctuary would not be without parallel, as we shall see. Personifications of time and worship of temporal units are well known from across the broader Mediterranean and Hellenistic world, including within the Seleucid empire (see Chapter 2). And, perhaps more germane, the copycat Arsacid and Kanishka Era counts each were sacralized with epoch sanctuaries, at Asaac in Turkmenistan and Rabatak in northern Afghanistan, respectively (see Chapter 3).

Back by the Mediterranean, in the emerging Seleucid heartland of northern Syria, the Seleucid Era epoch was similarly monumentalized. According to the account of John Malalas, the sixth-century CE chronicler of the great Seleucid colony of Antioch-by-Daphne: "In front of the city, on the far side of the river Orontes, Seleucus (I) set up another statue, this one of a horse's head and, beside it, a gilded helmet, with the inscription, 'Seleucus, fleeing from Antigonus upon this (horse), was saved; and, returning from there, he killed him (ἐφ᾽ οὗ φυγὼν ὁ Σέλευκος τὸν Ἀντίγονον διεσώθη· καὶ ὑποστρέψας ἐκεῖθεν ἀνεῖλεν αὐτόν).'"[79] The inscription's narrative comes remarkably close to that of the new cuneiform tablet discussed above. The chaotic events of the 310s BCE have been clarified into a flight from and return to

Babylon, bound together as a unified, single event. φυγὼν . . . ὑποστρέψας (having fled . . . having returned) almost reads as a translation of the tablet fragment's Sumerograms záh gur (he fled, he returned), displaying once again the Seleucid state's remarkable capacity to produce a panimperial discourse in local idiom. Moreover, as with the *akītu*-coordinated epoch, the Greek inscription seems to correlate Seleucus's heroic return with the defeat of his main rival.[80] Should this Byzantine chronicle be accurate[81]—and Malalas here made use of the civic history of the Imperial-period author Pausanias of Antioch[82]—this horse-head and helmet monument must be considered yet another public memorialization, perhaps even a sanctification, of the idealized origin and triumphant beginnings of the time of empire. It may be significant that Malalas's description of this statue group immediately follows his account of Seleucus I's adoption of the Babylonian automated calendar—"[Seleucus] ordered that the Syrian months be named after the Macedonian months (ἐκέλευσε δὲ ὁ αὐτὸς καὶ τοὺς μῆνας τῆς Συρίας κατὰ Μακεδόνας ὀνομάζεσθαι)"[83]—as if the calendrical directive and the monument of return coordinated the empire's new architecture of time, with respect to months and years. The statue group likely gave its name to the suburb of Antioch known as Hippocephalus (horse head).[84]

If Babylon's Temple of Day One and the horse head on the Orontes's bank represent orchestrated and public celebrations of the Seleucid Era epoch, so, pari passu, the enemies of the Seleucid house recognized and undermined this chronographic origin. Clearest of all is the third-century chronicle known as the Parian Marble, a two-meter-high stele from the Aegean island of Paros inscribed with a continual history of the world, extending from the mythical Athenian king Cecrops down to the second quarter of the third century BCE.[85] This remarkable work serialized its entries according to a double dating formula: first, a numericalized countdown to a year 1, a kind of reverse era system, that culminated in 264/263 BCE, the presumed date of composition;[86] and second, the name of the Athenian annual eponym (or king) in genitive absolute—for instance, "From when Homer the poet appeared, 643 years, Diognetus being king of Athens,"[87] or "From the death of Alexander and Ptolemy's mastery of Egypt, 60 years, Hegesias being archon at Athens."[88]

The Parian Marble's late fourth- and third-century BCE entries display an evident bias toward the Ptolemaic house: for instance, Ptolemy I's elevation to kingship is mentioned, but not those of the other Successors, and the birth of

Ptolemy II appears, but not those of his peer monarchs.[89] Accordingly, it has been suggested that the Parian Marble was composed to mark either the alliance of Athens and Alexandria in the Chremonidean War against Antigonid Macedonia[90] or the introduction of a hypothesized, though probably illusory, Ptolemaic "Soter Era."[91] Whatever the chronicle's teleological implications, the Seleucid dynasty is mentioned in only one entry: "From the time the sun disappeared and Ptolemy prevailed over Demetrius in Gaza and dispatched Seleucus to Babylon, [4]8 years, Po[le]mon being archon at Athens (ἀφ' οὗ ὁ ἥλιος ἐξέλιπεν, καὶ Πτολεμαῖος Δημήτριον ἐνίκα ἐν | Γάζει καὶ Σέλευκον ἀπέστειλεν εἰς Βαβυλῶνα, ἔτη ΔΔ[Δ]ΔΓΙΙΙ, ἄρχοντος Ἀθήνησιν Πο[λέμ]ωνος)."[92] The glorious epoch of the Seleucid empire—the battle of Gaza and Seleucus I's return to Babylon—has been reconfigured both politically and temporally. Far from being the autonomous act of independence asserted by Seleucid sources, cult, and chronography, here it is Ptolemy of Egypt, the lemma's nominative subject, who wins at Gaza and then dispatches Seleucus to Babylon; it is the instruction of a superior to an inferior. Furthermore, we know from both Diodorus Siculus and modern astronomical calculation that the eclipse of the sun reported in this entry in fact occurred two years later, in the Athenian archonship of Hieromnemon.[93] The Parian Marble's author has relocated this ominous event—the single instance of an eclipse in his entire millennium-plus account—to coincide with Ptolemy's victory at Gaza and, no doubt more significantly, the foundation of the Seleucid empire. Finally, the Seleucid Era's epoch is here redated by the chronicle's own directionless eponym system and numerical countdown: this Ptolemaic-affiliated inscription has relegated the Seleucid empire's miraculous and sanctified Year 1 to a passing episode in Athens's sequence of archons and as forty-eight years before the Parian Marble's own event horizon. This is a chronography of insult.

In another case, the eleventh chapter of the book of Daniel delivers a pseudoprophetic account of the Syrian Wars between the Seleucid and Ptolemaic empires for control of the southern Levant. A full analysis of this predictive history will be offered in Chapter 5, but the key point here is that this fiercely anti-Seleucid work again reconfigures the Seleucid Era epoch. Daniel 11:5 reports that, after the death of Alexander the Great and the fracturing of his kingdom, "the king of the south (Ptolemy I) will be strong, but one of his princes (Seleucus I) will be strong over him (ûmin-śārāyw wᵉyêḥᵉzaq 'ālāyw) and will rule and his rule will be a great rule." As in the Parian Marble, Seleucus I appears here as one of the subordinates of Ptolemy

43

I, as the prince (*śar*) of the king (*melek*) in the south.[94] As we shall see, the Judean author(s) of the book of Daniel have fixed the source of the Seleucid-Ptolemaic wars that so devastated their homeland in the Seleucid Era epoch, recoded as a subordinate's rebellion against his overlord.

The recognized significance of the Seleucid Era epoch, even beyond the dynasty's reach, can be detected as much in these undoings as in its commemorations.[95] I am reminded of Hayden White's truism, that "in order to qualify as historical, an event must be susceptible to at least two narrations of its occurrence."[96] The epoch needed to be reckoned with.

Conclusion

The Seleucid Era was nothing like the nineteenth-century CE imperial imposition of modern, historical, European temporalities on a supposedly timeless and archaic East or South. Quite the reverse: the Era, both in its birthplace and in its chronographic principles, was a conjunctural phenomenon, located in historical processes that brought hitherto relatively isolated societies into contact with each other. We have, on the one hand, Seleucid statecraft's particular tendency toward design and abstraction, and on the other hand, the prestige, scientific perfection, and automated progression of Babylon's luni-solar calendar and the theology of first creation outlined in its New Year festival.

The new year count selected and promoted a discrete historical event as the birth point of empire, as a Year 1. It is hard to think of any other premodern state that narrowed down so precisely on a date of emergence, on a clarifying punctuation in time. Such concentration of historical significance, gathered up by succession around an origin, has something of the sacred about it. It may be no accident that the eleventh-century CE Chorasmian polymath al-Biruni, the twelfth-century CE Sephardi astronomer Savasorda (Abraham bar Hiyya), and the eighteenth-century CE Talmudist Eliyahu ben Shlomo Zalman (the Gaon of Vilna) each placed this Seleucid Era epoch a thousand years after the theophany to Moses on Mount Sinai.[97] Whatever calculation lies behind their reconstructions, they share a sensitivity to the epoch's punctiform, momentous, and ultimately numinous quality. This was a fresh surging forth of time, the invention of a new beginning to the world.

CHAPTER 2

A Government of Dating

The Seleucid Era epoch was a function of royal charisma. The inaugural moment of empire was organized around the moving body of Seleucus I and its transformative effects. But the Era that it unfurled—the onward and regular passage of numbered years—routinized this charismatic origin. This chapter will explore the place of Era dating in Seleucid imperial governance and its role in molding the practices, ideas, and even sensibilities of the empire's diverse communities.

For our understanding of this question, two characteristics of the Seleucid Era are fundamental: first, the form in which the Era met the world; and second, its frequency. The Hellenistic period witnessed the emergence of a new visual culture of materialized time. Novel personifications of time were discovered and worshipped.[1] Sundials were cast across cityscapes and sanctuaries.[2] Hellenistic monarchs paraded and clothed themselves in natural times and seasonality: Demetrius Poliorcetes wore a cloak woven with the celestial zodiac;[3] the citizens of Teos crowned their cult statue of Antiochus III with the first fruits of each season;[4] the two great royal processions described by Athenaeus (of Ptolemy II Philadelphus at Alexandria and of Antiochus IV Epiphanes at Daphne) incorporated into their rolling tableaux representations of the morning and evening stars, the natural year (a six-foot-tall man wearing a tragic mask and carrying a cornucopia), the four seasons, night and day, dawn and noon.[5] All these figurations—and there were many more—expressed either the recurring cycles of nature and the heavens or the momentous and divine moment.[6] But the Seleucid Era, as a linearized, open, and historical temporality, resisted depiction in these allegorical or imagistic modes. It was only and exclusively materialized as number.

In contrast to the culturally embedded logics of figural iconography, "number, by its nature, flattens idiosyncrasies."[7] That is, in whatever script the Seleucid Era number was recorded—and we have it attested in the Greek,

TABLE 4
Greek alphabetic numbers

α'	1	ι'	10	ρ'	100
β'	2	κ'	20	ϲ'	200
γ'	3	λ'	30	τ'	300
δ'	4	μ'	40	υ'	400
ε'	5	ν'	50	φ'	500
ϛ'	6	ξ'	60	χ'	600
ζ'	7	ο'	70	ψ'	700
η'	8	π'	80	ω'	800
ϑ'	9	ϙ'	90	ϡ'	900

Akkadian, Phoenician, Aramaic, and Hasaitic counting systems—the year's numerical value was universally stable. As we shall see, within the extraordinary diversity of the imperial territories, the Seleucid Era, as a regular and homogeneous chronographic system, seems to have achieved a regulating and homogenizing force.

More particularly, the Greek alphabetic numbering system, by which the Era's year date was expressed in most court and administrative business, was given a unique ordering. The Greek numerical system is less well known but far simpler than the Latin. In short, the twenty-seven letters of an extended Greek alphabet were divided into three groups of nine letters each: the first nine letters of the alphabet, α' to ϑ', represented the units 1–9; the next nine letters, ι' to ϙ' (the letter *qoppa*), 10–90; and the final set, ρ' to ϡ' (the letter *sampi*), 100–900, as shown in Table 4.[8]

In every form of numbering across the Greek world except for the Seleucid Era year date, these numbers were arranged in backward alphabetic order, from larger to smaller—our "123," for instance, would be written ρκγ'. But the Seleucid Era count reversed this standard, increasing from units to tens to hundreds—123 SE would be written γκρ'.[9] The Era year date was the only numerical unit within the broader ancient world, within the Seleucid empire itself, and even in individual Seleucid documents, to be arranged in this way. For instance, an inscribed letter of Antiochus III, appointing a high priest of the Apollo and Artemis sanctuaries at Daphne, closed with the dating formula "δκρ' Δίου ιδ'" ("124 [SE], Dios 14");[10] and a letter of Antiochus VIII or IX, proclaiming the freedom of Seleucia-in-Pieria, ended with the date "γσ' Γορπιαίου κθ'" ("203 [SE], Gorpiaeus 29").[11] In each case, the entirely standard larger-to-smaller (reverse alphabetical) arrangement of the month's day numbers (ιδ' and κθ') directly follows the exceptional smaller-to-larger arrangement of the Seleucid Era year dates (δκρ' and γσ'). This peculiar ordering of the Era number was also a spoken phenomenon. An inscription from Euromus in Caria gives the Seleucid Era numerals as ει' καὶ ρ' ("115 [SE]"), where the καὶ ("and") reproduces the oral form.[12] 1 Maccabees, the Hasmonean chronicle of the Maccabean revolt, wrote out in full the Seleucid Era dates in their unique smaller-to-larger arrangement.[13] Josephus, synchronizing these Seleucid Era dates with Greek Olympiads, clarifies by comparison: the rededication of the Jerusalem Temple, for example, is dated "ἔτει γὰρ πέμπτῳ καὶ τεσσαρακοστῷ καὶ ἑκατοστῷ" ("in year five and forty and one hundred") and "ὀλυμπιάδι ἑκατοστῇ καὶ πεντηκοστῇ καὶ τρίτῃ" ("in the one hundred and fifty third Olympiad"):[14] the Seleucid Era date is fully written out in its reverse order of units, tens, and hundreds, its Olympiad equivalent in the standard hundreds, tens, units. To return to our example of 123: if this were, say, a quantity of grain we would read it "one hundred and twenty-three"; if it were the Seleucid Era year date, we would say, uniquely, "three and twenty and one hundred."

The reverse numericalization of the Seleucid Era date was a brilliant invention, for it served as an immediate diagnostic marker of the imperial chronography while requiring of readers and listeners no additional or specific skills. Seleucid Era year numbers could stand alone and be recognized for what they were; by contrast, the numerals of Olympiad or month dating (see above), as of everything else countable, required explicit designation.[15] This smaller-to-larger order of the Seleucid Era numerals made an argument about the texture and nature of time. Saying "three and twenty and one hundred" rather than "one hundred and twenty-three," for instance, produced a sense of an ever-accumulating temporal depth, of time moving increasingly away from its epoch. It manifested in written and verbal modes the Era's underlying chronographic principle of unbroken, irreversible linearity. It went in alphabetical order. Its absolute status, free of identifying or qualifying nouns, proposed a monopoly of political time and, as we have already seen on the Saros Tablet (discussed in Chapter 1), a temporal transcendence. All this requires a considered decision by the imperial center, presumably shortly after the Era's introduction, to develop and deploy this unique numerical form, offering yet further evidence for the early Seleucid court's politicization of and self-conscious reflection upon temporality. The scheme's adoption throughout the empire by both officials and subject communities (see below) attests the determined articulation of a differentiated temporal landscape, although we know nothing of the specific mechanisms that enabled or enforced such conformity.[16]

The second characteristic is the pervasiveness of dating. Seleucid Era year numbers were marked onto an unprecedented range of public, private, and mobile platforms. Era dates were affixed to market weights[17] and amphora handles,[18] royal and local coinage,[19] building constructions[20] and votive offerings,[21] seal rings and silver bowls,[22] royal letters[23] and civic decrees,[24] tombstones[25] and tax receipts, priest lists[26] and boundary markers,[27] astronomical reports[28] and personal horoscopes,[29] marriage contracts,[30] manumissions,[31] and much more.[32] Jewish, Babylonian, and presumably the lost Seleucid historiography dated events by the Era year. In our world full of date marks it is easy to underestimate the sheer novelty, and so historical significance, of this mass year marking. But, to the best of my knowledge, this was without precedent or parallel. In no other state in the ancient Mediterranean or west Asia did rulers and subjects inhabit spaces that were so comprehensively and consistently dated. Such pervasiveness was enabled but not, of course, required by the Era's materialization as number.

The Era's display on innumerable quotidian and public objects carried the imperial year count, and the whole political system it represented, deep into private spaces and, undoubtedly, personal thoughts with a penetration unavailable to any other imperial technology. The Era date would have been visualized, read aloud, repeated, internalized, and made entirely natural for the daily life of imperial subjects. Such domestication of the Era, through both state coordination and undirected routine, must have produced something approaching an intuitive sense of temporal location and a panimperial synchronicity, much as our own Anno Domini or Common Era dating has become international second nature.[33] To a degree not previously possible, the Seleucid Era count—irreversibly linearized, exclusively numericalized, and extensively displayed—asserted a temporal framework into which people, objects, and even the world at large could be fitted, sequenced, and related to one another. Think of Anton Chigurh in Cormac McCarthy's *No Country for Old Men:* "You know what the date is on this coin?" "No." "It's nineteen fifty-eight. It's been traveling twenty-two years to get here. And now it's here. And I'm here."[34] Everything dated in this way, from market weights to manumissions, had object histories—that is, an immediately legible and inherently comparative temporal depth. Indeed, if we wished to narrow down on "the difference that makes the difference," it would be this material, and thus social, pervasiveness.

Outside the infrequent but spectacular interventions of warfare, royal procession, benefaction, and so forth, one of the dominant quotidian experiences of empire would have been of this transcendent and linear time regime. These Seleucid Era markings were especially associated with those spaces and activities in which an instrumental rationality and a focus on the near future were dominant: fiscality (taxes, archives, and their authorizing procedures), trade (coinage, weights and measures, and the agora), and imperial communications of all kinds. The countersites—the places and practices saturated with alternative and nonlinear or nontranscendent time regimes—would include the countryside, the ruin, the shrine, the internal periphery (mountains, deserts, and so on), and all those spaces and behaviors characterized by women and the impoverished. Such zones of greater and lesser imperial time consciousness offer a strong visibility gradient for the density of state interventions. As we shall see, the Era functions as a synecdoche of the empire, its claims, and its demands. The following sections will discuss the cultural significance and logic of Seleucid Era dating in each of these fiscal, trading, and epigraphic spaces.

The Archive

The controlled preservation of administrative data had been characteristic of the complex states of the Near East and the eastern Mediterranean since the Bronze Age. But Seleucid archives stand out for their institutionalization of the serialized temporality as the functional logic of administration. Seleucid state or civic archives are attested in one way or another at Sardis in western Asia Minor, Antioch-by-Daphne in the Syrian Seleucis, Kedesh in northern Israel, Dura-Europus on the middle Euphrates, Seleucia-on-the-Tigris, Babylon, Uruk, and Nippur in Babylonia, and Susa in Elymaïs; domestic archives are known from Seleucia-on-the-Tigris and Uruk.[35] This broad geographical and typological spread suggests, as we would expect, a fairly dense matrix of public and private archives across the kingdom; doubtless, there are many more of each category waiting to be found. The coordination of administrative practices to the Seleucid Era can be seen most clearly from two kinds of data—the clay seal impressions, or *bullae*, that bound the now disintegrated documents once housed in the archives, and the architecture of the new archive buildings.

Date Keepers

Only a handful of papyrus or parchment documents have survived from the Seleucid empire, but the impressed clay sealings, by which such papers were bound, have been excavated in their thousands from Seleucid archives in Babylonia. Each of these *bullae*, as the clay rings or globules are known, bears on its surface the civic and imperial seals of the public administration alongside the personal seals of the private individuals engaged in the economic activity once documented within. The number of separate seal impressions on a single *bulla* ranges from two to forty. In other words, these *bullae* trace the institutionalized encounters between local and imperial economic agendas, as between personal iconographic identities and the standardized character of the archival administration. They are an exceptional resource for our understanding of the temporalized bureaucracy, on account of both the extraordinary size of the corpus and their close and exclusive identification with the Seleucid regime: for the *bullae's* hybrid technology—combining ancient Mesopotamian clay-based sealing practices with the demands of a Seleucid "paper" administration—emerged in the early third century, soon after the consolidation of Seleucid rule in Babylonia, and disappeared with the Parthian takeover in 141 BCE.[36]

Over half of the *bullae* (around sixteen thousand) that have been excavated from both state and domestic archives[37] prominently bear the standardized impression of a Seleucid Era year date. The shape, size, and content of these seals are uniform across Babylonia: ovoid impressions, with straight short sides and slightly curving longer edges, carrying a three- or four-line Greek inscription that named a fiscal department, regulated interest, or archival procedure (having to do with salt, slave, or grain sales, river travel and harbor dues, archive registration, or tax exemption); the Seleucid Era date in Greek alphabetic numerals; and the city (Seleucia, Uruk, or Nippur) or individual responsible for administration.

Figure 3 is a representative example from the public archive of Seleucia-on-the-Tigris. The stamp's top line identifies the sealed document as relating to the salt tax (ἁλικῆς). The bottom line names the fiscal authority, the royal

Figure 3. Tax *bulla* from Seleucia-on-the-Tigris (S9–602)

city of Seleucia-on-the-Tigris (Σελευκείας). In a much larger script at the center of the impression is the Seleucid Era year date, ηξ' (68 SE), organized in the uniquely Seleucid smaller-to-larger numerical order ("eight and sixty"). Typically, year dates are followed by a couple of small icons of still uncertain function, such as a half-anchor or a column.[38]

The Seleucid Era seals, small enough to compel a prioritization of display, privilege the imperial year number as their dominant and most visible element, ensuring the date's legibility even if, as in several extant cases, the *bulla* was not fully impressed or had its edges broken.[39] Furthermore, unlike the seals of both private individuals and archival officials, these date stamps favor epigraphic content over royal or personal portraits, imperial symbols, and figuration in general. The Seleucid Era year date sufficiently represented the imperial authority to make redundant or relegate to a minor position in the visual field other markers of the administration. Needless to say, the date consists only of the numericalized Seleucid Era year, never the month or day:

even if this represents an annual record-taking practice, the omission of any subyear divisions reproduces the empire's administrative and ideological concentration on the year unit.

These thousands of dated *bullae* permit the reconstruction of something approaching a bureaucratic sociology of Era-time. A recently arrived Thracian immigrant to Seleucia-on-the-Tigris; a proud Urukean descendant of Sîn-lēqi-unninni, the supposed editor of the Gilgamesh epic; a Tyrian from the Galilee; and, presumably, the merchant and landowning families of Elymaïs, Lydia, the Syrian Seleucis, and so on—all were equally required to bring their documents to the date stamp for guarantee, preservation, and filing into an ever accumulating fiscal library. We must imagine an annually repeated and, perhaps, emotionally charged act of impressing seals onto soft clay— imperial dates side by side with personal portraits or heirloom stones—that necessarily pulled all participants into the standardizing, linearized logic of the imperial time system.[40] These dated documents, with their affective penumbra of professional sensibilities, familial identities, and the *Zukunfts-angst* that accompanies episodic encounters with bureaucracy, became vehicles of imperial incorporation.

The uniformity and number of these date seals indicate that they were centrally engraved and distributed to archival officials. While the allocation of standardized seal rings was anticipated in the Neo-Assyrian imperial administration[41] and paralleled in the Ptolemaic dynastic cult,[42] by necessity the Seleucid distribution was regularized as an annual procedure. Through some still unknown, perhaps even ritualized, process, Seleucid and civic administrators received and then wore on their persons the new stamp for the current Seleucid Era year. Presumably, the previous year's seals were pulled off, invalidated, destroyed, or in some other way taken out of circulation, physically enacting the Era's irreversible and regular progress.[43] These archival officials were just as much marked with the Era date as the documents they impressed, populating their civic and domestic landscapes with mobile, breathing year numbers.[44]

Such a physical embodying of the imperial time system belongs to the metaphorical, even ontological, exchangeability of seal rings and bodies. This was a recurring motif throughout antiquity, expressing a range of relational associations, from religious and political delegation (such as the prophets Jeremiah and Haggai figuring the last Davidic kings as God's signet ring[45]) to epistemology (in Aristotelian and Stoic discussions of sight and sense perception[46]).

Importantly, an early episode in the encomiastic biographical tradition about Seleucus I, sometimes called the Seleucus Romance, shows that the association was celebrated at the imperial court. The Romance tradition adduces, as proof that Seleucus I Nicator had been sired by Apollo (see Chapter 3), the discovery of a seal ring, bearing the device of an anchor, in his mother's bed on the morning after her divine impregnation. We are told that, when Seleucus was born nine months later, he bore an anchor-shaped birthmark on his thigh, and that this birthmark passed as a bodily inheritance down the dynastic line.[47] The tale not only equates Seleucus's royal person with his seal ring; it also transforms every subsequent impression of the anchor ring into a replication of the imperial founder's body. While this is the transparent function of the birthmark inherited by every one of his genetic descendants, the story's power also lies in the anchor's employment as the dominant icon on the seals of imperial archival officials, including, as we have seen, the Era date stamps (see Figure 3). A rectangular seal engraved with a horizontal anchor above a lion, used by an imperial official at both Seleucia-on-the-Tigris and Uruk, is even captioned in Akkadian "sa-um-bu-lu šá lugal" ("the symbolon of the king"), borrowing the Greek term for a token, guarantee, and identity mark.[48] In other words, the Romance tradition used the anchor-engraved seal ring and the activities that inhered within it—impressing, authorizing, entrusting, passing on—as an archetype for the acts of delegating and reproducing monarchic power; we can read it as a founding myth for the imperial bureaucracy. When set within the material context of Seleucid statecraft, the seal-ring tale almost imagines the registered parchment documents and the bodies of the archival officials as extensions of the king's skin, an epidermis of empire, stretched out across the political landscape.[49] At its maximal interpretation, then, the archival officials' use and wearing of annually distributed Era rings simultaneously reproduced the king's authority-bearing body and the empire's temporality.

Timelines

If each stamping of a document marked a single bureaucratic encounter, its intelligibility came from the full "set," for it was the archive's systematized accumulation of date-marked documents, the consigning through a repeated technique, that established a relational temporal order. Our best way to access the meaning of this serialization is from the architectural remains of the newly constructed archive buildings. Unlike the inherited and reused administrative buildings at Uruk and Kedesh, the public archives in the new Seleucid foundations of Dura-Europus and Seleucia-on-the-Tigris were expressly designed to meet the requirements of the imperial bureaucracy. While these two archives may not seem much to go on, in fact they represent the best surviving evidence for ancient archival practice from the ancient world as a whole. We will see that each of the buildings materializes in stone and mud brick the basic principles of the Era count.

The excavated archive at Dura-Europus (Figure 4) was constructed as part of the mid-second century BCE expansion of the Euphratene military settlement.[50] It is situated at the middle of the northern edge of the colony's orthogonal agora (Unit A of Block G), in the very center of the enlarged town. It is the only space with civic functions so far uncovered in the public square. The large room A2 was likely used as the office of the archival administrators and

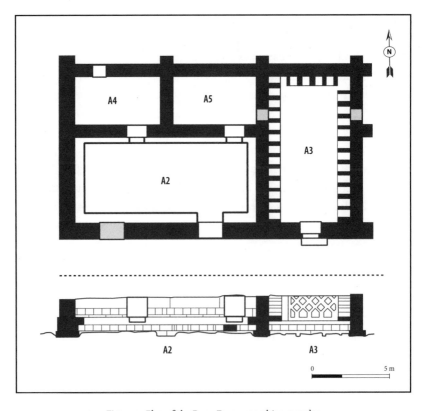

Figure 4. Plan of the Dura-Europus archive complex

Figure 5. Excavated depot (Room A3) of the Dura-Europus archive

for the registering of documents. It is likely that a statue of Seleucus Nicator, the colony's and empire's founder, was erected on a base against the room's northern wall, presiding over the impressing of seals. The registered documents were then removed for storage in the neighboring room A3 (ten by six meters in size), the walls of which consisted of banks of lozenge-shaped pigeonholes (Figure 5).[51] Cut into the plaster to the left of each niche were Seleucid Era year dates, written out as Greek alphabetic numbers in the specifically Seleucid smaller-to-larger order. The dates spiral in chronological order around the room and up the walls, continuing without break into the Parthian and Roman regimes.

Seleucia-on-the-Tigris's public archive was constructed in the first half of the third century BCE.[52] It is the largest archive building known from the ancient world, at 140 meters long (north-south) by 6 meters wide (east-west), extending along the entire western side of the city's main agora (Figure 6). The building had a modular layout of fourteen identical rooms, linked to one another by doorways. A wall between the central modules divided the complex into two identical suites of seven rooms each, requiring independent access. Excavation did not uncover any doorjambs in the eastern or western walls, suggesting that access was through the archive's narrow north and

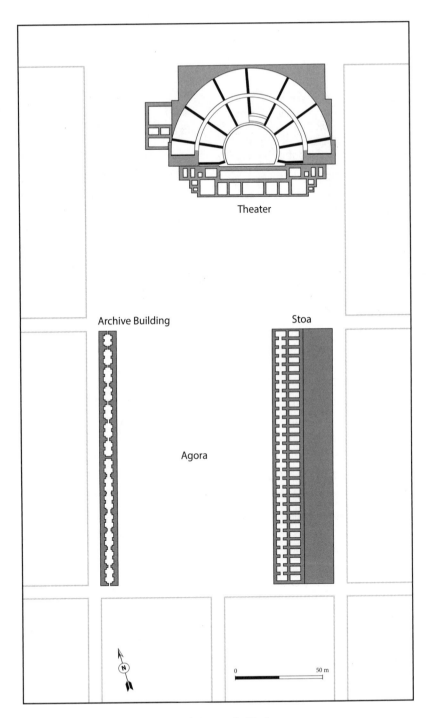

Figure 6. Seleucia-on-the-Tigris agora

south ends. The more than 25,000 *bullae*, discussed above, were found on the floors of these modular rooms following a devastating conflagration that burned the parchment and papyrus documents some decades after the Parthian conquest. Archaeologists have confidently restored a regularized system of wooden shelving for each of these rooms, between and beside their pilasters and doorways, with the registered documents kept in boxes or woven baskets. It is hard to doubt that some form of linearized ordering was at work, at least in origin; certainly, the high proportion of *bullae* with Seleucid Era dates suggests that this time system formed an organizing axis.[53]

At their different scales, appropriate to a military settlement on the middle Euphrates and the empire's "city of kingship" in Babylonia, the two archive buildings represent a streamlined governmental knowledge system: a uniform, predictable temporality extended over a defined, linearized space. The combination of date-wearing officials, date-stamped documents, and dated built infrastructure—the correlation of where and when—represents a fundamental simplification of statecraft. If I may reverse Marx's famous dictum, it is as if these archives were constructed by the bee and not the architect in their direct rendering into mud-brick cells of the Seleucid Era's underlying structure: a single idea, unity in construction, and the linear repetition of an elementary shape. An illuminating contrast may be found in the *bullae* of the Ptolemaic archive at Nea Paphos on Cyprus, excavated from beneath the mosaic floor of the House of Dionysus. Unlike the Seleucia data, only about 3 percent of these Ptolemaic *bullae* (around 330 of more than 11,000) are inscribed, and far fewer are dated. The stamped dates, expressed in the standard, non-Seleucid, greater-to-smaller alphabetic numerical system, cover years 1 to 36.[54] Because the Ptolemies employed regnal counting rather than era dating, it is impossible to supply a temporal referent: for instance, documents sealed with dates κ' to κϑ' (20 to 29) could refer to the regnal years of Ptolemy I, Ptolemy II, Ptolemy III, Ptolemy V, Ptolemy VI, Ptolemy VIII, Ptolemy X, or Cleopatra VII, and the problem is only more acute for the smaller year numbers. The historian's difficulty is also an ancient one: in contrast to the Seleucid system, Ptolemaic archival dating was a regime of temporal discontinuity and so could not be materialized into a regular form with a visible logic of recall. This applies to all things dated by regnal number.[55]

Near Eastern public archives were typically embedded within and so gained their authorizing character from an umbrella administrative complex, such as a palace or a sanctuary.[56] But the archive buildings at Dura-Europus and

Seleucia-on-the-Tigris were noncultic, freestanding, and set against their colonies' agoras;[57] indeed, they seem to have been original elements in their cities' master plans. We shall see below that the market activities of the abutting public squares were themselves strongly characterized by the imperial time regime. To contemporary inhabitants congregating in the agora and going about their urban business, the archive must have been a symbol of state as well as its instrument. For reasons of both security and mystery, the rooms we have been examining would have been closed repositories, bureaucratic recesses patrolled by date-wielding mandarins, into which fiscal and personal affairs were alienated. It is no accident that archival administrative titles—*bibliophylax* and *chreophylax*—were built from the sentinel verb φυλάσσω ("I guard"); Jacques Derrida speaks of archived documents as being under house arrest.[58] The archive building at Seleucia-on-the-Tigris suggests an especially powerful exterior experience. By delimiting the entire western side of the city agora for 140 meters, it presented an impermeable walled barrier, obliging movement along its length, quite unlike the similarly scaled stoa opposite, with its perforated facade,[59] or the urban *insulae*, with their internal passageways and shortcuts. Whatever the archive's interior organization, its external aspect was of an absolute, unrelenting linearity. Indeed, the building's very shape—long, document-shelved corridors without real depth to speak of—comes as close as an ancient monumental building could to a simple line, as if hypostatizing in its public architecture the serial, irreversible structure of the Era count.

We might be tempted to think of these Seleucid archive buildings as institutions oriented toward the past, preserving, accumulating, and even arresting temporality. But they were also, and in more significant ways, directed forward, toward an open future. The registration of documents was a responsibility and a promise, designed to anticipate the needs of prospective fiscal and legal scenarios.[60] More fundamentally, the archive buildings materialized a confidence in the continuity of the Seleucid bureaucracy and, therefore, of the imperial project itself. Document depots that must have remained mostly empty for many decades after their construction expressed the inevitability of a knowable, secure, and stable future. Put another way, the verbal tense enunciated—or, perhaps better, aspired to—by the Dura-Europus and Seleucia-on-the-Tigris archive buildings was the future perfect: this will have been. The unparalleled enormity of Seleucia's archive, in particular, was a translation into time of the capital city's spatial gigantism and all that this proclaimed about the Seleucid regime.

In recent decades, cultural theory and postcolonial history have employed the archive (or "Archive") as an organizing metaphor for any corpus of interested, selective remembrances and forgettings.[61] While these discussions tend to allegorize the archive and so to lead elsewhere, their insights can be turned back to the ancient institution. Especially illuminating is Derrida's *Le mal d'archive*, "Archive Fever," which coordinates a conversation between Sigmund Freud and Yosef Yerushalmi to draw attention to the ideological potency of the archive as a site and process of idealization, of "a spectral messianicity."[62] Accordingly, I would suggest that the Seleucid archive, in addition to being a depot for documents, constituted something approaching a utopia or, if I may, a uchronia.[63] For the arrangement of fiscal life into an annually serialized grid established an ordered geometry of imperial time at the heart of the city's public life, a locus of maximal temporal conformity and rationalization that idealized and euphemized the messiness of the greater societal picture.

Endings and Undoings

Indirect confirmation of this imperial "archive effect" comes from the extraordinary and gruesome violence it seems to have provoked. The evidence is found in the remains of minor archives we have not yet discussed: the domestic document collections of a wealthy salt-trading family or civic official from Seleucia-on-the-Tigris, and the administrative archive in the minor regional center of Kedesh, in northern Israel.

Concurrent with the large public archive at Seleucia-on-the-Tigris were a pair of smaller, domestic collections, designated by their excavators as Archives A and B, uncovered in Level IV of the so-called Great House in sector G6 of the city.[64] Archive A stored date-stamped fiscal documents from the 230s BCE down to the mid-second century, while Archive B contained material from the 180s BCE to the same period. Fragments of iron nails and bronze straps suggest that the documents were housed in wooden chests or niches set upon platforms of beaten earth. Both collections were deliberately burned at the end of Seleucid rule in Babylonia, most probably at the time of the Parthian conquest in the late 140s BCE—preserving, as elsewhere, only the clay *bullae* with which the documents had been sealed. In Archive B, the flammable contents on the earthen platforms had been reduced to ash before the ceiling collapsed, indicating purposeful arson from within. The case of Archive A is more intriguing: over and above evidence of burning, several

Map 3. Southern Levant

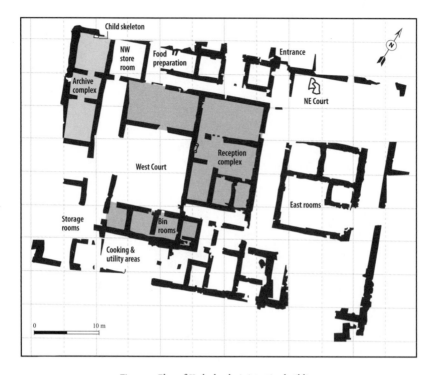

Figure 7. Plan of Kedesh administrative building

date-seal impressions were physically gouged out of the *bullae*. For their ex-cavator, "there can be no question but that the sealings and, presumably, the documents to which they had been attached were deliberately destroyed, not for purposes of cancellation but with hostile intent."[65] This was an effortful, malicious act that clearly went beyond what was necessary for the oblitera-tion of the parchment rolls.[66]

Our second case is the administrative center at Kedesh. Located in the Tyrian hinterland at the edge of the Hula basin in northern Israel (see Map 3), this monumental building had first been constructed in the early Achaemenid period.[67] It was repurposed by the Ptolemaic regime in the early third century BCE, following some decades of abandonment, as a more modest storage and distribution center for agricultural goods. The famous Ptolemaic adminis-trator Zenon twice visited it during his regional tour of 259 BCE, first to col-lect flour and then to take a bath.[68] Following Antiochus III's conquest and incorporation into the empire of the southern Levant at the very beginning of the second century, a small archive complex was inserted into the building's

Figure 8. Interred child, Kedesh archive

northwestern corner (Figure 7). It seems probable from the two thousand or
so *bullae* excavated from the room's floor that the administrators were local
Tyrian elites working on behalf of themselves and their Seleucid overlords. It
does not take too much imagination to reconstruct from the topography,
small finds, and architectural evidence a likely fiscal route of resentment:
local inhabitants were obliged to hike up the steep tel to the administrative
center on its ridge to deposit taxes in kind into plastered collection bins. On
entering the site through its narrow, northern doorway, taxpayers passed
by a lavishly decorated suite of reception rooms, outfitted with mosaic floors,
imported glass and ceramic wares, and painted, molded stucco, newly

constructed under the Seleucid regime. Records of payments and debts were then recorded on-site, sealed, and deposited in the archive.

Excavation revealed that this administrative building fell victim to two independent disasters in short succession. First, the site was suddenly abandoned during or shortly after 145 BCE, most likely as a result of Jonathan Maccabee's pursuit of Demetrius II's army to the immediate vicinity, as reported in 1 Maccabees.[69] Then, the archive, alone of the entire administrative center, was deliberately burned; so intense was the fire that the stones of the interior walls cracked from the heat. Before the conflagration, the room's doorway had been sealed up with stones and two infants buried within. One burial had been too disturbed for examination, but the other was sealed against the northern wall beneath a layer of burned brick debris. This child had been interred without grave goods and may have had its hands and feet amputated (Figure 8). Even when squatters briefly reoccupied the site, this archive room was never used again.

The burning of archives was a recurring if not frequent event in antiquity.[70] Where we have textual testimony on such destructions (for Dyme in contemporary Achaea and Jerusalem during the First Revolt), it is evident that the violence was as much directed at the political regime materialized through the archival procedures as the documents registered within.[71] But in our two Seleucid cases, both probably dating to the 140s BCE, the destruction seems to indicate an additional and rather shocking assault. I would suggest that the probable gouging out of Era dates before the burning of the Seleucia domestic archive points to an attempted symbolic annihilation of the archival administration, its date rings, and the temporal system of which these were manifestations. The motivations behind the deposition and possible amputations of the child at Kedesh remain obscure. Perhaps we are dealing with a magical defiling or purging—even a cosmological unworking—of the imperial archive:[72] certainly, we have seen that the Era year date was worn on officials' hands and that bodies were numericalized through digit counting. Perhaps it was an archaizing, antiquarian act, reviving a defunct religious tradition in the thrill of a newly won autonomy—in 141 BCE Tyre was declared *hiera* and *asylos*—much as the Maccabean wars renewed or were held to renew forms of violence from the Israelites' conquest of Canaan and cultic centralization (see Chapter 6).[73] I am reminded of the fifteenth historical thesis of Walter Benjamin, in which he recounts "new Joshuas" firing on the dials of clock towers during Paris's July Revolution ("De nouveaux Josués au pied de chaque tour / Tiraient sur les cadrans pour arrêter le jour"); or, with Michael Löwy's commentary on Benjamin's thesis, the report that in April 2000 members of an indigenous opposition movement shot arrows at the celebratory clock counting down to the five hundredth anniversary of the Portuguese "discovery" of Brazil.[74]

The destruction of these Seleucid archives seems as much about enacting a temporal ending as inciting bureaucratic amnesia. To anticipate a little what is to come, these symbolic, on-the-ground gestures of termination correspond to contemporary apocalyptic eschatology. For each, the empire's timeline—its linearized, open-ended, and institutionalized Era system—was a provocation to finitude. Indeed, as we will see in Chapter 5, the book of Daniel and 1 Enoch's Apocalypse of Animals adopt the archival imagery of sealing and filing away as their framing metaphor, thereby displacing sovereignty from the Seleucid king and administration to God and His angels.[75] This cleft door of resistance—a physically violent, this-worldly confrontation with empire; the theological fantasy of God's exclusive kingship on earth—will recur throughout our story.

The Market

We now leave the patrolled, exclusive, and idealizing archives for the most unrestricted and kinetic space within Seleucid urban life: the market. The

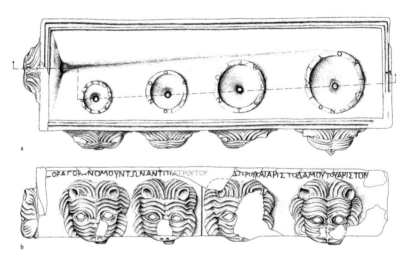

Figure 9. Stone *sekōma*, Maresha

volume and color of the ancient agora, from the haggling and stall displays to deft-fingered pickpockets and competing merchants, have been irrevocably lost to us. But the excavated, reusable devices of market trade permit at least the outlines of a hands-on reconstruction. The four artifacts introduced below can stand as representative of the enormous number of types and qualities of cultural technique by which the Seleucid Era pervaded commercial transactions.[76] All come from Levantine rather than Mesopotamian cities, on account of the intensity of archaeological excavation, the relative significance of market versus redistributive economies, and regional institutional reforms introduced by Antiochus IV.[77]

The first artifact is a stone *sēkōma* from a merchant's shop in the lower city of Maresha (Figure 9), the major Idumean settlement about forty kilometers southwest of Jerusalem.[78] This was a market device for measuring standard liquid volumes: four funnels of increasing size are labeled "sixteenth," "eighth," "quarter," and "double-*kotyle*," respectively; each is faced with a lion protome, to give the impression that the measured liquid poured from their mouths. A neatly carved Greek inscription, running in a line above this pride of protomes, gives the Seleucid Era year date in Greek alphabetic numerals, ordered in the peculiarly Seleucid smaller-to-larger system, and the names and patronymics of two *agoranomoi*, or market officials, in the genitive absolute: "[year] 170 (143/142 BCE), the *agoranomoi* being Antipa[ter, son of Helio]dorus and Aristodamus, son of Ariston[(L ορ' ἀγορανομούντων Ἀντιπά[τρου τοῦ Ἡλιο]δώρου καὶ Ἀριστοδάμου τοῦ Ἀριστον[)."

The second is the handle of a bag-shaped Hellenistic jar (Figure 10), discovered at Tel Aviv.[79] Incised on the handle, prior to firing in the kiln, was a small, roughly written, but easily legible two-line inscription in Greek, "L ηνρ' | με," which gives the Seleucid Era year date 158 (155/154 BCE), in its Seleucid ordering, possibly followed by the beginning of the name of the potter, trader, or *agoranomos* or the jar's contents.

The third is a round lead weight (Figure 11) from the still unlocated colony of Demetrias-by-the-Sea.[80] Arrayed around a dolphin and caduceus—iconographic markers of Apollo and Hermes, dynastic and mercantile patrons, respectively—is the inscription "ϑνρ', Δημητριάδος τῆς πρὸς ϑαλάσσῃ, ἡμιμναῖον," giving in Greek alphabetic numbers the Seleucid Era year date 159 (154/153 BCE), the name of the city where this weight was manufactured and used, and the half-mina unit that it guaranteed.[81]

Figure 10. Handle of a jar, Tel Aviv

Figure 11. Lead weight, Demetrias-by-the-Sea

And the fourth is a bronze coin of Apamea-on-the-Axios (Figure 12), bearing on the obverse the diademed portrait head of Alexander I Balas and on the reverse the image of Zeus holding a spear or scepter and a crested Corinthian helmet. The reverse carries the ethnic legend "ΑΠΑΜΕΩΝ" ("of the Apameans") and the Seleucid Era date "γξρ'" for year 163 (150/149 BCE).[82]

Figure 12. Bronze coin of Alexander I Balas, Apamea-on-the-Axios

What do these Seleucid Era dates do? The adversarial nature and information asymmetries of the buyer-seller relationship fixed distrust into the marrow of ancient commerce.[83] Indeed, marketplace suspicions may have been especially acute in the newly founded and ethnically hierarchized settlements of the Seleucid world, where customary sanctions presumably had a reduced capacity to regulate cheating. Accordingly, throughout the east Mediterranean and west Asia it had long been the responsibility of civic and imperial overseers to control and authorize weights, measures, and other devices of trade.[84] This vesting of integrity in the impersonal mechanisms of commerce, rather than in the face-to-face interactions of fellow traders, was achieved through visible guarantees marked onto the surfaces of market standards. These stamped symbols, countermarks, or names of civic officials indicated to participants in economic activities that a regulated and unitary system was at work. For instance, Athenian weights and measures bear on their surface images of the head of Athena, a double-bodied owl, wheels, turtles, crescents, cornucopias, knucklebones, *boukrania,* and so forth; several also carry the inscription "ΔΕΜΟ" or "ΔΗΜΟ" (an abbreviation of δημόσιον), indicating the sovereign civic authority.[85] Hellenistic amphorae from Rhodes, far and frequently traded in this period, carried the stamped name of an annual magistrate.[86] Recently discovered grain measures from Hieron II's Sicilian kingdom were impressed with the name of a government official along with a participial form of the verb ἀκριβάζω, "the one who measures precisely."[87]

Within the Seleucid empire, as our four artifacts demonstrate, the guarantee mark was the Seleucid Era year number, either standing alone or associated with imperial icons or local magistrates' names. In other words, the precise volume of a liquid, the carrying capacity of an amphora, the correct weight on a scale, and the supposed worth of a coin were all authorized by the imperial year date. This number, in its peculiar ordering, functioned as a shorthand for both the entire system of certifying bureaucratic institutions and the categories of calculable and exchangeable units. It is perhaps the first example of numericalized time serving as a commercial assurance or, to borrow Anthony Giddens's helpful phrase, a "faceless commitment."[88] These market devices, guaranteed by the visible Seleucid year date, either would themselves be regularly employed or would set the standard for shopkeepers' measures. We can even imagine these Era-marked weights and measures called upon as the unimpeachable signs of fairness to arbitrate conflicts. Note that, given the Era count's irreversibility, all objects guaranteed in this way were visibly self-aging, incorporating market activities into a historicizing depth of imperial time. In these ways, the imperial time system became the locus of the conventions and categories of fair exchange, the criteria of reliability and credence, and, fundamentally, political predictability and accredited knowledge.

Each of these market devices juxtaposed two different measures, the Seleucid Era date and whatever commercial unit it guaranteed (double-*kotyle*, mina, tetradrachm, and so on). Some kind of analogy seems to be at work here between a temporal regime that, for the first time, was systemized as a regular, numerical, and countable abstraction and those sites and activities where consistency and precision of units were required. Such quantification of prices, weights, volumes, and now years was a shared technique of simplification, a transformation of the world into a problem of arithmetic so that it could be solved.

This analogous relationship may be illuminated by the quality of "portability," as outlined by Paul Kockelman and Anya Bernstein: this property of portability describes the degree to which a semiotic technology is widely applicable and contextually independent, permitting one domain to be treated as if it were constituted by the same principles, or contained the same features, as another.[89] Strikingly, recognition of such ontological transposability between the Era count and standardized trading units can be found in contemporary apocalyptic writings. The book of Daniel, to which we shall return in far greater detail in Chapter 5, responds with enormous sensitivity to the often subtle ways of Seleucid power. Two passages explore the relationship between the imperial time system and the activities of market commerce.

In the book's second chapter, a sleep-troubled king Nebuchadnezzar summons his dream interpreters, exorcists, sorcerers, and Chaldeans to demand that they not only interpret his dream but first describe its details. When the wise men of Babylon protest that nobody could hope to meet such a challenge—although Daniel, of course, will succeed—the king interjects: "I know for a fact that you are buying time (*'iddānā' 'antûn zoḇnin*)."[90] The Aramaic Pe'al participle *zoḇnin* ("buying"), cognate with Akkadian *zibānītu* ("weighing scales"), is rendered in the Greek translations as "ἐξαγοράζετε," a denominative verb derived from the market square; the "time" that the wise men are supposedly purchasing is the definitive Aramaic *'iddānā'* and Greek καιρός.[91] This passage represents, to the best of my knowledge, the earliest attestation of the "buying time" idiom in all ancient literature and the source text for its employment in the New Testament.[92] As Anathea Portier-Young has observed, the king's threat "evinces a view of time as commodity that can be bought or gained at will,"[93] entirely at odds with the divinely controlled "full" temporality that the apocalypse will go on to unveil. Such commodification of time, in the voice of Nebuchadnezzar II, enunciates well the new logic of the dated Seleucid agora.[94]

In the similarly plotted court tale of king Belshazzar's feast, in chapter 5 of the book of Daniel, the Chaldean sages are again confounded and the Judean exile triumphant. While the king of Babylon and his thousand nobles and his wives and his concubines were making merry with the sacred loot from the first Jerusalem Temple, a disembodied hand suddenly appeared and inscribed mysterious words on the plaster wall of the palace. Only Daniel is able to read them—"*m^ene' m^ene' t^eqel û)arsin*," according to the Masoretic vowel pointing—and interprets them as follows: "God has numbered (*m^enâ*) your kingdom and brought it to an end; you have been weighed (*t^qil^etâ*) in the scales and found wanting; your kingdom has been divided [or assessed][95] (*p^erisat*) and given to the Medes and the Persians."[96] That very night, the chapter concludes, king Belshazzar was killed and the kingdom handed over to Darius the Mede, clocking forward from the first to the second of the four world empires.

The intricate graffito riddle embraces the polysemy that unpointed Aramaic permits. Modern exegesis, emboldened by the obvious inconsistency between Daniel's reading and the number and quantity of the written terms, has proposed additional or alternative interpretations.[97] For our purposes, it is key that the words, when first voiced as nouns, give the technical names of market weights: *m^ene'* for a mina, *t^qel* for a sheqel, and *p^eres* (here in the plural or dual) for a half-mina.[98] Daniel's own explanation of *m^ene'* directly links this

69

weight unit to the enumeration of time and, according to the apocalypse's theology, to the completion of the Neo-Babylonian empire's predetermined chronological allowance.[99] His interpretation of t^eqel adopts the imagery of a pair of weighing scales to visualize God's judgment.[100] Like a shopkeeper's faulty weights in a Seleucid market, Belshazzar and Babylon have been tested and found wanting. Indeed, it is likely that Daniel's interpretation plays on the Aramaic verb $tiqqal$—the second person Pe'al imperfect of q^elal meaning "you are underweight," in the sense of cheating the scales.[101] Finally, parsîn ("half-minas") is a transparent wordplay on the ethnicity of Babylon's conquerors; if it is the dual, "two half-minas," it neatly captures the yoke-team of the Medes and Persians. Additionally, the tale's market-weight terminology implies a comparative evaluation, presumably of the sequence of empires. Throughout the Hebrew Bible the language of weightiness and lightness expresses judgments of personal character, and we even find the specific terminology of market weights employed in the Babylonian Talmud to valorize academic lineage: "It is more proper that a mina, son of a half-mina, should go (for study) to a mina, son of a mina, than that a mina, son of a mina, should go to a mina, son of a half-mina."[102]

In sum, the famous writing on the wall pulls into a concentrated knot of mutual and intertwined identifications market weight units, the enumeration of time, and the fate of empires. This semantic portability, first attested in the Seleucid administrative practice of this period, is even more starkly expressed in a passage from 4 Ezra, an apocalyptic text first composed in Hebrew during the reign of the emperor Domitian.[103] Here, quoting a now lost, earlier text, the archangel Jeremiel answers the righteous souls who urgently inquire when the promised Eschaton will arrive: "When the number of those like you is completed, for He has weighed the age in the balance (in statera ponderavit saeculum), and measured the times by measure (mensura mensuravit tempora), and numbered the times by number (numero numeravit tempora)."[104]

The book of Daniel dramatically undoes the presumed imperial control of weighing, measuring, and the market, as it does the Seleucid archive and its authorizing procedures (see above). It is not simply that the language of Hellenistic government is displaced to God.[105] Rather, it is the specific material expressions of the Seleucid Era, and the calculative, instrumental, authority-vested time system these enunciate, that God's true sovereignty and His underlying historical plan expose as hollow and transitory.[106]

Inscriptions

Epigraphy forms our third and final field in which the imperial time system was made most visible and socially significant. The subjects of Seleucid Asia inhabited cityscapes and sanctuaries that were filled as never before with publicly inscribed writing—decrees, remembrances, agreements of alliance, poems, manumissions, letters, and so forth by private, civic, or imperial authors, located at temple entrances, on public buildings, in market squares, citadels, and graveyards, and beside roadways. This epigraphic habit, as it is known, though densest along the Aegean coasts and hinterlands of Asia Minor, operated throughout the entire imperial landscape and with the explicit promotion of the Seleucid authorities.[107] Scholarly attention rightly has been attracted more to the inscriptions' substantive contents than their dating formulae, which have been exploited only for their contributions to prosopography, political history, and reconstructions of bureaucratic practice.[108] But these publicly inscribed and prominently displayed Seleucid Era dates were among the most monumental and prestigious bearers of the imperial time system.

For our purposes, it is helpful to distinguish between two formal categories of epigraphic dating: lengthy, compound formulae at the beginning of inscriptions and curt, numerical displays at their end. The first type was used in civic decrees, alliances, manumissions, and other communal or personal documents, almost all authored by the empire's subjects.[109] The temporal location of these inscribed stones was fixed by the names of the imperial authorities and the Seleucid Era year number, often with the secondary addition of the holders of local priesthoods or magistracies. An inscribed decree from Xanthus, an ancient Lycian city on the southwestern coast of Asia Minor (see Map 4), can be taken as representative. It begins:

βασιλευόντων Ἀντιόχου καὶ Ἀντιό[χου]
τοῦ υἱοῦ, ϛιρ', μηνὸς Ὑπερβερεταίου
ἐπ' ἀρχιερέως Νικάνορος ἐν δὲ Ξάνθ[ωι]
4   ἐφ' ἱερέως τῶμ μὲν βασιλέων <Π>ρασί-
[δ]ου τοῦ Νικοστράτου πρὸ πόλεως δὲ
Τληπολέμου τοῦ Ἀρταπάτου· ἐκκλη-
σίας οὔσης κυρίας, ἔδοξεν Ξανθίων
8   τῆι πόλει καὶ τοῖς ἄρχουσιν . . .

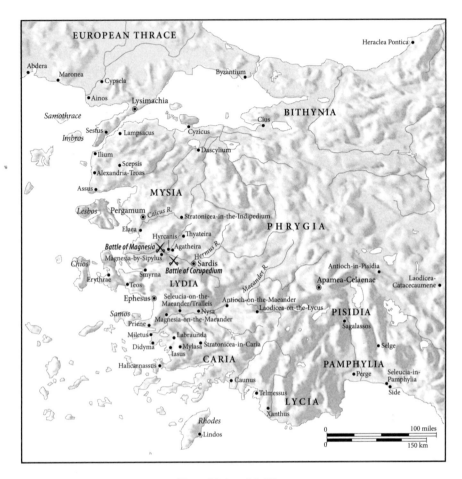

Map 4. Western Asia Minor

Antiochus (III) and Antiochus the son being kings, 116, month Hyperberetaeus, in the high-priesthood of Nicanor; in Xanthus, in the tenure of
Prasidas, son of Nicostratus, as priest of the kings, and of Tlepolemus, son
of Artapates, as priest *pro poleos*; in plenary assembly, it seemed good to
the city and magistrates of the Xanthians . . ."[110]

The inscription goes on to record the Xanthians' praise and honors for a
certain Themistocles, a talented orator from Ilium. The decree represents an
interaction between one *polis* and an individual from another *polis* that is in
no way mediated by the institutions, mechanisms, or agendas of the Seleucid

empire. Yet the dating formula—first, the supralocal world of kings, Seleucid Era year, Macedonian month, and provincial high priesthood, then the civic horizon of eponymous priests—locates this political procedure within the imperial time system. In other words, the date formula was a monumentalized acknowledgment of the community's subordination to the Seleucid kings and their world.[111] Note, moreover, that the compound formula juxtaposes the transcendence of the Seleucid Era—appearing here as the unqualified alphabetic number ϛιρ' (116)—with the bounded and socially embedded eponyms. While such compound formulae were common to the epigraphy of all the Hellenistic kingdoms,[112] the Seleucid Era number generated a uniquely deep and unified impression of empire. Compare, for instance, a manumission inscription from Seleucia-on-the-Eulaeus (formerly Susa) with an opening date formula of "βασιλεύοντος Σελεύκου, ἔτους λρ'" ("Seleucus being king, year 130")[113] to a dedication from the Egyptian Fayum on behalf of Cleopatra III and Ptolemy X, with the shallow and fractured regnal dates "L ιγ' τὸ καὶ ι'" ("[year] 13, which is also 10)."[114]

The second dating type is found on communications emanating from the royal court: imperial letters, responses to petitions, and commands of various kinds.[115] These inscriptions monumentalized on the surface of a single stone stele a series of epistolographic dispatches and supporting documents. Whether these instructions moved down the successive administrative cataracts or plunged straight to the most relevant on-the-ground official,[116] the missives' valedictions were typically followed by the date of their composition, expressed in the simple, standard imperial calendar: the alphabetically numbered Seleucid Era year (with its peculiar ordering), the Macedonian month name, and the day. The organizing function of these dates is easily apparent from even a brief glance at these dossiers (see, for example, the Heliodorus stele, Figure 13). Concluding each communication, set off to the right-hand border, often assigned its own line, and sometimes in a larger script, these dates gave immediate visual separation to the dossiers' constituent documents, much as our own e-mail platforms apply color changes or insets to a chain of messages. Additionally, for dossiers arranged according to a hierarchical rather than chronological logic, the engraved dates permitted reconstruction of the procedural sequence.

If our first epigraphic group (the compound formulae of civic and personal epigraphy) employed the imperial time system as a marker of political incorporation, the second type (the simplex dates of an imperial document

Figure 13. Heliodorus inscription, Maresha

chain) gave a symbolic unity to the spatially and chronologically distributed acts of empire. Each ordering event gained its meaning not from itself alone, but as part of a succession in a temporal continuum. These imperial dossiers, erected by command or permission in the most visible locations across the political landscape, can almost be seen as the temporal equivalents of milestones, both in demonstrating the existence of a coherent, extensive field of time and in identifying within it legible points of orientation. Moreover, the incorporation of the Seleucid Era dates into the authorized register of instructing, judging, decision making, and efficacious power-speech in general reasserted to what the time system alluded and from where it gathered its prestige. In contrast to the homogenizing and authorless discourse of civic decrees, these imperial communications, with their first-person voice, familiar forms of address, and courtly discretion, presented themselves as reproducing the personal will and oral speech of the king.[117] This is most apparent in certain document dossiers that give a monarchic monopoly over dating. For instance, the most complex Seleucid inscription of which we know is a stele, found at Hefzibah in northern Israel, that treats the territorial possessions of Ptolemaeus, first governor and high priest of the newly conquered province of Coele Syria and Phoenicia. This dossier consists of nine abbreviated documents and communications between kings Antiochus III and Antiochus the son, governor Ptolemaeus, and the provincial administrators Cleon, Heliodorus, Marsyas, Lysanias, Leon, and Dionicus, covering the five years from 201/200 to 196/195 BCE (Seleucid Era dates βιρ' to ζιρ').[118] Copies of the stelae were erected with explicit royal permission in the villages under Ptolemaeus's control.[119] Of all the Hefzibah dossier's documents, only those authored by the kings were dated.[120] That is, above the polyphony of local administrative voices we hear the distant king speaking the date: a visual and rhetorical reminder, reaching even to the humble hamlets of the Jezreel valley, of the time system's source and character: the year of our lord.[121]

Conclusion

The Seleucid empire was not timeless. It did not fashion itself as archetypal or self-mythologizing. It did not precede or transcend history. Far from it: more than its predecessor, peer, or even successor regimes, Seleucid kingship and governmental practice employed and encouraged the dating of the world. The Era was everywhere visible as a basic and necessary technology of administrative knowledge and urban life, well adapted to the enactment of concrete actions, from the archive to the agora, in civic decrees and royal speech.

It is tempting to frame this phenomenon within a history of abstraction, of the growing power of human reason to order and take the measure of the world. Certainly, the Era seems to have allowed an unprecedented quantification, neutralization, and even commodification of time. Characteristics quite unusual in the ancient world point toward what has often been considered diagnostic of a temporal modernity: the indifference and departicularizing of temporal referents, the disciplinary use of time, and so forth.[122] Yet the meaning, as well as the facticity, of this time regime is important. For its potency came from the combination of the categorical with the political: by the later third and second centuries BCE, at least within the kingdom's core, the Era had become a naturalized and inevitable way of understanding the world; concurrently, its royal origins and imperial coloring were everywhere reasserted.

The Seleucid Era, by being exclusively numericalized—that is, by avoiding eponyms or regnal identifiers—marked dates, not events. By structuring the kingdom's visibility and institutional practices around a continuous, irreversible, and predictable accumulation of years, it made the empire historical in a radically new sense, perhaps even the first truly historical state.

CHAPTER 3

Dynastic Time

We have explored the material presence of Seleucid Era numbers in three especially visible and significant domains. This final chapter of the book's first part will move from these quantitative aspects of the Era count to their qualitative transformations. It will argue that the Seleucid Era, in alignment with other expressions of imperial temporality, produced a distinctively Seleucid historicity, a specific system of thought and politics in which the Era was both representative and constitutive of a temporal regime that downplayed monarchic individuality and advanced a restorative, intergenerational, even protomessianic dynastic project.

A structural tension of "the king's two bodies," to employ Ernst Kantorowicz's influential formula, can be identified in most monarchic states:[1] between the individuality of the reigning monarch and the dynastic corporation of the kingdom; between the complexities of private personhood and the harmonies of abstracted office; and, in temporal terms, between the punctuated life cycle of royal biography and the dynastic linearity of inherited rule. These alternative orientations are woven through the practices and discourses of imperial administrations, court lifestyles, and royal ideologies, their paradoxical union brought most obviously under strain by the king's mortality—an inescapable problem of ideology and anthropology.

In various and predictable ways the Seleucid empire, much like its peer, predecessor, and successor kingdoms, conformed to the life-cycle tempo of its individual rulers. It is for some good historical reasons, as well for the metonymic spotlighting of our surviving evidence, that "personal monarchy" has been identified as the gravitational center and legitimizing principle of the Hellenistic kingdoms, with all that follows for the character of court and provincial interactions. Thus, the birthdays of Seleucid kings and queens were formally celebrated and their deaths mourned by the empire's subject communities. Coin portraits and royal epithets visually or verbally distinguished between homonymous monarchs. Kings' journeys functioned as chronological reference points. Royal accession could trigger a kaleidoscope-shaking detonation of violence, something between a test by ordeal of the new king's worth and his immediate exercise (characteristic of installation ceremonies everywhere) of the particular rights and capacities of the recently occupied office: the removal by assassination of potential challengers, the elevation to new vacancies of court favorites, provincial revolts, external invasions, new military campaigns.

These were the universal biorhythms of ancient monarchy. But where the Seleucid empire stands apart, and so where we must muster our historical attention, is in its relative downplaying of the death-accession moment and, indeed, of monarchic individuality itself.[2] An ineluctable effect of the Seleucid Era's unbroken linearity was an overriding emphasis on the dynastic continuum. The accession of a new king could never be chronographically celebrated, as it was in other kingdoms, as a new beginning. This was not simply the ongoing consequence of Antiochus I's decision to continue his father's year count. Even in origin, the Seleucid Era had been disassociated from the coronation event: as we saw in Chapter 1, Seleucus I chose to backdate the Era epoch from his assumption of the diadem to his heroic return to Babylon. Year 1 had never represented an enthronement.

This chronographic underplaying of death and accession was reinforced by the practice, again introduced by Seleucus I, of coregentship: the formal elevation of the living king's son or brother to monarchic status, often associated with command of a supraprovincial territorial block. This succession strategy of overlapping kingships—Antiochus I under and after Seleucus I, a Seleucus and then Antiochus II under and after Antiochus I, perhaps Antiochus Hierax under Seleucus II, Antiochus III under and after Seleucus III, Antiochus the son and then Seleucus IV under and after Antiochus III, Antiochus V under and briefly after Antiochus IV—eliminated all temporal gaps, both the "big interregnum" between monarchs and the "little interregnum" between the predecessor's death and the successor's full legitimization by installation. Such predeath baton passing, unstitching the royal life cycle at each end, contributed to the unusual dominance of the dynastic temporality: not the monopolized throne of "the king is dead, long live the king," but the shared scepter of "the kingship never dies."[3] Indeed, we may wonder whether this trivializing of accession and formalized multiplication of monarchs encouraged the dynastic chaos of the empire's closing decades, when brothers, cousins, and pretenders simultaneously grasped at a kingdom none could possess or embody.

The relative devaluing of monarchic individuality had three significant consequences for the empire's ideology of rule: first, it uncovered a kind of political abstraction previously unseen; second, it identified new moments of historical significance; and third, it emptied the kingdom of occasions for renewal.

The Seleucids' dynastic temporality allowed their kingdom to be imagined, even to its contemporaries, as a coherent political-temporal unit, an isolated block of continuous, homogeneous time. If we look back to the Saros Tablet of eclipse dates (Figure 2), discussed in Chapter 1, we can see that the Seleucid empire manifests itself quite unlike the Neo-Babylonian, Achaemenid, and postconquest Macedonian kingdoms: while these polities required the naming of the reigning monarch alongside his regnal year, the Seleucid empire is identified only by the syllable, prefixed with the male-determinative mark, ^{I}*Se*. This does not abbreviate the name of the enthroned monarch, as it is repeated without change through successive reigns. Rather, the tablet's author is groping toward a new possibility of political location: to represent a kingdom without naming the king, *basileia* without *basileus*.[4] In a curious way, it is the passage of time that has become the site of the immutable.

The book of Daniel, with its simultaneous sensitivity to imperial symbolism and capacity for ideological critique (see Chapter 5), gives a powerful image to this new experience of overarching dynastic time. In its seventh chapter, the crucial pivot from Aramaic court tales to Hebrew visions of the end-times, Daniel records seeing four beasts emerging from the great cosmic sea, of which the last is the Seleucid empire:

> (7) And after this I was watching in visions of the night and—behold!—the fourth beast, dreadful and terrible and exceedingly strong; it had great iron teeth; eating and smashing and trampling the remnant with its feet; it was different from all the other beasts that were before it; it had ten horns (*w^{e}qarnayîn ^{ca}śar lah*). (8) I was contemplating the horns and—behold!—another horn, a little one, came up in their midst; and three of the former horns were uprooted before it.[5]

Transparently, each horn represents a Seleucid monarch, ordered as a king-list down the beast's trunk like the plates of some Jurassic monster. The vision effectively resolves the two-body ambivalence of fragmented-monarchic and whole-dynastic temporalities: Daniel figures the Seleucid empire as a unitary body from which individual monarchs sprout, overlap, compete, and perish.

The second consequence is that this dynastic temporality opened a historical extension beyond the easy reach of other kingdoms. The Era's and empire's irreversible time system recognized high-range dynastic dates in addition to or in preference to the monarchic life cycle, making possible and significant new moments of imperial celebration. Three examples, in chronological order, should suffice.

First, the famous Borsippa Cylinder of Antiochus I is a cuneiform clay foundation text recording, in the king's own voice, his reconstruction of the Ezida temple of the god Nabû in the city of Borsippa, close to Babylon. The clay inscription dates itself to "^{iti}še ud.20.kám mu.43.kám," or 20th Addaru, year 43 (se) — that is, March 27, 268 bce — the very last days of the Babylonian calendar year.[6] The incorporation of a month-day-year date, itself a break with the Babylonian scribal tradition,[7] pulls this ritual of religious renewal into the kingdom's formal temporal framework; it belongs within the general Seleucid culture of precise and public dating. But the year number 43 also had special significance in local Babylonian history: the reign of Nebuchadnezzar II, the most glorious and best remembered of native kings, lasted forty-three years. The number was well known, being attested in the autobiographical inscription of Adad-guppi, mother of Nabonidus[8] (see Chapter 1) and, for the Hellenistic period, in the *Uruk King List*[9] and Berossus's *Babyloniaca*.[10] Antiochus I's reconstruction of the Ezida — a ritual and inscription that made deliberate allusions to Nebuchadnezzar II and his building foundation texts — in Seleucid Era year 43 marks a deliberate engagement, even competition, with the potent legacy of the long-reigning local hero.[11] Only ten days later, at the Babylonian New Year *akītu* festival in which Antiochus I surely participated, the Seleucid empire would clock forward into year 44. Needless to say, the significant date is the cumulative Seleucid Era number, not the regnal count of Antiochus I.

Second, the greatest of Seleucus Nicator's successors, king Antiochus III, spent seven years restoring Seleucid authority in the Upper Satrapies and the Persian Gulf (his so-called second *anabasis*) and then arrived at Babylon to join the city's New Year *akītu* in Nisannu 107 se (205 bce), as we have seen in Chapter 1. This is our only explicitly attested episode of a Seleucid monarch personally participating in the festival. Since ascending the throne, Antiochus

III's political actions, legitimizing discourse, and even numismatic types had been directed toward reclaiming the provincial landscape and repeating the foundational acts of his great-great-grandfather, Seleucus I. Nisannu 205 BCE marked the precise centennial of the real-world inauguration of the Seleucid Era count and Seleucus I's self-coronation as *basileus*. In other words, Antiochus III incorporated himself into the New Year ritual paradigm at the precise moment when the empire and Era had been born one hundred years earlier (see Chapter 1). His participation should be understood as an ideologically motivated reenactment and rejuvenation, in much the same way as his recently forged peace with the Indian king Sophagasenus repeated and renewed Seleucus I's treaty with Chandragupta Maurya.[12] Indeed, it is worth considering the possibility—the chronology permits this, but the relevant Astronomical Diary is still unknown—that Antiochus III initiated his eastern expedition by participating in the Babylonian *akītu* of Nisannu 101 SE (211 BCE), the centenary of Seleucus I's inaugural return to the city and the idealized epoch of the Era.[13]

Third, as we have seen in Chapter 2, Seleucid Era alphabetic numbering in Greek was distinguished by its unique smaller-to-larger ordering. This regular, imperial arrangement is attested on objects of market trade at the Idumean city of Maresha, with the exception of six weights, all guaranteed by the year mark 205 SE (108/107 BCE). For this year alone—when Maresha, long a Seleucid stronghold, was under the immediate threat of Hasmonean conquest[14]—the alphabetic numbering was reversed: what would standardly have been represented as εσ' appears as σε'.[15] Additionally, each of these six weights has as obverse decoration the distinctive outline of a Macedonian shield.[16] I would suggest that the Seleucid alphabetic numbering was reversed so as to allow the Era date σε' (*se*) simultaneously to represent the year number and, by its abbreviation of the imperial dynasty's name—identical to that on the Saros Tablet and elsewhere—its ultimate and original political referent.[17] In Maresha's last days this reverse *gematria*, together with the weights' ethnically marked blazon, asserted a Seleucid, Macedonian, and stridently non-Judean identity. Indeed, there may be some confirmation of this concentrated ideological force in the deliberate breaking and shearing through of one of these *se*-dated shield weights, presumably soon after the Hasmonean takeover (see Figure 14). Rather than being cut for recycling, this appears to be evidence of a will to erasure,[18] a kind of violence against temporal symbols we have already seen in the Kedesh and Seleucia-on-the-Tigris archives (Chapter 2).

Figure 14. Cut-up σε' lead weight, Maresha

In other words, the Seleucid Era, alongside its denotational function as a quantitative technology of dating, with the year number as a position in an accumulating sequence, could also be a connotational marker of historical meaning, with the year number as an event. This was a numerology of empire that identified thresholds, fateful episodes, and historical rhymes. Dynastic time brought to the surface the punctuated moments of imperial significance that regnal dating would have left invisible.

The third and most profound consequence is that the relative undervaluing of the monarchic life cycle effectively eliminated the opportunity for the empire's periodic re-creation in ritual. Historians, anthropologists, and theorists of religion working in the traditions of James Frazer and A. M. Hocart have identified in the cyclical temporality of monarchic death-accession the reconciliation of social community to nature's rhythms of growth and decay. The prosperity of a people required the king to die, it is argued, for the very rituals of coronation that elevated his successor to the royal office would also bring about, in some stronger or weaker sense, as ideological fiction or political reality, the springtime of the state.[19] In Mircea Eliade's famous formula, the enthronement of a monarch was a "myth of the eternal return," an antihistorical annulment of time that sought to restore—if only momentarily—the dawn of first ages: "Almost everywhere a new reign has been regarded as a regeneration of the history of the people or even of universal history. With each new sovereign, insignificant as he may be, a 'new era' began."[20] However idealizing this statement, however dubious its agenda,[21] it succeeds in capturing something of the political theology of the Seleucids' neighboring regimes. Most visibly in the hybrid kingship of the Ptolemies, the return of Year 1, in all its rosary-bead cyclicality, achieved some kind of restoration of relations and re-emergence of original states. The trilingual priestly decrees configure this ideology in the traditional mode of a Horus-pharaoh reconfirming the divinely sanctioned order. Hence, according to the Rosetta Stone, at a moment of severe crisis, Ptolemy V "restored to order whatever things were neglected in former times"; "he took [Lycopolis] by storm and destroyed all the impious men in it, just as Hermes and Horus, the sons of Isis and Osiris, subdued formerly those who had rebelled in the same places"; "and of the revenues and dues he receives from Egypt some he has completely remitted and others he has reduced . . . and he has freed those who were in the prisons and who were under accusation for a long time from the charges against them."[22] This ideology of cyclical renewal even breaks the surface of Polybius's pragmatic history, whose account of the Majority of Ptolemy V reechoes with such expectations: "The palace officials . . . at once began to occupy themselves with the celebration of the king's Proclamation. Although his age was not such as to make it pressing, they thought that the state would achieve a certain restoration (νομίζοντες δὲ λήψεσθαί τινα τὰ πράγματα κατάστασιν) and that it would be, once again (καὶ πάλιν), the beginning of a change for the better (ἀρχὴν τῆς ἐπὶ τὸ βέλτιον προκοπῆς)."[23] The same historian judged king Perseus, the last of the Antigonids, to be "truly royal (ἐπέφαινε . . . τὸ τῆς βασιλείας ἀξίωμα)," for canceling all debts, releasing prisoners, summoning back exiles, and restoring their original property immediately on his accession in 179 BCE.[24]

All this is absent from the Seleucid empire. While the fragile and fragmentary state of our evidence demands the utmost caution, nothing in extant royal inscriptions, civic decrees, iconography, archaeology, or Greek, Latin, Jewish, and Babylonian historiography (much of which engages with the empire's fraught succession moments) suggests a cyclical, typological coronation reset of this kind. The overriding impression from all attested Seleucid investitures is of the arbitrariness of timing, the variety of location, the insignificance of ritual, and the lack of communal participation or witnessing.[25] As far as can be reconstructed, Seleucid history and empire ran only forward in an accumulation of irreversible actions that, even when regretted, could never be symbolically undone.[26] As an example, the well-attested attempts by mid-second-century BCE kings to restore Seleucid dominion over Judea, reversing the cultic persecution, are represented in preserved royal letters, the historiographic record, and its apocalyptic transposition as fundamentally cumulative and historic. The narratives of 1 and 2 Maccabees, dated by the ongoing Seleucid Era, are in no way organized or segmented by the accession of new kings.[27] This emptying from the kingdom of prescribed and institutionalized occasions for restoration, rebalance, and forgiveness—a social therapy for which the subject populations must have yearned: recall the destroyed fiscal archives of Chapter 2—may go some way to explaining the forceful reemergence, displaced to the kingship of God, of a Year 1 ideology in the prospective visions of second-century BCE apocalypticism.

A second, related absence—again visible only when seen in the context of other ancient monarchies—was heroic priority. Egyptian, Mesopotamian, Achaemenid, and other Hellenistic kings regularly proclaimed their unparalleled surpassing of predecessors' achievements, of course with profoundly different emphases and agendas between and within dynasties, from the Neo-Assyrian program of "enlarging the borders"[28] to Ptolemaic claims of unprecedented elephant hunting[29] or Antigonid assaults on Olympus.[30] This was not present within Seleucid royal fashioning. Instead, we find an ever-unfulfilled aspiration: to strive to seek to match the imperial achievements and conquests of the kingdom's founder, Seleucus I Nicator.

The Royal Mission

The primary focus of the Seleucid kings lay in reasserting and confirming what had originally been won. The unbroken dynastic temporality produced an intergenerational royal mission, constructed out of an internal Seleucid historical framework and analysis. The strange texture of Seleucid temporality demarcated a closed historical field: despite its fragmentary state of survival, official memory seems to present a fairly coherent boundary of recollection, beyond which it did not reach. This temporal field, generated at court, propagated throughout the empire, and reenunciated by subject communities, took as the limit point of appropriate historical reference the Seleucid Era epoch or, as prologue, the early life of the kingdom's founder, Seleucus I Nicator. Such a "Seleucid horizon" generated an exclusive, totalizing, and homogeneous present of empire, an enclosed, self-referencing temporality that blocked out other viewpoints and constructed for itself a recursively linked past and future. We can see it at work in various official forms of Seleucid memory: royal communications, court historiography, and dynastic cult.

The Seleucid language of precedent, whether deployed to justify conquest or to negotiate status, was exclusively restricted to the ruling dynasty. In extant royal inscriptions and the historiographic record of diplomatic argument, the reigning king's forebears were collectively invoked as πρόγονοι (ancestors)[31] and πατέρες (fathers),[32] or specifically identified by the kinship terms πατήρ (father),[33] πάππος (grandfather),[34] or ἀδελφός (brother).[35] So, in a recently published, very fragmentary royal letter from Drangiane (modern Seistan, on the Iranian-Pakistani-Afghan frontier), king Seleucus II invoked the precedent of his grandfather with respect to certain decisions and rescripts.[36] In a better-known inscription from Asia Minor, Antiochus III used identical language in establishing a high priest over the cis-Tauric sanctuaries.[37] In each case the emulated πάππος (grandfather), respectively Antiochus I and Antiochus II, was left unidentified, coloring provincial interventions with the closeness of a familial frame and the anonymized, "steady-state" achronicity of biological regenesis.[38]

Cities and other incorporated communities picked up this language of dynastic precedent, expressing gratitude or encouragement for kings acting in accordance with the generous example of their ancestors.[39] Of course, the experience of ancient or recent non-Seleucid benefaction was relevant for many of these populations, especially on the much-disputed far western frontier, from which the bulk of our epigraphic evidence comes. But even in these limit cases, where subject communities invoked a pre-Seleucid past, the discursive strategies of the Seleucid horizon can be seen to be at work. Take, for instance, the negotiations between the ancient Ionian *polis* of Erythrae and king Antiochus II, as preserved in a royal letter to the city. The Seleucid monarch confirmed a range of benefactions largely on the basis of historical precedent. In an extended motivation clause, Antiochus declared:

And since Tharsynon and Pythes and Bottas (the envoys from Erythrae) pointed out that your city, under Alexander and Antigonus, was autonomous and free from tribute (ἀπέφαινον διότι ἐπί τε Ἀλεξάνδρου καὶ Ἀντιγόνου αὐτό[ν]ομος ἦν καὶ ἀφορολόγητος ἡ πόλις ὑμῶν), and (since) our ancestors were always zealous on its behalf (καὶ οἱ ἡμέτεροι πρόγο[νοι] ἔσπευδον ἀεί ποτε περὶ αὐτῆς), we, observing that they determined justly and not wishing to fall short in benefactions (θεωροῦ<ν>τες τούτους τε κρί[ναν]τας δικαίως καὶ αὐτοὶ βουλόμενοι μὴ λείπεσθαι ταῖς εὐερ[γεσ]ίαις), shall assist in preserving your autonomy and tax-exemption.[40]

The inscription establishes an antithetical contrast between not the policies, but the temporal fields of, on the one hand, Alexander the Great and Antigonus Monophthalmus and, on the other hand, of Antiochus II and his *progonoi*.[41] (Lysimachus, who ruled Erythrae for the two decades between Antigonus and Seleucus I, goes unmentioned.) Here, in the voice of Antiochus II, the policy of Alexander and Antigonus toward Erythrae is simply a piece of testimony, local evidence brought before the king by the city's ambassadors, external to Antiochus II's own memory and concern. By contrast, the Seleucid king could confidently declare how his own ancestors had acted; in a knowledgeable and approving autohistoriographic stance, Antiochus II gazes upon and evaluates his dynasty's past. Note that the contrasting verbal subjects are "οἱ ἡμέτεροι πρόγο[νοι]" ("our ancestors") and, not the earlier Macedonian conquerors, but "ἡ πόλις ὑμῶν" ("your city"); Alexander and Antigonus are denied the royal title and demoted to an adverbial clause.[42] Such strategies of distinction—syntactic imbalance and omitted titles—sunder the pre-Seleucid past from the temporal isolate of the dynastic present. This Seleucid historical horizon would have been reinforced further by the Seleucid Era date with which the king's letter would have concluded.

Antiochus II's autohistoriographic gaze is paralleled in a number of sources, especially those dealing with the thickly attested Jewish-Seleucid interactions. According to 2 Maccabees, Antiochus IV explained in his final letter to the Jews of Judea that the elevation of his son to coregent status was motivated by the example of Antiochus III: "seeing that my father (θεωρῶν δὲ ὅτι καὶ ὁ πατήρ), at the time he marched into the Upper Satrapies, appointed a successor."[43] Even if not genuine, this valedictory epistle figures the reigning king looking backward upon and then emulating a specific decision of his (unnamed) predecessor. In a letter of Antiochus III to Zeuxis, his general-commander of cis-Tauric Asia Minor, the king proclaimed his wish to transfer two thousand Babylonian and Mesopotamian Jewish families to the troubled regions of Lydia and Phrygia. Antiochus writes, "I am persuaded (πέπεισμαι)" that the Jews would make loyal guardians of Seleucid interests on account of, first, their piety toward God and, second, because "I know that they have had the testimony of my ancestors to their trustworthiness and eagerness to do as they are asked (μαρτυρουμένους δ' αὐτοὺς ὑπὸ τῶν προγόνων εἰς πίστιν οἶδα καὶ προθυμίαν εἰς ἃ παρακαλοῦνται)."[44] When the hopes of Antiochus III were disappointed and Jewish piety turned against the empire, dynastic history was again invoked. The Friends of Antiochus VII Sidetes, who was conducting a

siege of Jerusalem in 134 BCE, "reminded him of the hatred felt by his ancestors toward this people (ὑπέμνησαν δὲ αὐτὸν καὶ περὶ τοῦ προγενομένου
μίσους τοῖς προγόνοις πρὸς τοῦτο τὸ ἔθνος)" and "related in detail (διεξιόντες)"
Antiochus IV's persecution in order to encourage its repetition.[45] The verb
διέξειμι is characteristic of historiographic expounding.[46] Similarly, according
to Polybius, Antiochus III's battlefield exhortation to his assembled troops at
Raphia, in 219 BCE, dwelt on the glory and achievements of his ancestors ("τῆς
δὲ τῶν προγόνων δόξης καὶ τῶν ἐκείνοις πεπραγμένων ἀναμιμνήσκοντες").[47]
We have here the faint outlines of a particular royal pose: the king and his
court publicly engaged in a dynastic remembering—contemplating their own
past, trusting their ancestors' testimony, and emulating their good example.

This intradynastic interest is confirmed from the few surviving scraps of
court-sponsored historiography. The literary output of the Seleucid empire's
pioneering phase, in the early decades of the third century BCE, had been
characterized by geographic and ethnographic writing, as Megasthenes, Pa-
trocles, Demodamas, and others attempted to forge a legible and meaningful
imperial landscape.[48] But by the high empire of the later third century BCE,
court literature had taken a turn toward historiography.[49] Like their royal patrons, these court historians seem to have fixed their gaze resolutely within
the Seleucid horizon, narrating the origins of the dynasty, the heroism of its
earlier kings, and thereby the precedents for current restoration. Thus, in the
court of Antiochus III, a certain Simonides, an epic poet from the Seleucid
colony of Magnesia-by-Sipylus, sung of Antiochus I's victory over the Galatians, perhaps in the famed Elephant Battle described in Lucian's *Zeuxis*.[50] Hegesianax of Alexandria Troas, appearing in Polybius and Livy as Antiochus III's
trusted Friend and ambassador to the Romans,[51] wrote of the crossing of the
Galatians from Europe in the aftermath of Seleucus I's death.[52] This Galatian
invasion of Asia Minor was narrated in thirteen books by Demetrius of Byz-
antium, who went on to compose, perhaps in continuation, a history titled
On Antiochus and Ptolemy and Their Settlement of Libya that dealt with Antiochus I Soter's support of Magas during the First Syrian War.[53] While we
know nothing more of Demetrius, his historical focus suggests a Seleucid affiliation.[54] Mnesiptolemus, another historian associated with Antiochus III,[55]
who loyally named his son Seleucus, incorporated into his *Histories* personal
autopsy of king Seleucus II's or III's drinking habits, indicating an earlier presence at court.[56] Protagorides of Cyzicus described the extravagant downriver
voyage of a king Antiochus, probably Antiochus IV Epiphanes during the

Sixth Syrian War, and the great festivities at Daphne near Antioch.[57] An otherwise unknown Timochares treated, in his *On Antiochus*, the defensive topography of Jerusalem. Together with the work's title, his exaggerations of the city's size (forty stades in circumference) and natural situation—"it is difficult to conquer (δυσάλωτον), since it is surrounded on all sides by precipitous ravines (πάντοθεν ἀπορρῶξι περικλειομένην φάραγξι)"—indicate an encomiastic biography of either Antiochus IV, who plundered the Judean capital, or, more likely, Antiochus VII, who besieged it.[58] Additionally, the surviving fragments of official city-foundation narratives, centered entirely on the potent image of the builder-king, seem to have omitted all pre-Seleucid precedent.[59] The evidence is undeniably scanty, but no Seleucid court histories of Alexander the Great or Asia's pre-Hellenistic past are known.[60]

Finally, this coherent and exclusive block of Seleucid time was given religious formulation. The political myth of Seleucus I's divine paternity, already operative in the early third century,[61] functioned as a mechanism of cultural uprooting and historical erasure: the founder king was granted not a Macedonian and mortal father but the divine and universal Apollo. As we cannot stare behind the sun, so we cannot see beyond a god: there was no history before the founder's birth. Kingship descended from heaven. Similarly, the ancestor cult, introduced at a state level by Antiochus III[62] and attested in its civic form at Antioch-in-Persis, Seleucia-on-the-Tigris, Seleucia-in-Pieria, Dura-Europus, Scythopolis, Samaria, and Teos,[63] did not reach back before Seleucus I.[64] These cults of divine ancestors, into which the reigning monarch was sometimes incorporated, honored an ordered succession of god-kings. For instance, a priest list from Seleucia-in-Pieria, dating to the reign of Seleucus IV, names a certain [- -]ogenes, son of Artemon, as priest of "Seleucus (I) Zeus Nicator and Antiochus (I) Apollo Soter and Antiochus (II) Theos and Seleucus (II) Callinicus and Seleucus (III) Soter and Antiochus (the son) and Antiochus (III) Megas."[65] The cultic activities, the incorporation of local elites as its priestly agents, and the unifying king-list structure of the cult recipients, opening with Seleucus I and omitting all predecessors, pretenders, or regents, formed a kind of authorized historical review in religious mode, a closed series of the imperial past.

It is worth pausing to recognize how strange and exceptional was this dynastic here and now, walled off from past regimes or precedent and absorbed in its own reflection: the Seleucids' peer Hellenistic kingdoms contrast with its every element. Space permits only the briefest of surveys. The perennial rival, the Ptolemaic empire, publicly embraced both its Argead and Nectanebid heritages. Ptolemy I himself composed a history of Alexander the Great's *anabasis*, seized and buried the conqueror's body, incorporated him into a state cult, and established his own genealogical link to the Argead house and perhaps even to Philip II.[66] The god Dionysus, at the dynasty's head, was relegated far back to the primordial Macedonian forests, some twenty-two generations before the son of Lagos took the throne.[67] Similarly, the sarcophagus of Nectanebo II, the last Egyptian pharaoh, was moved to the Ptolemaic dynastic mausoleum in Alexandria;[68] the Alexander Romance, which may go back to the third century BCE, insists on the Macedonian conqueror's filiation from this Nectanebo;[69] and the cult of Nectanebo-the-Falcon grew to prominence under the Ptolemaic monarchs.[70] Up in Macedonia, in the Antigonid kingdom, Antigonus Gonatas rebuilt the great Argead tumulus at Aegae, which had been plundered by Pyrrhus's Galatian garrison in the mid-270s BCE,[71] and established on Delos an ancestor monument that lined up in genealogical order over twenty statues, reaching to an ancient divine or heroic progenitor.[72] Philip V arranged a digest of those passages of Theopompus's history that treated Alexander's father, Philip II,[73] and explored through his own antiquarian research ("κατὰ τὴν ἱστορίαν") the original costume of the ephebic "hunters of Heracles."[74] The Attalids of Pergamum promoted the myths of Telephus, Pergamus, and Heracles to give deep territorial legitimacy to their imperial claims.[75] A statue group at Delos set Attalus I and Eumenes I among the local heroes and personified places of their capital region, each with a lengthy genealogical inscription.[76] Their Galatian victories were cast as but the latest of Graeco-barbarian confrontations: for instance, the Attalid dedication on the Athenian acropolis, a kind of universal myth-history in bronze, celebrated in a series of statue groups the Gigantomachy, Amazonomachy, Persian Wars, and Galatian Wars.[77] The minor post-Seleucid kingdom of Commagene also established a deep past for itself: the *hierothesion* of Antiochus I of Commagene, at the top of Nemrud Dağı, was another *progonoi* monument, displaying the king's fifteen paternal Persian ancestors (beginning with Darius I) and seventeen maternal Macedonian ancestors (opening with Alexander the Great and continuing down the line of Seleucid kings to Antiochus VIII Grypus and his daughter Laodice Thea Philadelphus).[78] Even the pedigree coinage of the Graeco-Bactrian king Agathocles, a remarkable series that reproduced in his own name the portraits and reverse types of the region's previous Hellenistic rulers, incorporated Alexander the Great into the line of honored kings.[79]

By contrast, the Seleucids' pervasive and exclusive orientation to their dynastic present meant that they owed no historical or moral debt to their predecessors. The empire was grounded only in itself, implying not so much its illegitimacy as the impossibility of its delegitimization. In this way, Seleucid temporality comes to resemble significant and well-studied features of Islamic time: the total discontinuity of a new beginning and the consigning of all pre-epochal phenomena to the historical basement of a *Jahiliyya* or "age of ignorance."[80]

The Seleucid imperial temporality, constantly turning back to its origin, required an accounting in historical terms of the kingdom's transformation, of the changes wrought by time. For the bizarre situation of the early Hellenistic period gave the Seleucid state a trajectory of territorial shrinkage from the first, a decline almost coeval with its foundation. In the empire's opening three decades Seleucus Nicator carved a political landscape out of the maximal Macedonian conquests of Alexander, from Central Asia to European Thrace, but his successors, despite ephemeral expansions and short-lived reconquests, inherited a progressively diminishing sovereign space. Indeed, the empire's linear time indexed its own deflation with unforgiving clarity, identifying as the point of retrospective comparison not the individual reigns of preceding rulers but the empire's original and early condition.

Greek historiography has preserved traces of the emic, internal Seleucid explanation for the empire's unfortunate trajectory. Polybius, Appian, Diodorus Siculus, and Strabo account for the breaking away or loss to rivals of European Thrace, the Chersonese, Cappadocia, Commagene, Hyrcania, Parthia, Bactria, and the eastern provinces in general as the result of royal distraction. In a fairly coherent vocabulary, we are repeatedly told that the kings were occupied elsewhere.[81] Especially prominent are the verb περισπάω, in the passive voice ("to be distracted" or "to be drawn away"), and its derivative noun περισπασμός ("distracting circumstances").[82] Diodorus Siculus, for instance, recounts that Ptolemaeus, the Seleucid governor of Commagene, could revolt in the mid-second century without fear of the Syrian kings "because they were distracted with their own affairs (διὰ τοὺς ἰδίους ἐκείνων περισπασμούς)."[83] Antiochus III, according to Polybius, responded to the Roman L. Cornelius Lentulus's querying the legitimacy of the king's conquests in Asia Minor and European Thrace with the argument that these cities or provinces had forcibly been wrested away by Ptolemy III and Philip V ("πρῶτον μὲν Πτολεμαῖον παρασπασάμενον σφετερίσασθαι τοὺς τόπους τούτους, δεύτερον δὲ Φίλιππον") because of "the distractions faced by his own ancestors (κατὰ δὲ τοὺς τῶν αὐτοῦ προγόνων περισπασμούς)."[84] The jingle of the Ptolemaic and Antigonid παρασπασάμενον ("wrested away," from παρασπάω) with the Seleucid περισπασμούς ("distractions," from περισπάω) suggests the formulation's lexical significance. So pervasive was this mode of explanation that it was even deployed against the Seleucid king Antiochus VII by the rebel Simon Maccabee.[85]

This Seleucid theory of their own retreat stands apart from external evaluations. Since the *polis* paradigm of ancient political theory was inadequate to the territorial disintegration of grand empires, external or retrospective analysis of Seleucid decline was channeled instead toward a tribunal-like moral censure. Take Strabo's thundering denunciation of the rise of piracy in late Seleucid Cilicia: the geographer-historian inveighs against, in succession, the worthlessness of the kings ("ἡ τῶν βασιλέων οὐδένεια"), the conflicts between the dynasty's senior and cadet lines ("διχοστατοῦντές τε ἀδελφοὶ πρὸς ἀλλήλους"), the wickedness of the administrators ("κακίᾳ τῶν ἀρχόντων"), and the rebellious activity of Diodotus Tryphon and others ("τῷ γὰρ ἐκείνου νεωτερισμῷ συνενεωτέρισαν καὶ ἄλλοι").[86] This accounting by decadence was, transparently, not the story the Seleucids told themselves about themselves. Rather, the Seleucid *explanans* of royal distraction euphemized the imperial trauma: its generic, anonymous language ascribed no blame and recognized no rivals; it refused either to look history in the eye or to give it shape; it projected decline as a temporary aberration; and so it allowed for and encouraged its reverse.

For all the reasons outlined so far—from the historical model of distraction to the absence of the coronation reset, from the Seleucid memory horizon to the downplaying of monarchic individuality—the guiding program of Seleucid monarchy, and an ever heavier charge to pass down the generations, was the reestablishment of the political landscape closed by Seleucus I and Antiochus I. This agenda had already found explicit formulation under Antiochus I Soter, shoring up the kingdom after his father's assassination at the Hellespont: an honorific inscription from Ilium, reproducing the language of the court, praised the king because "he sought to restore the cities throughout the Seleucis . . . to peace and their former prosperity (ἐζήτησε τὰς μὲν πόλεις τὰς κα(τὰ) τὴν Σελευκίδα . . . εἰς εἰρήνην καὶ τὴν ἀρχαίαν εὐδαιμονίαν) . . . and to recover his paternal dominion (ἀνακτήσασθαι τὴμ πατρῴιαν ἀρχήν)."[87] According to the local historian Nymphis of Heraclea Pontica, Antiochus I "in many wars, though with difficulty and not in its entirety (εἰ καὶ μόλις καὶ οὐδὲ πᾶσαν), restored his paternal kingdom (ἄνασωσάμενος τὴν πατρῴαν ἀρχήν)."[88] Subsequently, wherever historiographical or epigraphic data permit access to the royal voice of Seleucid policy, we find motivation by this master program. It is abundantly and volubly demonstrated for the career of Antiochus III.[89] Similarly, Antiochus VII Sidetes, according to 1 Maccabees, expressed a wish to lay claim to his kingdom "so that I will restore it as it formerly was (ὅπως ἀποκαταστήσω αὐτὴν ὡς ἦν τὸ πρότερον)."[90] It is the most plausible guide for the military campaigns of, at the very least, Antiochus I, Seleucus II, Antiochus III, Antiochus IV, Demetrius I, Demetrius II, and Antiochus VII. These attempted restorations of the original imperial territory amounted to an ideal of kingship: an intergenerational mission that reconciled a royal legitimacy located in both traditional and charismatic elements—that is, in the inheritance of territory, lineal descent from Seleucus I, and respect for the empire's memory traditions with the victorious conquests and personal divinity of a warrior on horseback. Understood this way, all Seleucid campaigning was a sort of reenactment or historical echo of the empire's foundation, looping back to and reaching for its grandest beginnings. Vincent Gabrielsen has effectively characterized this as "the perpetuation of 'creation' of the empire by means of its recurrent 're-creation.'"[91] The crushing of revolts and the reintegration of territories allowed the Seleucid monarch to act, at one and the same time, as both protector of a legacy and its conqueror.

A kingdom in decline; a dynastic temporality permitting neither pause nor restart; a self-enclosed, self-referencing historical field; a privileging of the Golden Age of imperial foundation; an emic accounting of territorial shrinkage; and a retrospective, inherited, and ultimately unattainable royal mission of restoring the original dispensation: the empire's linearized, forward-rolling temporality seems to have provoked in its rulers, just as it would in its subject populations (see Part II), the concomitant desire to turn back the clock. Like Walter Benjamin's first historical thesis—the famous parable of the little hunchback hiding beneath the automaton's chess table[92]—Seleucid dynastic time, alongside its rationalizing, neutralizing quality explored in the previous chapter, approaches if not quite messianism, then the messianic situation.

Confirmation by Contradiction

The invention and comprehensive employment of the Seleucid Era had re-vealed to the ancient world, or at least to a good part of it, a new discovery: on the one hand, the simple effectiveness of continuous, transcendent time and countable, open futurity for process and thinking; on the other hand, a hitherto underexplored medium of political legitimization. This strange double identity of the Seleucid Era—as both a naturalized chronographic technology and an assertion of an imperial worldview—underlies the modal-ities by which the empire's rivals and rebels attempted to cut it short or to transpose and adapt it.

Take the case of Diodotus Tryphon, the Seleucid army commander who, after championing for two years the cause of king Alexander I Balas's son, Antiochus VI, in 141 BCE committed his unfortunate royal ward to the surgeon's table and took the throne in his own name.[93] In contrast to most other rebels from within, who anxiously sought legitimacy in dynastic terms, Diodotus attempted to pull down the very pillars of the kingdom's temporal edifice.[94] His monarchic self-styling represented an openly pug-nacious break with the preexisting imperial structures of authority. Di-odotus did not pretend to Seleucid kinship or take a dynastic throne name. He added the unprecedented epithet *autokratōr* to the royal *basileus* title, appearing on his coinage as Diodotus Tryphon, "the self-empowered" king.[95] These coins introduced a novel iconography—a cavalry helmet, from which sprouted a single large goat horn (Figure 15)—that engaged with the symbolism of divine kingship and, perhaps, the apocalyptic im-agery circulating in the contemporary Levant.[96] Most germane for our pur-poses, Diodotus abandoned Seleucid Era dating in favor of his own regnal count. Reworking the imperial date-world of pervasive Era-stamped ob-jects (see Chapter 2), Diodotus's regnal numbers were inscribed on market and military objects and struck onto the coins of his Phoenician and Coele Syrian mints. At his base of Dora (Tel Dor on the northern coast of Israel, near modern Zichron Ya'akov), Levantine storage jars of local fabric bear the inscription "L β'" on their handles, marking the second year of Di-odotus Tryphon's usurpatory reign;[97] only a few years earlier the city had used market weights guaranteed by the Seleucid Era date.[98] Remarkably, a lead slingshot from the same site, specifically molded to target the be-

Figure 15. Tetradrachm of Diodotus Tryphon

sieging Seleucid army of king Antiochus VII, bears the Greek-Phoenician inscription "For the victory of Tryphon! D[ora?]. Year 5. Of the city of the Dorians. Taste sumac!"⁹⁹ The gustatory insult is paralleled on other ancient slingshots, but the incorporation of a year date is, to my knowledge, unique.¹⁰⁰ It represents, in the most literal sense possible, the weaponization of time against the Seleucid empire: the army of Diodotus, a usurper monarch powerfully fashioning himself as a new beginning, was firing his regnal dates at the forces of Antiochus VII, a dynastically aware king eager to restore the empire's once great territories. Indeed, the ideological potency of Diodotus's new dating is highlighted by the way in which his rival, Antiochus VII Sidetes, is introduced in the Hasmonean chronicle, 1 Maccabees 15:10: "In year four and seventy and one hundred Antiochus came forth to the land of his fathers (ἔτους τετάρτου καὶ ἑβδομηκοστοῦ καὶ ἑκατοστοῦ ἐξῆλθεν Ἀντίοχος εἰς τὴν γῆν τῶν πατέρων αὐτοῦ)"—dynastic time, ancestral space.

Around the very same time that Diodotus Tryphon did away with both Antiochus VI and the Seleucid Era, an analogous chronographic development was taking place in neighboring Judea. 1 Maccabees reports:

In year seventy and one hundred (ἔτους ἑβδομηκοστοῦ καὶ ἑκατοστοῦ) the yoke of the nations was lifted from Israel (ἤρθη ὁ ζυγὸς τῶν ἐθνῶν ἀπὸ τοῦ Ισραηλ), and the people began to write in their contracts and transactions, "In Year 1, under Simon, great High Priest, general, and leader of the Jews" (ἔτους πρώτου ἐπὶ Σιμωνος ἀρχιερέως μεγάλου¹⁰¹ καὶ στρατηγοῦ καὶ ἡγουμένου Ιουδαίων).¹⁰²

1 Maccabees here makes an explicit identification between Israel's political independence and the introduction of a new system for dating years. Year 1 of Simon—the gentile yoke was lifted from Israel. The chronographic change marks a fundamentally new period of Jewish history. This new Big Bang, like the clarifying Seleucid Era epoch discussed in Chapter 1, idealizes and makes absolute what was in fact an ongoing, contested, and fluctuating historical process of increasing Judean autonomy. The popular decree in honor of Simon, reproduced in the next chapter of 1 Maccabees, confirms the obligation of this high-priestly dating formula: "all contracts in the land should be written in his name."[103] Indeed, there may be a reference to the chronographic reform in the *Megillat Ta'anit*, a Second Temple-period list of thirty-five calendar days on which mourning was forbidden because of "miracles performed for Israel." Alongside other well-known episodes from the Maccabean struggle—23rd Iyyar, the men of the Acra left Jerusalem; 27th Iyyar, the crown tax was revoked; 25th Kislev, the Temple was rededicated (Ḥanukkah); 28th Shevat, king Antiochus (IV or, more likely, VII) left Jerusalem—the *Megillah* prohibits fasting and eulogizing on 3rd Tishri because "the mention was removed from documents."[104] The lemma is curt and mysterious, but we now know from the recently discovered inscription of Seleucus IV, found at Maresha, appointing Olympiodorus as high priest of Coele Syria and Phoenicia (Figure 13) that the province's contracts required a dating formula incorporating the name, and so respecting the legitimacy, of the empire's regional high priest.[105] If the objectionable "mention," removed from legal, fiscal, and administrative documents on 3rd Tishri, were indeed the full Seleucid dating formula, we have here witness to an institutionalized and annual celebration of Simon's time-keeping reform.[106]

More is at stake: the concatenation of Seleucid Era date, Simon's autonomous date, and biblical allusion in this passage is crafted so as to imply the fulfillment of prophecy.[107] From Nebuchadnezzar II's destruction of the Jerusalem Temple and dethroning of the Davidic monarchy in the early sixth century BCE down to this very year—170 SE, Year 1 of Simon—Judea had remained under the control of dominating, foreign kings. Now, as framed by 1 Maccabees and formalized by the new indigenous dating system, "the yoke of the nations was lifted from Israel." The agricultural idiom of yoking conquered populations was a widespread metaphor of imperial dominance in the ancient world and so, conversely, the raising or smashing of a yoke was an image of liberation from oppressive vassalage.[108] 1 Maccabees here makes undoubted reference to certain prophecies of Isaiah, Nahum, Jeremiah, and

Ezekiel, which used the current imposing and promised lifting of the yoke to figure, first, the present period of Israel's defeat, exile, and divine punishment and, then, a future return, restoration, and glorious vindication.[109] Jeremiah, for instance, was instructed by God "to make a yoke out of straps and crossbars and put it on your neck"[110] as the physical symbol of inescapable subjugation to the Neo-Babylonian empire.[111] Then, in a moment of prospective hope, Jeremiah consoles: "In that day—declares the Lord of Hosts—I will break the yoke from off your neck and I will rip off your bonds. Strangers shall no longer make slaves of them."[112] 1 Maccabees, an authorized Hasmonean chronicle, deliberately echoes this prophetic formulation to imply that the long-heralded day, when God's wrath will complete itself, has been reached in Year 1 of Simon; indeed, the historian will go on to color Simon's rule with the language of realized eschatology.[113]

This momentous Year 1 is synchronized with 170 SE. While Seleucid Era dating was employed throughout our Maccabean narrative accounts, probably reflecting a broader if all but lost chronographic practice for intra-imperial historiography,[114] it was not applied thoughtlessly. For instance, 2 Maccabees suppresses the Era date for the rededication of the Jerusalem Temple—a ritual declaration of God's, not the king's, sovereignty over Judea—noting only that the rites were restarted after a two-year interval ("μετὰ διετῆ χρόνον").[115] As we have seen in Chapter 2, Seleucid Era numbers were uniquely ordered from smaller to larger, so that 170 SE appears as "year seventy and one hundred" ("ἔτους ἑβδομηκοστοῦ καὶ ἑκατοστοῦ").[116] This year date, especially in the literary and historical context just outlined, hints at the completion of another of Jeremiah's prophecies—the much-cited seventy years of Judah's servitude and Jerusalem's desolation.[117] The evident failure of the Achaemenid restoration to bring to a close Israel's sufferings had demanded creative reinterpretations of this seventy-year period, rising to an exegetical, numerological obsession during the national and theological crisis of the Maccabean period: the book of Daniel, which presents its pseudonymous author reading and pondering this prophecy (see Chapter 4), extended it to seventy weeks of years;[118] the Enochic Apocalypse of Animals transformed it into the successive rule over Israel of seventy wicked shepherd-angels[119] (see Chapter 5); and it was reworked in various ways in other apocalyptic fragments from Qumran.[120] Just as the Seleucids themselves could recognize and play with the connotational function of specific Era numbers (see above), it is probable that the prophetic significance of the "year seventy" was recognized in Israel. In other words, the introduction of an autonomous year count in 170 SE—a deliberate choice and a contemporary fact—turned the imperial chronography into a sign of its own undoing.

We do not know when Seleucid Era dating was entirely abandoned in Judea. Certainly, the only Hasmonean coins to incorporate year dates—for the twenty-fifth year of king Alexander Jannaeus—employed not only regnal counting but also the standard, larger-to-smaller, non-Seleucid ordering of numbers ("L κε'").[121] In this light, it is noteworthy that shortly before Gadara (the Decapolis city to the east of the Jordan river) was conquered by Alexander Jannaeus after a ten-month siege,[122] the city displayed facing outward on its acropolis defensive wall the inscription "L̦ ηκσ' Φιλώτας | καὶ Σελε[υκέ]ων | τῶν ἐν Μεσ[- -] | ἡ πόλι[ς]," combining the Seleucid Era date "eight and twenty and two hundred" (85/84 BCE) with the city's dynastic name, "Seleucia-in-Mes[-?-]," and its Greek constitutional status.[123] As with Diodotus Tryphon at Dora, besieged by Antiochus VII, so at Gadara, besieged by Alexander Jannaeus, we have the symbolic opposition of time systems, marked, in each case, onto the very materials of military confrontation.

The Attalid and Artaxiad stories appear similar, with the new adoption of regnal counting as a sign of political autonomy and a traditionally monarchic biorhythm.[124] In each of these situations, as is particularly well-delineated in the Hasmonean, we see at work the explosive and long denied potential of the Year 1 ideology.

While regnal dating was reintroduced to the empire's western hemisphere by rebel monarchs and warrior-priests, the Seleucid Era technology was adopted and repurposed in two geographic areas and in two distinct ways: by the independent city-states of the Levantine and southern Anatolian seaboard, as part of the irrepressible self-assertions of *poleis* against empire, and by the post-Seleucid monarchies of the Upper Satrapies and beyond, as chronographic claims to imperial succession.

Along the east Mediterranean Seleucid seaboard the grant of autonomy to a city, whether as the strategic benefaction of a potent center or as a grudging confirmation of imperial weakness, was often celebrated with the inauguration of a new civic era count: on local coins, this new autonomous year dating replaced the Seleucid Era number, just as images and legends of local reference supplanted royal portraits and imperial icons. The pattern was set by Aradus, on the northern Phoenician coast, and Perge, in Pamphylia in southern Asia Minor. The situation of the former city, both as a major naval power and the main third-century BCE buffer between the Seleucid and Ptolemaic territories, has long explained its precocious tendency to anticipate by several generations emancipating developments elsewhere. Its autonomous era, with an epoch of 259/258 BCE, first attested on coins in 243/242 BCE, probably marked a grant of independence by Antiochus II at the start of the Second Syrian War.[125] Similarly, Perge's autonomous count most likely commemorates a strategic grant of emancipation at the opening of Seleucus III's ill-fated Asia Minor campaign.[126] Such civic eras of autonomy and freedom proliferated over the suppurating Seleucid polity of the later second and early first centuries BCE: Tyre in 126/125 BCE, Sidon in 112/111 BCE, Seleucia-in-Pieria in 109/108 BCE, Ascalon in 104/103 BCE, Tripolis circa 100 BCE, and Laodicea-on-the-Sea in 81/80 BCE.[127] The overwhelming majority of these cities adopted the era count immediately or very shortly after the autonomy grant that the epoch celebrated: they were contemporary assertions, not retrospective recognitions, of historical rupture.[128] Whereas the civic eras later introduced under Rome could mark conquest, liberation, provincialization, city foundation, or battlefield victories,[129] the city eras that proliferated within the Seleucid Levant and southern Asia Minor celebrated only and exclusively autonomy from empire.[130] Even if arrived at through royal grant or negotiation, these city times were inescapably polemical, orienting all future events around the local memory of Seleucid withdrawal. Imperial time had engendered its antithesis—free time.

The great successor empires of the east demonstrate a differently oriented reworking. The Arsacid dynasty of the Parthians, who would come to absorb the bulk of the Seleucids' trans-Euphratene territories, introduced its own imperial era count, based on an epoch of 1 Nisannu 247 BCE. These Arsacid Era year dates are attested from the second century BCE onward in the clay, stone, and parchment record.[131] This new beginning, precisely sixty-four years after the idealized Seleucid Era epoch, similarly opened the empire around a clarifying moment of foundation—most likely, the formal recognition of Arsaces I as king of the nomadic Parni.[132] Isidore of Charax recounts in his itinerary of Parthian territories that Arsaces was first proclaimed king in the city of Asaac ("πόλις δὲ Ἀσαάκ, ἐν ᾗ Ἀρσάκης πρῶτος βασιλεὺς ἀπεδείχθη") and that, in consequence, the city housed an eternal flame ("καὶ φυλάττεται ἐνταῦθα πῦρ ἀθάνατον").[133] The location of Asaac in Astauene, modern Ostova, to the north of the Kopet Dagh mountains and between the river Atrek and the Caspian Sea, is good evidence that the Parni had not yet descended into the province of

Parthia to face down the Seleucids.[134] Their epoch marked not an outward-facing freedom but an internal political reorganization that would give birth to empire. Indeed, the Arsacid Era's dynastic temporality would be rendered into nomenclature, with all subsequent kings taking the throne name Arsaces.[135]

Farther east, in the riverine oases of Bactria and the Punjab floodplains, chronographic systems modeled on the Seleucid Era were introduced by the Graeco-Bactrian, Indo-Greek, Indo-Scythian, Indo-Parthian or Gondopharid, and Kushan kingdoms.[136] The era systems of this most complex and politically fraught region have been reconstructed from a handful of administrative documents, royal inscriptions, Buddhist reliquaries, and Sanskrit treatises in an overawingly erudite reconciliation of numismatic evidence, astronomical data, calendrical systems, and classical, early Christian, Indian, and Chinese historiography. A single new discovery may well blow the whole house down. We do not know whether or in what way Diodotus, the rebel satrap and first independent monarch of Graeco-Bactria, broke with Seleucid dating; his carefully graduated self-emancipation may suggest the Seleucid Era's retention, as was the case under Eumenes I of Pergamum.[137] But the publication of a new parchment document demonstrates the existence of a Euthydemid Era, likely descending from Antiochus III's in-person recognition, during his second *anabasis* in 206 BCE, of the Bactrian king Euthydemus's sovereign independence and his son Demetrius's comonarchic status.[138] A second parchment document, a tax receipt dated to the fourth year of kings Theos Antimachus (I), Eumenes, and Antimachus (II), demonstrates king Antimachus I's abandonment of this Euthydemid Era and introduction of a new count.[139] It is possible that Antimachus I's newly inaugurated dating system is what was later known as the Yona or Yavana (Greek) Era, with an epoch of 175/174 BCE; this Greek Era was employed with its Macedonian calendar to north and south of the Hindu Kush into the second century CE.[140] As in the Seleucid empire, Yona or Yavana Era dates were struck onto the coins of Indo-Greek kings—years 47 (μζ′) and 48 (μη′) for king Plato, years 57 (νζ′) and 83 (πγ′) for king Heliocles I—perhaps offering a residue of unity for a fracturing and chaotic inheritance.[141] It may be significant, perhaps as an inheritance of the Euthydemid Era, that the Seleucid numerical ordering was reversed.[142]

A spherical reliquary, dedicated by a certain Apakhraka, son of Heliophilus, at the Buddhist Traṣaka stupa and inscribed with multiple, simultaneous era dates, has shown that the Indo-Scythian king Azes introduced his own era count in 47/46 BCE; it is surely not accidental that this was the second centennial of the Arsacid Era (201 AE = year 1 in the Azes Era).[143] Similarly, in 127/128 CE the great Kushan emperor Kanishka I introduced his own era count on the third centennial of the Yona or Yavana Era (301 Greek Era = year 1 in the Kanishka Era). Kanishka discussed at length the inauguration of his own era in a splendid temple foundation text inscribed in Bactrian on a stone plaque, discovered at Rabatak in northern Afghanistan in 1993. The Kushan monarch explicitly connected his "inauguration of the year one (ιωγο χþονο νοβαστο)" with his conquest of India, the establishment of a new sanctuary—"the temple (βαγολαγγο) was founded in the year one"—and the replacement of Greek with Bactrian ("Aryan") as the language of administration.[144] According to his inscription, this year one temple would house statues of Zoroastrian deities, king Kanishka, and his royal predecessors, beginning three generations back with his great-grandfather Kujula Kadphises. Whether in emulation of the Seleucids' Temple of Day One at Babylon (see Chapter 1) or as a cultic reflex of an inherent temporal logic, these eastern epochs, where we have evidence, were sacralized by the new establishment or repurposing of a sanctuary—Arsaces I's undying fire of Asaac and Kanishka's year one temple at Rabatak. Indeed, it is tempting to identify in these Seleucid, Parthian, and Kushan epoch-sanctuaries a coherent type of political religion, focused on the commemoration and celebration of imperial and dynastic origins.

These new era systems, whether operating in the territorial empires of the Hellenistic far east or among the city-states of the Mediterranean littoral, follow the logic of the Seleucid Era even as they undo it. Their epochs, commemorating a newly won political autonomy, elevation to kingship, or conquest of a legitimizing space, betray the deep penetration of Seleucid temporal thought by replicating its foundational terms. Indeed, even where new time systems were introduced, the Seleucid Era could be retained as an overarching, umbrella referent, with documents double-dated by the unqualified, absolute Seleucid Era number alongside, say, the high-priestly year of Simon Maccabee or the Arsacid Era year.[145] This was different from the apparently similar phenomena of bilingual inscriptions, in *koinē* Greek and local tongues, or semiautonomous coinage, which could oppose imperial Seleucid and pre-Hellenistic local iconographies. For the temporal *diglossia* distinguished between a supraregional, unmarked standard—the Seleucid Era—and a local, marked dating system that, while often expressive of an ancient indigenous identity, was by necessity forever younger and globally embedded.[146] In short, the Seleucid Era, by illustrating the power of uniform time, multiplied it. The empire's territorial retreat and political decline was also a protracted process of pluralization in an already variegated temporal landscape.

Conclusion

The Seleucid empire, as expressed through the temporal regime it invented and promoted, constituted a *Weltanschauung*, a total worldview—a mode of thinking about power, duty, origins, historical development, and the future that was both relatively coherent and fundamentally different from all that had come before. The Seleucids' linearization and absolutization of time was not only rationalizing and "empty," characterized by measure and number. Their temporal regime, perhaps even despite itself, was also "full," generating an interpretation of experience, a horizon of contemporary memory, a political responsibility, a pessimism about the direction of history, and, ultimately, the necessity of redemption.

PART TWO

Indigenous Past and Future

CHAPTER 4

Total History 1:
Rupture and Historiography

The first part of this book has argued that the Seleucid empire proposed, for itself as for its subjects, an extended and exclusive imperial present. Here and in the following chapter I will suggest that this temporal orientation created the conditions for new attitudes to the preconquest world. Within several of the empire's indigenous communities we can detect the emergence of a distinctive historical perspective, characterized by the distantiation, numericalization, and segmentation of the past—or, in the terms of intellectual history rather than phenomenological affect, total history, chronography, and periodization. In addition to conveying an argument about imperial time and local pasts, it is my hope that these two chapters can begin to constitute a coherent library—or at least a shelf—of Seleucid indigenous histories. As a corpus they reveal that the Seleucid empire and its temporal regime prompted within at least some parts of its subject communities reflections of remarkable profundity on the texture, structure, and meaning of history.

An anecdote in Plutarch's *Life of Alexander* maintains that the Macedonian conqueror, taking possession of the Achaemenid palace complex at Persepolis in 330 BCE, happened upon a statue of the Persian king Xerxes that had toppled to the floor. "Shall I pass on and leave you lying there, because of your expedition against the Greeks?" Alexander inquired of this fallen great king. "Or shall I set you up again, on account of your magnanimity and excellence?"[1] Alexander's choice captures not only the central paradox of his own kingship but also a continuing concern of historical scholarship: the extent to which the Macedonian *anabasis*, the generation-long civil war over Alexander's inheritance that followed, and the survivors' sedimentation of armies into kingdoms represented a significant rupture from the Near East's Iron Age and Achaemenid past.

The conviction that the Macedonian conquest opened a deep historical fissure has a pedigree as old as Gaugamela. The unpredicted and swift fall of the dominant, two-century-old world power, trampled underfoot and scattered to the winds, was for many contemporaries as much an emotional disturbance and intellectual perception as a political event. With Darius III, last of the Achaemenids, being hunted through the Iranian uplands, the orator Aeschines wondered before the Athenian democratic assembly: "Well, then, what strange and unexpected thing has not taken place in our time! For we have not lived the life of normal men, but were born to be a marvel to posterity. Is not the king of the Persians, he who channeled mount Athos, bridged the Hellespont, demanded earth and water from the Greeks . . . is he not now struggling, no longer even for lordship over others, but already for his own life?"[2] From a different tradition, the late-prophetic ninth chapter of Zechariah pronounced the destruction, in a north to south order, of Hadrach, Damascus, Hamath, Tyre, Sidon, Ashkelon, Gaza, Ekron, and Ashdod as the prelude to the arrival of the messianic king, the battle of "your sons, Zion, against your sons, Greece (bānayîk ṣiôn ʿal-bānayîk yāwān)," and God's self-disclosure to His people. This divine warrior hymn, with its fearful expectation of a Greek invasion, eschatological undertones, and orthographic pun (ṣywn-ywn), has been persuasively characterized as a response to Alexander's post-Issus campaign down the Levantine coast and the collapse of Achaemenid domination.[3] Most far-sighted of all, the Peripatetic philosopher-tyrant of Athens, Demetrius of Phalerum, reflected in his treatise on *Tychē* (Fate), "Do you think if, fifty years ago, one of the gods had told either the Persians and their kings or the Macedonians and their kings what the future would bring, they would ever have believed that by this present time nothing would remain even of the name of the Persians, who were the rulers of all the *oikoumenē,* and that the Macedonians, who before were nameless, now rule over all of it? But fate (τύχη) . . . having established the Macedonians in the fortune that used to belong to the Persians, has lent these blessings to them, until it should determine for them something different (ἕως <ἂν> ἄλλο τι βουλεύσηται περὶ αὐτῶν)."[4] In Jerusalem, in Athens, and presumably elsewhere,[5] the fall of Persia thundered, prompting immediate reflections on the mutability of fate, the transitoriness of earthly power, and the unpredictability of the future.

From these contemporary observers in the late fourth century BCE right up until recent decades, the watershed quality of the Macedonian conquest has been assumed and adopted into the very shaping of our academic disciplines. Undergraduate textbooks and introductory survey courses for both Near Eastern and Greek history—the reproductive distillations of our academic fields' arguments and reflexes—typically conclude with the arrival or death of Alexander.[6] However, three developments since the Second World War have worked to undermine this historical periodization: first, the better integration of Near Eastern sources, with all they reveal of the vitality, deep-rootedness, and permanence of indigenous cultural traditions across the imperial sequence;[7] second, archaeological and historical turns toward landscape survey and the *longue durée* that have deemphasized political agency and downplayed dynastically articulated change;[8] and third, if only barely perceptible, a shift in scholarly self-identification away from both Greeks and rulers of empire.[9] Scholars arguing for the absence or reduced significance of an Achaemenid-Hellenistic rupture have focused on imperial structure, state machinery, and the modes and relations of production.[10] And in these areas it cannot be doubted that Hellenistic technologies of administration and surplus extraction were deeply indebted to Achaemenid and earlier precedents. It comes as little surprise, for instance, that late-fourth-century BCE Aramaic documents from Bactria and Idumea in the southern Levant, two ends of empire, betray little immediate change between the Persian and early Hellenistic administrations.[11] The model of the royal economy laid out in the second book of the pseudo-Aristotelian tract *Oeconomica* applies as much to the Achaemenid kingdom as to those of Alexander and his successors.[12] Similarly, it has now been amply demonstrated that, long before the Macedonian takeover, the Anatolian peninsula and the coastlands of the eastern Mediterranean had been variously impressed by "glocalizations" of Hellenic culture.[13]

These rejoinders to the overstatements and imperial overenthusiasms of the past are salutary, capturing essential and long-overlooked aspects of the early Hellenistic world.[14] But two things need to be emphasized: differentiation and design. The impact of the Macedonian conquest had wildly divergent historical trajectories at distinct locales and scales; we are dealing with a variegated landscape, discrepant experiences, and an inconsistent tempo of change. Compare, for instance, the battle-free renewal enjoyed in Egypt with the major dislocations of Babylonia, respective heartlands of the Ptolemaic and early Seleucid kingdoms. The former, only recently reconquered by the Persians, enjoyed a peaceful "liberation" by Alexander, easily resisted an invasion by the Macedonian regent Perdiccas, and then settled into renewed quasipharaonic rule under the Ptolemaic house. Babylonia, by contrast, was the site of repeated and devastating warfare: the succession crisis in 323 BCE; Eumenes's conflict with Seleucus in 317 BCE;[15] and above all the three-year war between Seleucus I and Antigonus Monophthalmus, from 311 to 308 BCE.[16] The cuneiform record of this last, bitter conflict laments plundered cities, burned sanctuaries, and a ravaged countryside, with repeated reference to "weeping and mourning in the land (bikīt u sipdu ina māti)."[17] Food prices skyrocketed: one sheqel of silver, which would normally have purchased between 100 and 200 liters of barley in the Babylonian marketplace, bought a mere 7.5 liters in 309 BCE, the lowest volume ever recorded in several centuries of monthly records.[18] Western Iran and the northern and southern Levant offer similar stories of continual and devastating conflict, albeit with little documentary evidence.

Second, the question of continuity with or separation from the Persian imperial past was visible to the Macedonian conquerors and open to deliberate engineering, as the anecdote of Xerxes's statue indicates—if not so much in the Near East's embedded political and economic structures, then in imperial policy, monarchic ideology, cultural epiphenomena, and emic interpretations of history, all of which were more immediately visible to ancient populations. We have seen in Chapter 3, for example, that Seleucid court historiography, dynastic cult, and appeals to precedent took the life of Seleucus I as the horizon of appropriate recollection, sundered from all that had come before. Similarly, the extensive Seleucid program of colonization was a massive, deliberate, and visible break from the earlier political landscape: fixing entirely new imperial centers in northern Syria and the middle Tigris; downgrading earlier settlements and relocating their populations; replicating a rigidly geometric, distinctively "modern" urban orthogonality, in architectural contrast to preexisting towns; employing colonial names and foundation narratives that suppressed the historical agency of all but the ruling dynasty.[19] Above all, as we have seen throughout the book's first part, the Seleucid Era count, running only forward, had a periodizing effect. All of the objects it dated (numbering in the many hundreds of thousands, at the very least) had an immediately legible depth with reference to its Year 1, a horizon reaffirmed in its every use. BSE, Before the Seleucid Era, was unthinkable in its own terms.[20] It would be as if we employed Common Era dating without any Before the Common Era enumeration—the existence of a prior world would in no way be doubted, but it could not be as easily grasped or placed within the same temporal understanding.[21] Indeed, several comparative studies, ranging from the Solomon Islands to Revolutionary France, suggest that the introduction of new calendrical or dating systems sharply differentiates a "time before" from "this time," giving substantive consequence to a chronographic distinction.[22]

All in all, it is within the core regions of the Seleucid empire that the noticeable signs and felt experience of historical discontinuity were both most profound and most vigorously evoked. When the waters receded at the end of the wars of succession, the political landscape of these territories had been fundamentally reshaped and its inhabitants made acutely conscious of this. Indeed, the fact that modern textbook surveys of the Near East tend to close with Alexander the Great and that those of Egypt continue into late antiquity confirms, and may be a much-mediated reflex of, a fundamentally different historicality for the Seleucid and Ptolemaic kingdoms.[23]

Accordingly, this chapter will propose that a sense of temporal distantiation, of a past that has been superseded and a present that owes little to it, can be observed in various kinds of historical writing produced by the Seleucids' indigenous subjects. These works—"writings from Year 1," to paraphrase Carlo Ginzburg[24]—are characterized by an attitude of externality toward a pre-Hellenistic local past that could now be objectified, totalized, classified, and sealed. The witnesses to this sense of belatedness, distance, and closure will, by necessity, be few and selective, taken from the rich seams of third- and second-century BCE Babylonia and second-century BCE Judea. For our purposes, these works of historiography stand not only as a proxy for the experience of rupture or distance that operates internally, in personal, psychological spaces, but also as a sophisticated medium for the clarification, reproduction, and social dissemination of such a perspective on the world.

Babylonia

A helpful and early witness to this new historical positioning is the *Babyloniaca* of Berossus, one of the more intriguing of all the ancient world's historiographic texts. This was a three-book account of Babylonian culture and civilization, composed in Greek in the third century BCE by the cuneiform-literate Babylonian priest Berossus (from the Akkadian Bēl-re'ušunu, meaning "the Lord is their shepherd"). Unfortunately, inevitably, the *Babyloniaca* has not survived the centuries, failing to pass the classical assault course of stylistic standards and narrow historical qualifications. But a first-century BCE epitome by the Greek prisoner of war Alexander Polyhistor was extensively cited in the apologetic works of Jewish and Christian scholars—the first-century CE Jewish historian Josephus, the otherwise unknown Imperial-period Abydenus, the third–fourth-century CE Christian chronographer Eusebius of Caesarea, and the ninth-century CE Byzantine churchman Syncellus—and so gives us a reliable impression of the work's content and texture. In brief, the *Babyloniaca*'s first book described Babylonia's geography, flora, fauna, cultural practices, and astronomical knowledge before narrating, in the voice of the primeval sage figure Oannes, a reworked version of the Babylonian genesis account, the *Enūma eliš*. The second book gave an account of the antediluvian kings, the great flood myth, and then a king-list-like succession of postdiluvian rulers up to the reign of Nabonassar (or Nabû-nāṣir) in the eighth century BCE. Berossus's third and final book recounted, in a thicker narrative derived from royal inscriptions and other local historiography, the succession of Neo-Assyrian, Neo-Babylonian, and Achaemenid rulers of Babylon. In its entirety and in part, the *Babyloniaca* has received much recent and sophisticated attention, with treatments orbiting a supposedly ancient but decidedly modern contest of genealogies—Berossus's relative debts to Greek and Babylonian historiography, his advocacy of local paradigms of kingship and championing of Babylonian heroes, and his political stance vis-à-vis the Seleucid regime.[25] Here, my focus will be on Berossus's fundamental reworking of the historical time of the Babylonian past.

Berossus opened his *Babyloniaca* by establishing his own temporal location. His proemial comments at the beginning of book one indicated first, that he was in some sense contemporary with the coming of Alexander the Great—"κατ᾽ Ἀλέξανδρον γεγονώς" can indicate either his birth or his adulthood at the time of the conquest—and second, that he dedicated his three-book history to a king Antiochus.[26] This Antiochus is identified either as the third ruler after Alexander the Great,[27] in which case, with some wriggling, as Antiochus I Soter, or as the third ruler after Seleucus I Nicator,[28] in which case, with equal wriggling, as Antiochus II Theos. While most scholars have preferred the earlier king, it is at least suggestive that, during the reign of Antiochus II, the *šatammu* (high priest) of Babylon's Esagil temple, and so the most senior indigenous official in the city, was a certain Bēl-re'ušunu.[29] In either scenario, Berossus chose in his introductory comments to closely situate his biography and literary product within the Macedonian conquest and the subsequent Seleucid regime: he is of the new world.

Given his explicit self-positioning with reference to the Graeco-Macedonian imperial environment, it is all the more significant that Berossus's history ran, as a continuous and unbroken narrative, from the beginnings of organized life up to the conquests of Alexander the Great.[30] This was a totalizing and unbroken account of Babylonian culture and civilization from Alorus, the first king, until the fall of the Achaemenid empire; and then it stopped. Depending on the date of dedication (to Soter or Theos) this left a textual no-man's-land, a historical interstice of between five and nine decades, between the closing point of Berossus's history and the Seleucid context of its composition—approximately the duration of the entire Neo-Babylonian dynasty. As far as we can tell, Berossus's contemporary world had no place within his history; his textual content came to a close at the Seleucid horizon. It is noteworthy that a text of close comparison and possible influence, the *Indica* of Megasthenes (a Seleucid-affiliated three-book ethnographic history of Gangetic India composed in the early third century BCE), ran from the primordial cultural heroism of the god Dionysus up to and including the contemporary Mauryan state, beyond the boundaries of the Seleucid empire, to which Megasthenes had been dispatched as Seleucus I's envoy. Whereas Chandragupta Maurya's peer kingdom is the culmination of the logics of Indian history,[31] Antiochus's imperial formation stands beyond Babylon's.

The overriding impression given by Berossus and his *Babyloniaca* is of externality: the Seleucid present is an observation point for gazing back upon and evaluating all the formations that have preceded it. Even the autoethnography of Babylonia's landscape at the opening of the first book—a description of the region's productive and infertile soils, its grains, fruits, and roots, together with glosses for Greek readers—displays a remarkable capacity for distance and disembeddedness; its self-description is entirely without parallel in any cuneiform text.[32] More fundamentally, Berossus's project constituted an unprecedented unification into a single, coherent, linked-up narrative account of Mesopotamian myths, dynastic histories, local cultural practices, and other elements of scribal science and local lore. This emptying of the cuneiform drawers, their ordering by chronology, and the translation of their contents comes close to a universal library or total history of pre-Seleucid Babylonia: "the totality of the past . . . meant something in and of itself, distinct from the series of discrete events that constituted the histor[y]."[33] The *Babyloniaca* achieved this unity and completeness of its history in two ways: first, it emphasized the sealed totality of knowledge; and second, it established a single, coherent timeline.

In his first book, Berossus reported that a certain Oannes, the earliest of Babylonia's culture-heroes (Akkadian *apkallū*), squelched out of the Persian Gulf to educate humanity in all that constituted an ordered existence:

> In Babylon there was a great crowd of men, who had settled in Chaldea. They lived without order like wild animals (ζῆν δὲ αὐτοὺς ἀτάκτως ὥσπερ τὰ θηρία). In the first year (ἐν δὲ τῷ πρώτῳ ἐνιαυτῷ) there appeared out of the Erythraean Sea, at the place bordering Babylonia (κατὰ τὸν ὁμοροῦντα τόπον τῇ Βαβυλωνίᾳ), a sentient creature by the name Oannes . . . having a body entirely like a fish, but underneath his head another head which had grown under the head of the fish, and feet similarly of a man which had grown out from the tail of the fish; it possessed a human voice, and the image of him is preserved even now (τὴν δὲ εἰκόνα αὐτοῦ ἔτι καὶ νῦν διαφυλάσσεσθαι). [Berossus] says that this creature passed the days with the humans, not taking any food, but giving to the humans knowledge of writing and learning and crafts of all sorts; and he was teaching also the founding of cities, the building of temples, the introduction of laws and geometry; he was revealing planting and the harvesting of crops, and in sum all the matters that pertain to the amelioration of life he was handing over to men (παραδιδόναι τοῖς ἀνθρώποις). From that time, nothing more in addition has been discovered (ἀπὸ δὲ τοῦ χρόνου ἐκείνου οὐδὲν ἄλλο περισσὸν εὑρεθῆναι).[34]

Berossus goes on to describe how Oannes, after civilizing humanity, composed and handed over a written account of creation.

> And Oannes wrote about origins and government (τὸν δὲ Ὠάννην περὶ γενεᾶς καὶ πολιτείας γράψαι) and handed over this account to mankind (καὶ παραδοῦναι τόνδε τὸν λόγον τοῖς ἀνθρώποις). "There was a time," he says, "when all was darkness and water (γενέσθαι φησὶ χρόνον, ἐν ᾧ τὸ πᾶν σκότος καὶ ὕδωρ εἶναι) . . ."[35]

Berossus then reproduces, in the embedded voice of Oannes, a subtly reworked version of the *Enūma eliš*, the great myth of Marduk's victory over the chaotic Tiamat and the cosmogonic shaping of the world from her carcass.

This inaugural moment of Babylonian civilization was also its completion—since Oannes, humanity has discovered nothing. To Berossus's evident argument for Babylonian cultural primacy, presumably directed as much toward other Near Eastern populations as to the youthful Greeks and Macedonians, he has added the claim of a singular, momentous, and unchanging cultural package, of—to glance forward a little—a canon.[36] As expressed by Stanley Burstein, "the beginning of history was also its end since everything thereafter could only be, and quite explicitly was, preservation, exegesis and application of that initial revelation to life."[37] While a sense of antediluvian "classic" texts certainly existed earlier than the Seleucids, such an all-embracing model of civilizational, textual origins and subsequent historical stasis is not found outside Berossus.

This textual totality appears once more in Berossus's retelling of the deluge myth in his second book. His account combines traditional and archaizing elements—for example, the flood hero is the Sumerian Xisouthros (Ziusudra), not the Standard Babylonian Utnapištim or Atrahasis[38]—with a new emphasis on the survival and redistribution of antediluvian cuneiform knowledge. The god Kronos, visiting Xisouthros in a dream to warn of the impending cataclysm, charged him to bury "the beginnings, middles, and endings of all writings (γραμμάτων πάντων ἀρχὰς καὶ μέσα καὶ τελευτάς)" in the city of Sippar and then construct his ark. After the deluge, when the flood waters had withdrawn and the survivors dared to disembark from their ark, now stranded in the Armenian mountains, a voice from heaven instructed them to return to Babylon, excavate the writings deposited at Sippar, and pass them on to mankind ("εἶπέ τε αὐτοῖς, ὅτι ἐλεύσονται πάλιν εἰς Βαβυλῶνα, καί ὡς εἵμαρται αὐτοῖς, ἐκ Σι[σ]πάρων ἀνελομένοις τὰ γράμματα διαδοῦναι τοῖς ἀνθρώποις").[39] The fact that Sippar, city of the sun-god Šamaš, escaped the cataclysm is already found in the *Erra Epic* from the early first millennium,[40] but the preservation there of all writings is otherwise unattested.[41] The *Babyloniaca*, therefore, transforms the ancient flood myth into an iteration of the original total revelation, a now human-directed repetition of Oannes's wisdom from the waters.[42]

It is tempting to detect in these two episodes an internal model for Berossus's own project: that what the merman-sage had imparted to the first of men and the flood survivors to the postdiluvian world, Berossus is now directing to the Macedonian king.[43] If so, this third revelation projects a periodizing distinction, a closure and new beginning quite as profound as the flood and requiring a similar operation of cultural salvage. Indeed, Johannes Haubold has brilliantly suggested that Berossus's phrase for the completeness of the rescued knowledge—"the beginnings, middles, and endings of all writings"—precisely maps the tripartite division and totalizing embrace of his own *Babyloniaca*,[44] from cosmogony and civilizational foundations, through a postdiluvian line of kings, to the succession of territorial empires up to Alexander. Needless to say, "endings" ("τελευτάς") implies completion, the closure of the textual archive.[45]

The second way that the *Babyloniaca* achieved a sense of historical unity and completeness was by establishing a single, foundational timeline in which myths were embedded, empires sequenced, and events related to one another. Babylonia's historical beginning is established in the first book, at the opening of the civilizational narrative. As we have seen, in the passage cited above, Berossus started his historical account not with the creation of the world, but with the appearance and cultural heroism of the fish-man Oannes. The *Babyloniaca*'s account of the world's genesis is an embedded, retrospective, written text delivered by Oannes to the first humans only after they have become civilized and literate. This exegetical orientation to earliest history—that is, the scribal quotation of an already existing textual canon—is crucial to the chronological shaping and historical positioning of Berossus's entire work; it is as distinctive as Hellenistic rewritings of the Hebrew Bible that open not with the Genesis account, but with the giving of the Torah at Sinai and Moses's retrospective reading of the written record of creation.[46] Berossus dates this episode of Oannes's cultural heroism and recital of creation "in the first year (ἐν δὲ τῷ πρώτῳ ἐνιαυτῷ)." Except for its geographic and ethnographic elements, the narrative of the first book is entirely devoted to this Babylonian year.[47] Such a numericalized opening of time has no precedent or background in the extant cuneiform record;[48] as far as we can tell, no first year or month or day had ever been marked or discussed. Rather, we can discern behind Berossus's dated beginning an echo of the idealized Seleucid Era epoch, as outlined in Chapter 1. Berossus's Year 1 is both subsequent to the uncountable *illud tempus* of the created but unordered world[49] and immediately anterior to the creation of kingship. As John Dillery has emphasized, Oannes's appearance is not assigned to the regnal date "Year 1 of king Alorus," the first antediluvian ruler in Berossus's king list: ordered government had not yet been revealed, so there was no throne for Alorus to ascend.[50] Instead, it is a freestanding, absolute time marker, a directional origin for all subsequent narrative, establishing the foundational point of a linear, forward-moving temporality. It is, in essence, the epoch of Babylonian history. In these respects, it corresponds to 1 SE, which both acknowledged a preexisting, chaotic world and was distinct from Seleucus I's subsequent self-coronation. Moreover, the content and location of Oannes's teaching—a version of the *Enūma eliš* at newly founded Babylon[51]—suggest that what is being presented at this moment is the first and archetypal *akītu* festival. As we have seen in Chapter 1, Babylon's New Year opened with the programmatic recitation of this cosmogonic text by the Esagil temple's *šešgallu* priest. Pulling all this together, it seems that the *Babyloniaca*'s establishment of civilization and recitation of the *Enūma eliš* account in a transcendent "first year" precisely reproduce, albeit in a mythic and local frame, the origin of the imperial temporality employed by its Seleucid dedicatee.

Just as creation is seen retrospectively, from the perspective of a Year 1, so mythic timelessness in general—the bottomless, Eliadian *illud tempus*— is evacuated from the *Babyloniaca*. Berossus presents a newly historicized, thoroughly numericalized past: all events, even those of legend and myth, are assigned a specific temporal location on a continuous, regularly unfolding timeline. Thus, when the god Kronos visits Xisouthros in a dream, "he said that humanity would be destroyed by a deluge on 15th Daisius (φάναι μηνὸς Δαισίου πέμπτῃ καὶ δεκάτῃ τοὺς ἀνθρώπους ὑπὸ κατακλυσμοῦ διαφθαρήσεσθαι)."[52] The significance of this passage lies not only in Berossus's use of the Seleucid calendar (Daisius is the eighth Macedonian month) with all that this implies for his intended readership, but also in the fact that the precise dating of the flood was unparalleled in the main cuneiform versions.[53] When Seneca attributes to Berossus a precise, calendrical prognosis of a future cataclysm, the distant future and deep past have the same, numericalized temporal texture.[54] Note that, if the citation is accurate, Berossus here emulates the action of Kronos. The Macedonian calendar appears once more in Berossus's account of the contemporary *Sakaia* festival, a Saturnalia-like carnival of inverted hierarchy, fixed to begin on 16th Loos, the tenth Macedonian month.[55] With so little of the *Babyloniaca* surviving we must be cautious, but the use of the Macedonian calendrical framework for both the great flood in the second book and the ethnography in the first suggests that it would have been employed throughout, unifying this total account of Babylon's past by the regular time of its Seleucid present.

Berossus's account descends vertically in a continuous line from the transcendent Year 1 of Babylonian civilization to the Macedonian conquest. His second book generates a chronographic backbone from king Alorus to king Nabonassar, counting years either by the regnal lengths of successive kings or by coherent, often dynastic, and somewhat era-like groupings. Thus, Berossus calculated the total chronological periods of the ten antediluvian monarchs, from Alorus to Xisouthros, as 432,000 years; then, after the flood, the next eighty-three kings from Xisouthros to the Gutians ("Medes"), followed by eight Gutian rulers, forty-nine Chaldean kings, nine Arabian kings, and forty-five Assyrian kings—all of whose reigns together total something more than 34,592 years (only some numbers are extant and none can be trusted).[56] The third book continues the count, from the Neo-Assyrian Sennacherib through the Neo-Babylonian and Achaemenid dynasties, again as individual monarchic reigns or grouped into periods.[57]

All historical episodes or *logoi* are subordinated to the directional time frame in which they are anchored; the *Babyloniaca* takes its structure and meaning not from a Herodotus-like accumulation and juxtaposition of episodic narratives but from this underlying, unbroken, and now complete temporal axis.[58] This chronologizing, a repeated totting up of dates for Babylonia's deep history, was recognized by Berossus's readers as a distinctive characteristic of his work: for Josephus, Berossus "catalogues the descendants of Noah (i.e., Xisouthros) and appends their dates (τοὺς ἀπὸ Νώχου καταλέγων καὶ τοὺς χρόνους αὐτοῖς προστιθείς)";[59] for Eusebius, "in [his books] he narrates the counting of the eras."[60] Accordingly, Berossus provided for Babylonia an uncomplicated, linear chronology from Oannes's original revelation to the coming of Alexander. Indeed, the "total duration" of Babylonian clay records, as given in several classical sources, likely derives from this Berossean sum. According to Diodorus Siculus, for example, the Chaldeans "count up (καταριθμοῦσιν) that it has been 473,000 years from when they began to make their observations of the stars in early times (ἀφ' ὅτου τὸ παλαιὸν ἤρξαντο τῶν ἄστρων τὰς παρατηρήσεις ποιεῖσθαι) until Alexander's crossing over into Asia (εἰς τὴν Ἀλεξάνδρου διάβασιν)."[61] Pre-Hellenistic Babylon appears as an objectified, closed, and numericalized unit.[62]

If the *Babyloniaca* as a whole is a product of Hellenistic rupture, one fragment in particular suggests Berossus's direct reflection on the Seleucid Era and horizon. The great second-century CE astronomer and geographer Claudius Ptolemy had opened the dates of his *Royal Canon* in 747 BCE, the accession year of the Babylonian king Nabonassar.[63] Syncellus, citing Berossus, explains this choice as the result of king Nabonassar's own self-periodizing ambition:

> The Chaldeans kept accurate records of the times of the movements of the stars from Nabonassar, and the Greek astronomers derived these from the Chaldeans. As Alexander and Berossus, who treated Chaldean ancient history, report, the reason is that Nabonassar collected and destroyed the deeds of the kings who preceded him (Ναβονάσαρος συναγαγὼν τὰς πράξεις τῶν πρὸ αὐτοῦ βασιλέων ἠφάνισεν) in order that the enumeration of the kings of the Chaldeans start from him (ὅπως ἀπ' αὐτοῦ ἡ καταρίθμησις γίνεται τῶν Χαλδαίων βασιλέων).[64]

The Alexander of this passage is Alexander Polyhistor, the epitomator of the *Babyloniaca*, so our Babylonian priest is the single and original source. The most likely context for this passage is the king list–like second book, which came to a close with Nabonassar's reign. Berossus has a curious and not at all simple relationship to this episode. On the one hand, the destruction of records is accepted not only as an apology for the spare, detail-free texture of the *Babyloniaca*'s "middles,"[65] but also as the fundamental organizing principle of its postdiluvian history—the second book runs from the flood up to Nabonassar ("in the second book he described the kings, one after another, until he says 'Nabonassar was king'"[66]) and then the third, with more narrative detail, from Sennacherib to Alexander.[67] On the other hand, Berossus's first and second books assert the capacity of a scholar-historian to resist the erasure of all before the king. Berossus, like the survivors of the flood who would rescue antediluvian knowledge, has redeemed from Nabonassar's biblioclasm the names, sequence, and year lengths of his predecessor monarchs. It is difficult not to see behind this attitude to Nabonassar's destructive hope that "the enumeration of kings be from him (ἀπ' αὐτοῦ ἡ καταρίθμησις γίνεται τῶν . . . βασιλέων)" a more general response to the periodizing distinction generated by the Seleucid Era count, the Temple of Day One, and the empire's horizon of historical reference.

The potency of such annihilation of the textualized past, as well as its scribal response, can be confirmed by the only parallel in ancient west Asia or the Mediterranean.[68] In 167 BCE, Antiochus IV, great-great-grandson or great-grandson of the *Babyloniaca*'s dedicatee, incorporated historical erasure into the cruelties of his Jewish persecution: "Whatever scrolls of the Law (τὰ βιβλία τοῦ νόμου) they found, they tore up and burned and whoever was found with a scroll of the covenant (βιβλίον διαθήκης) in his possession or showed his love of the Law (τῷ νόμῳ), the king's decree put him to death."[69] As we shall shortly see, the Hebrew Bible was taking shape in the Hellenistic period as a "total history," so Antiochus IV's violent remolding of the Judean community was an assault not only on its prescribed lifestyle but also on the complete historical narrative in which its covenant and laws were embedded (see Chapter 6).[70] The overcoming of this Seleucid cataclysm is held to have prompted a record-collecting salvage operation, a reconstitution of the library: the Second Epistle at the opening of 2 Maccabees reports that, alongside the rededication of the Jerusalem Temple, "Judas gathered together [the books] lost on account of the war that had come upon us, and they are in our possession ('Ιούδας τὰ διαπεπτωκότα διὰ τὸν γεγονότα πόλεμον ἡμῖν ἐπισυνήγαγε πάντα, καὶ ἔστι παρ' ἡμῖν)."[71] As with Berossus vis-à-vis Nabonassar and, in some sense, Seleucus, so here in second-century BCE Judea the monarchic fantasy of absolute beginnings is the inversion or consummation or provocation of total history.

Berossus's *Babyloniaca* is a repeated drama of textual loss, historical closure, and scribal responsibility, of which it must be the culmination and end. At each stage, a temporal rupture has prompted the sealing and historiographical packaging of distinct *spatia historica*: fish-man Oannes for prehistoric, timeless creation; the flood survivors for antediluvian wisdom; and Berossus for the postdiluvian kings and, indeed, for all that preceded. Each re-presentation of the Babylonian world involved a distancing from it. Berossus's external positioning extends even to the *Babyloniaca*'s form. The history was not inscribed in Akkadian on clay tablets—the medium of his source texts that, as Berossus himself was aware, could survive cataclysm, biblioclasm, and the passing of millennia.[72] Instead, the *Babyloniaca* was composed on papyrus or parchment and in Greek—the prestige- and power-language of rulers who arrived on stage only after the curtain had come down. As Svetlana Boym has observed, some things can only be written in a foreign language: "they are not lost in translation, but conceived through it."[73]

We find this powerful sense of a deep historical continuity and its Seleucid supersession even in Akkadian cuneiform. The so-called Uruk List of Kings and Sages presents another scribal perspective on such distantiation.[74] The tablet in question was inscribed by the lamentation priest Anu-bēlšunu ("Anu is their lord") in 165 BCE, during the reign of Antiochus IV. It was excavated from the Bīt Rēš, the vast Hellenistic sanctuary complex of Anu, god of the sky, in the southern Babylonian city of Uruk.[75] Anu-bēlšunu is well-known, appearing in Urukean legal transactions of the late third and early second centuries BCE as a scribe, witness, and purchaser of temple offices.[76] His personal horoscope, a remarkable survival, provides his birthdate as 2nd Tebetu, 63 SE (December 29, 249 BCE) and, on the basis of astrological calculations, promises old age ("his days will be long [ud^{meš} gíd^{meš}]") and male descendants ("he will have sons [dumu^{meš} tuk]").[77] If the Uruk List of Kings and Sages was inscribed by this same Anu-bēlšunu, the stars were correctly read, for it would have been written in his eighty-third year; its melancholic retrospection may be as much a personal as a panimperial sensibility. The tablet must be reproduced in full, as part of its force is visual:

Obv.1 [In the ti]me of king Ayalu, U'an was sage *(apkallu)*.
[In the ti]me of king Alalgar, U'anduga was sage.
[In the time of] king Ameluana, Enmeduga was sage.
[In the time of] king Amegalana, Enmegala was sage.
[In the time of] king Enmeušumgalana, Enmebuluga was sage.
[In the time of] king Dumuzi, the shepherd, Anenlilda was sage.
[In the time of] king Enmeduranki, Utu'abzu was sage.

---

8 [After the flood,] in the time of king Enmerkar, Nungalpirigal was sage, [whom Ištar] brought down from heaven to Eana. [He made] the bronze lyre, [whose . . .] (were) lapis lazuli, according to the technique of Ninagal. The lyre was placed before Anu [. . . ,] the dwelling of (his) personal god.

12 [In the time of king Gilgam]esh, Sîn-lēqi-unninni was scholar *(ummânu)*.
[In the time of king Ibb]i-Sîn, Kabti-ili-Marduk was scholar.
[In the time of king Išbi]-Erra, Sidu, also known as Enlil-ibni, was scholar.
[In the time of king Abi-e]šuḫ, Gimil-Gula and Taqīš-Gula were scholars.
[In the time of king . . . ,] Esagil-kīn-apli was scholar.

Rev. 1 [In the time of] king Adad-apla-iddina, Esagil-kīn-ubba was scholar.
[In the time of] king Nebuchadnezzar, Esagil-kīn-ubba was scholar.
[In the time of] king Esarhaddon, Aba-Enlil-dari was scholar, whom
the Arameans call Aḥiqar.

5 [. . .] Nicarchus.

---

6 Tablet of Anu-bēlšunu, son of Nidintu-Anu, descendant of Sîn-lēqi-
unninni, the lamentation priest of Anu and Antu, an Urukean. By his
own hand.
Uruk, 10th Iyyar, 147 (SE), king Antiochus. The one who reveres Anu
will not carry it off.

The tablet comprises a historical review, organized into three sections. The
first part, lines 1 to 7 of its obverse side, lists in chronological order the names
of seven antediluvian kings, each paired with a fish-man sage (apkallu). As
has long been recognized, this accords with Berossus's presentation of primordial history in the first book of his Babyloniaca—the tablet's first yoke-
team of king Ayalu and apkallu U'anu is identical to king Alarus and fish-man
Oannes.[78] A double line, incised across the full breadth of the tablet's obverse,
represents the great deluge. The second section, from line 8 of the obverse side
to line 4 of the reverse side, continues its double list of kings and scholar-
advisers, from Enmerkar (Berossus's Euexios[79]) until the Neo-Assyrian king
Esarhaddon.[80] As we shall see, line 5 of the reverse side concludes the historical review in the third century BCE by breaking its pattern. Finally, following another incised line, the colophon provides Anu-bēlšunu's name and
genealogy and the place and date of inscription.

The double list incorporates a chronologically ordered selection of
southern Babylonian rulers, spaced at intervals of several centuries—
Enmerkar and Gilgamesh, the legendary, third-millennium BCE hero-kings
of first-dynasty Uruk;[81] Ibbi-Sîn and Išbi-Erra, respectively the last ruler of
the Ur-III empire and the founder of the Isin dynasty, in the early second millennium BCE; Abi-ešuḥ of the first dynasty of Babylon in the seventeenth
century BCE; Adad-apla-iddina and Nebuchadnezzar I, both of the late-
second-millennium BCE fourth dynasty of Babylon; and Esarhaddon, the
seventh-century BCE successor of Sennacherib.[82] The sages and scholars
(apkallū and ummânū) are known from first-millennium BCE library catalogues, tablet colophons, and various kinds of scribal listings as authors of

epics, incantation series, medical works, and so forth.[83] This tablet, synchronizing each celebrated ruler with his learned adviser, constructs an idealized image of a stable, millennia-long union of Mesopotamian political leadership and indigenous intellectual authority. It is a model that emerges at the first origins of civilization and continues after the flood, through military conquests, and over dynastic transitions.

This double review, unprecedented in its historical systematization and implicit claims, bears the imprint of the Seleucid temporal regime: it reworks the Babylonian literary past to the new imperial paradigm, and it demonstrates the closure of this indigenous model. Line 12 of the tablet's obverse side gives us the remarkable entry "at the time of king Gilgamesh, Sîn-lēqi-unninni was scholar ([*ina tar-ṣi* ᴵᵈ*bilga-m*]èš lugal? ᴵᵈ30.ti.ér ˡú*um-man-nu*)." The lemma is burnished with local and personal pride—Gilgamesh was Uruk's most heroic scion, and Sîn-lēqi-unninni ("Sîn, who receives supplication") was claimed as Anu-bēlšunu's own ancestor, as the tablet's colophon shows.[84] A text-author catalogue from the great seventh-century BCE library of Aššurbanipal demonstrates that this Sîn-lēqi-unninni, a scribal editor from, in all likelihood, the Kassite period in the second half of the second millennium BCE, was assigned responsibility for the Standard Babylonian twelve-tablet recension of the *Epic of Gilgamesh*.[85] Anu-bēlšunu's historical review has reshaped these traditional notions of the epic's authorship, origin, and political context: the historical Sîn-lēqi-unninni has been recast as the contemporary scholar-adviser of the legendary Gilgamesh. His inheritance and editing of the epic literary stream has been looped back as its original source.[86] Thus, most splendidly, the Standard Babylonian *Epic of Gilgamesh* has been recategorized as the contemporary record of a court historian. In Anu-bēlšunu's list, Sîn-lēqi-unninni has come to resemble the Seleucid writers of the late third and second centuries BCE, discussed in Chapter 3, a Timochares on king Antiochus or a Mnesiptolemus on king Seleucus, writing encomiastic accounts of their dynastic here and now.[87]

More significantly, Anu-bēlšunu's repetitive catalogue patterns a millennia-long paradigm of Babylonian history to demonstrate, in the final line, its Hellenistic undoing. The tablet, advancing several centuries from Esarhaddon, names as the tenth postdiluvian entry (reverse line 5) a certain Nicarchus (ᴵ*ni-q*(*a*)-*qu-ru-su*!-*ú*).[88] It is all but certain that this Nicarchus was a late contemporary of Berossus and the mid-third-century BCE *šaknu*, or governor, of Uruk, called at birth Anu-uballiṭ. From Anu-uballiṭ's foundation inscription at the Bīt Rēš, Uruk's main sanctuary, which he reconstructed and rededicated in Nisannu 68 SE (244 BCE), we learn that he had been awarded a Greek name by the Seleucid king, possibly Antiochus I, probably Antiochus II. The temple's dedicatory cylinder gives his full title as "Anu-uballiṭ, son of Anu-iqṣur, descendant of Aḫ'utu, šaknu of Uruk, whom Antiochus, king of lands, assigned as his other name Nicarchus (ša ˡan-ti-'-i-ku-su lugal kur.kur^meš ˡni-qí-qa-ar-qu-su mu-šú ša-nu-ú iš-kun-nu)."[89] Like Joseph at pharaoh's court or Daniel and his companions in Nebuchadnezzar's Babylon,[90] by a kind of linguistic baptism this local Urukean representative has been redesignated with a cognomen derived from the ruler's prestige tongue.[91] Anu-bēlšunu had opened his List of Kings and Sages at the world's first light with the pair Ayalu (or Alarus) and U'an (or Oannes); he closed it not with a Seleucid king and a Babylonian scholar, but with Uruk's governor at the time of his birth. Nicarchus's entry represents a distinct rupture from all that had come before. He stands alone, unpaired with a list mate. He is given no title, neither šarru (king) nor ummânu (sage). He is represented only by his Greek name, an award of the Seleucid monarch; moreover, in contrast to the šaknu's own practice, his birth name—the fine Babylonian theophoric, Anu-uballiṭ ("Anu causes to live")—is simply omitted. Indeed, alone of every word and name in the tablet, Nicarchus is meaningless cuneiform.[92] The entry's inscribed line length falls far short of all the others, creating at the list's conclusion, as in the English translation, a visual and logical emptiness. Furthermore, the line lacks the regnal dating formula ina tarṣi ("in the time of") employed in all the previous entries. Nicarchus is without explicit chronographic marking: perhaps his tenure as city governor could date itself; perhaps the Seleucid Era system, employed by Anu-bēlšunu in his colophon, is recognized as a distinct, nonregnal temporal order. Whatever the tablet's mood and purpose (nostalgia,[93] an appeal for scribal recognition,[94] or the construction of an intellectual genealogy[95]) the tenth and final postdiluvian entry shows the disintegration of its historical system.

The Uruk List of Kings and Sages is situated within the age-old Mesopotamian scribal tradition of Listenwissenschaft (list-thinking), which "takes as its prime intellectual activity the production and reflection on lists, catalogs, and classifications, [and] which progresses by establishing precedents, by observing patterns, similarities, and conjunctions, by noting repetitions."[96] These Babylonian scribal catalogues are pointed. They yield propositions. As Alan Lenzi has observed, "the culmination of an Akkadian list occurs in its final line where matters are summarized or its telos obtained."[97] The Uruk List of Kings and Sages characterizes several millennia of southern Babylonian history and then, in its closing, seals them off. The Seleucid present of Anu-bēlšunu is sundered from this totalized past by a linguistic, chronographic, and structural distinction.[98] Despite being written in cuneiform and by a scribe claiming an honorable, ancient genealogy, its periodization is as effective and final as Berossus's exterior positioning and use of Greek.

Judea

Judea provides our other rich seam of extant historical writings from among the empire's subject populations. Despite falling beneath the Seleucid scepter over a century later than Babylonia, after Antiochus III's great victory over the Ptolemies at Panium in the Golan, at the very beginning of the second century BCE, Judea's experience of historical distantiation closely parallels Babylonia's. The Seleucid incorporation of the formerly Ptolemaic southern Levant—"Coele Syria and Phoenicia," in imperial parlance—through institutional changes, colonial foundations, and political and economic realignments seems to have coincided with and catalyzed profound developments in ancient Jewish attitudes toward their own past. The Seleucid horizon was reproduced in the most basic theological tenets and historical conceptions of Hellenistic Judaism.

The canonical Hebrew Bible is divided into three parts: first, the Torah or Pentateuch, the five-book account attributed to Moses that proceeds from creation to the border of the promised land; second, the *Nevi'im* or Prophets, divided into the Former Prophets—that is, the historiographic accounts of the conquest of the land (Joshua and Judges) and the rise and fall of the house of David (1 and 2 Samuel and 1 and 2 Kings)—and the Latter Prophets (Isaiah, Jeremiah, Ezekiel, and the so-called Twelve: Hosea, Joel, Amos, Obadiah, Jonah, Micah, Nahum, Habakkuk, Zephaniah, Haggai, Zechariah, and Malachi); and third, the *Kethuvim* or Writings, a miscellany of literary and historical works including Psalms, Proverbs, Esther, Daniel, Ezra-Nehemiah, and 1 and 2 Chronicles. This scriptural collection is the end product of a lengthy, contested, and highly complex process of textualization, exclusion, and canonization that was still incomplete in the Hellenistic period. Recent work on the Septuagint (the Greek translation of the Bible), ever-greater attention to noncanonical texts, and, above all, the manuscript discoveries at the Qumran caves by the Dead Sea have demonstrated beyond all doubt the social and textual diversity of third- and second-century BCE Judaism:[99] while the Pentateuch and the Prophets were certainly established as an authoritative grouping by at least the early second century BCE,[100] there was no accepted delimitation of the wider scriptural corpus, let alone a fixing of a consonantal biblical text. Yet even if the boundaries of the scriptural embrace remained fluid, as textual traditions vied for wide acceptance, the Hellenistic period witnessed a fixing of the horizon of historical reference. We have not a textual, but a chronological, closure.

As far as we can reconstruct, all branches of Hellenistic Judaism (irrespective of the sharp political and theological controversies over Temple cult and calendar, Zadokite priesthood, evil, angels, Enoch, and end-times, and the Ptolemaic, Seleucid, and eventually Hasmonean kings) seem to have assumed a fundamental break at the fall of the Achaemenid empire. The frame of acceptable reference for biblical and parabiblical writings opened with the creation of the world and ended in the later fourth century BCE: cosmogony, the antediluvian period, the tales of the patriarchs, the exodus and wilderness narratives, the conquest of the promised land, the transition from judges to kings, the regnal histories of Israel and Judah, the Assyrian and Babylonian conquests and exiles, the Persian restoration, the refounding of God's House, and the reconstitution of a territorial community—any of these could be the setting or subject of historical, prophetic, or novelistic writings, many of which were composed or edited in the Hellenistic period.[101] The Pentateuch-Prophets pair, in particular, extended as a continuous, coherent, and chronologically ordered historical narrative, united as a chain of periods, a succession of heroes, a line of genealogies, and occasionally a count of years.[102] In short, the Hebrew Bible, pre- as well as postcanonization, constituted a Berossus-like total history that, as with the *Babyloniaca*, was not able to accommodate the Hellenistic period within its limits.

We must denaturalize a historical scope that centuries of liturgy and exegesis have made inevitable: why could the age after Alexander not be incorporated? The transitions from Achaemenid to Macedonian or from Ptolemaic to Seleucid imperial rule were as nothing beside the traumatic disorientations of the Babylonian conquest and Persian restoration, which the Deuteronomistic and late prophetic traditions had managed to integrate into the Hebrew Bible's single story. Yet despite the lively scribal cultures of Hellenistic Judea, the contemporary world seems to have constituted a discrete and inescapably postclassical temporal field.

Three interrelated phenomena suggest Jewish reflection on this chrono-logical closure, this terminus of revelation. First, several sources explicitly state or implicitly assume the end of prophecy, which encompasses not only the kinds of oracular pronouncement associated with an Isaiah or Hosea but divinely inspired writings of all kinds, including historiography.[103] The Hel-lenistic extension of the book of Zechariah, for instance, condemns in the voice of this Persian-period sage all future prophets as liars and announces their eradication: "They will not put on a prophet's garment of hair in order to deceive. Each will say, 'I am not a prophet. I am a farmer; the land has been my livelihood since youth.'"[104] Psalm 74, most likely a Maccabean-period com-position, deplores that "no signs appear for us; there is no longer any prophet ('en-'ôd nābī'); no one among us knows for how long."[105] 1 Maccabees laments that in the aftermath of Judas's death "great trouble came upon Israel such as had not come since the day when the prophets ceased to appear among them (ἀφ' ἧς ἡμέρας οὐκ ὤφθη προφήτης αὐτοῖς)."[106] The Prayer of Azariah, Dan-iel's companion cast by Nebuchadnezzar into the fiery furnace, incorporates into its plea for salvation a description of a humiliated Israel, left without pro-phetic guidance.[107] This pious embellishment of the book of Daniel, found in the Greek versions, likely dates to the period of Antiochus IV's persecution.[108]

This "end of prophecy" is presented as both a periodizing marker and an existential condition. Like all such closural declarations, it was both an at-tempt to schematize a complicated set of ongoing cultural transformations and a bid at establishing alternative sources of authority. Early rabbinic works correlated the withdrawal of prophetic inspiration with the Macedonian con-quest. For the *Seder 'Olam,* a work of harmonizing chronography, "Alexander of Macedon reigned for twelve years; until that time, prophets spoke proph-ecies through the holy spirit, but since that time, 'Incline your ear and listen to the words of the Sages.'"[109] Remarkably, some later sources directly iden-tify this retreat of divine inspiration with the introduction of the Seleucid Era count: "the Era of Contracts is the closing of prophecy."[110] Similarly, several other rabbinic texts determine the Achaemenid triad of Haggai, Zechariah, and Malachi as the final prophets of this world.[111] Malachi, a sort of prophetic last of the Mohicans, concludes the book of the Twelve minor prophets with the announcement of a future, eschatological figure to herald the end of history: "See, I will send you the prophet Elijah before the coming of the great and dreaded day of the Lord."[112] Until this still unborn end-time there is only a revelationless *Zwischenzeit,* a morbid no-man's-land. Indeed, the book of the Twelve, already recognized as a closed and numbered collection in early Seleucid Judea,[113] coordinated its oracular pronouncements to a post-Mosaic historical backbone: the chronological superscriptions to the Twelve's prophecies—Amos, "in the days of Uzziah, king of Judah, and Jeroboam, of Israel," for example, or Haggai, "in the second year of Darius"—ordered a succession of inspired prophets alongside the sequence of Neo-Assyrian, Neo-Babylonian, and Achaemenid empires.[114] Even the numbered title of this collection signifies in Second Temple Judaism a completed wholeness—twelve patriarchs, twelve tribes, twelve types of heroic ancestors, twelve hours, twelve months, twelve ages, twelve apostles, twelve gates of heavenly Jerusalem, twelve rungs of Jacob's ladder, and here twelve prophets.[115] Very much like the Uruk List of Kings and Sages discussed above, the book of the Twelve patterned a normative national institution—the joining of admonitory prophecy to the kings of the earth—that had run for centuries and across regimes only to run out in the Hellenistic period.

Second, the Seleucid empire could not go unnoticed by God, but the Syrian kings' treatment of Israel was turned into an object of historiographic *prediction*, viewed as if from the deep, pre-Hellenistic past. The second-century BCE apocalypses of Daniel and 1 Enoch, to be discussed in Chapter 5, though steeped in the bitter juices of contemporary events, enter history backward. They are framed as sealed and archived revelations, their authorship and embedded historical narratives attributed to wise, ancient seers—Daniel, the Judean exile in the Neo-Babylonian, Median, and Persian court; and Enoch, the great-grandfather of Noah, "who walked with God." While such pseudonymy had its own logics of identification, offering cause and consolation sanctioned by authority and guaranteed by predictive accuracy,[116] it fundamentally reinscribed the chronological horizon of acceptable representation. Hellenistic history fell beyond what could be written. To speak to the present as revelation, authors had to cast themselves in the past, a textual time traveling back to the lost world where God or his angels intervened.[117] To enter the kingdom of the living they were forced to mimic the dead. Indeed, as we shall see in the next chapter, these apocalyptic historical reviews, extending into and over the Seleucid world, nonetheless marked the Hellenistic age off from all earlier history as a qualitatively distinct temporal and spiritual landscape.

Third, and in consequence of these developments, the conditions of access to divine knowledge were in certain respects relocated from direct revelation to the exegesis of authoritative texts. Hellenistic Judaism, particularly after the Seleucid conquest, turned toward the study of already textualized prophecy. Unlike earlier, inner-biblical exegesis—annotations, adaptations, and harmonizations within the authoritative scriptural texts[118]—Hellenistic extrabiblical exegesis insisted on a chronological, as well as a typological, distinction between the Pentateuch and the Prophets, on the one hand, and their interpretation, on the other.

The banner case for this process is found in the ninth chapter of the book of Daniel, an unprecedented dramatization of reading-based inspiration. This episode's narrative setting is the first regnal year of Darius the Mede; Belshazzar's Neo-Babylonian kingdom has just fallen, awakening Jewish hopes of liberation from exile. Accordingly, "I, Daniel, observed in the books the number of years which were specified by the word of the Lord to Jeremiah the prophet for accomplishing the desolations of Jerusalem—seventy years (šibʿim šānâ)."[119] As we saw in Chapter 3, Jeremiah's prophecy of suffering servitude for only seventy years, delivered at the opening of the Babylonian conquest, was a provocation to intensive exegetical reworking in the second century BCE. Daniel, here pondering upon this written number, addresses to God a confession of Israel's sins and a plea for her forgiveness and is rewarded for his supplicating prayer with an unmasking of the prophecy's true meaning. The ninth chapter continues with the angel Gabriel suddenly appearing before Daniel "to give you insight, understanding (lᵉhaśkilᵉkā bînâ)."[120] This divine messenger then proceeded to multiply Jeremiah's seventy-year prophecy by seven, into seventy weeks of years—that is, four hundred and ninety years, or ten jubilee cycles.[121] Needless to say, the crisis of the Antiochid persecution, the date of this pseudonymous text's actual composition, falls in the final half-week of years: the *eschaton* is imminent.[122]

This mode of retrospective interpretation is thematized throughout the book of Daniel: again and again revealed knowledge consists of two sharply distinct, chronologically separated phases—first, the source matter, the *raz* (mystery), whether that is the prophecy of Jeremiah, the mysterious writing on Belshazzar's palace wall, or Nebuchadnezzar's dream; and second, a subsequent divine gift of interpretation to unveil the hidden inner meaning.[123] It has long been recognized that Daniel 9 closely parallels the *pesher* technique, or textual commentary, attested in the Dead Sea Scrolls. Like Daniel reading Jeremiah, these *pesharim* were eschatological interpretations of the Major and Minor Prophets.[124] They not only assumed and reconfirmed a periodizing distinction between a prior, pre-Hellenistic revelation and its secondary, contemporary study, but they also valorized the very belatedness of the interpreters. For these watchmen of the apocalypse, lashed to the crow's nest of history, it was their very proximity to the end-time that allowed them to recognize the once hidden inner truth of prophetic scripture.[125] The well-preserved *pesher* of Habakkuk, for instance, declares that the inspired prophet-author did not grasp the full significance of his own words—"[God] did not make known to him the fullness of that time *(w't pmr ḥqṣ lw' hwd'w)*"; it was necessary to await the Hellenistic-period exegete "to whom God has made known all the mysteries of His servants the prophets *('šr hwdy'w 'l 't kwl rzy dbry 'bdyw hnb'ym)*."[126] This is a transcendence of the prophetic that, at the same time, narrows its influence: as Gershom Scholem observed, with a degree of distaste, what once could hardly be proclaimed more widely or with sufficient volume has, in this late age, become a kind of esoteric and exclusive knowing.[127]

Judea's profound sense of chronological closure, coordinated to the arrival of the Macedonians in general and the Seleucids in particular, produced a new subgenre of retrospective reflection: the historical review. Throughout the wisdom, historiographic, and even novelistic literature of second-century BCE Judea we find embedded "potted histories" of pre-Hellenistic Israel, pocket versions of the scriptural narrative. Of course, remembrance of sacred history had always played a role in Israel's life as a nation. Creedal statements were historical in essence—"I am the Lord your God who brought you out of the land of Egypt, out of the house of bondage"[128]—and God could be known only insofar as He had revealed Himself historically.[129] For Yosef Yerushalmi, "in Israel and nowhere else is the injunction to remember felt as a religious imperative to an entire people."[130] But the historical reviews proffered in Hellenistic Jewish texts stand apart from earlier writings for their totalizing, systematic, and closed accounting of sacred history, from creation to the Achaemenids.[131] While each extant review is differently structured—selecting, balancing, and directing its logics for contextually appropriate exhortation, prayer, lament, or praise—all share three basic characteristics: they reproduce the full scope of Israelite-Jewish history, conceived as a sacred and organic whole, from origins to the end of prophecy; they are strictly chronological; and the contemporary world of the author or protagonist is sundered, explicitly or implicitly, from this coherent and ordered past. These reviews represent a codified package of an objectified and completed history.

To cite once more the insight of Dillery, discussing Berossus's *Babyloniaca*: "the totality of the past . . . meant something in and of itself, distinct from the series of discrete events that constituted the histor[y]."[132]

A couple of examples should sufficiently demonstrate these qualities. The work known as the Wisdom of Joshua ben Eleazar ben Sira or, more simply, Ben Sira or Sirach or Ecclesiasticus, was composed in Hebrew in Judea in the first quarter of the second century BCE. It is our best witness to a particular elite historical positioning in the very first generation of Judea's incorporation into the Seleucid empire.[133] Ben Sira self-consciously situated his book within the long-established and widespread genre of wisdom writing. These collections of ethical maxims and practical injunctions, a kind of scribal self-help literature, plotted routes of security, piety, and virtue through the multiple injustices and deathly terrors of the world. But in contrast to the cosmic, universalizing, timeless, and strictly ahistorical character of earlier Jewish wisdom writings, such as the books of Proverbs, Job, or even Qoheleth/Ecclesiastes (a third-century BCE work),[134] the Wisdom of Ben Sira culminated in an unprecedented and extravagant historical review. Immediately following praise of God's wondrous creation (42:15–43:33), the "Hymn in Honor of our Ancestors" (running for six full chapters [44–49]) reviewed in a sequence of poetic units over fifty biblical heroes (and some villains) and associated historical events. These extend from the antediluvian sage Enoch (44:16) or Noah (44:17)[135] to Nehemiah, agent of the Persian kings, "who raised up for us the walls that were fallen" (49:13).[136] The hymn is a strictly chronological parade, following and thematizing the scriptural order of the Torah and Prophets.[137] This earliest text from Seleucid Judea is also the first Jewish composition to take as its subject the praise of Israel's gallery of historical figures.[138] Importantly, the events, periods, and heroes are not taken as subjects in their own right but function as components articulating the aggregated and unitary historical past:[139] "All these were honored in their generations and were the glory of their times" (44:7).

Another example: the exhortatory speeches of the Maccabean heroes, as represented in our historiographic sources, incorporated historical reviews as encouragements to faithfulness amid unimaginable hardship. These *parakleseis*, to employ Polybius's rhetorical term for such addresses,[140] stand apart from the motivational harangues of earlier and contemporary Greek historiography in their historical focus and totalizing scope.[141] So, according to 1 Maccabees, the deathbed testament of Mattathias, father of the brothers

Maccabee, urged his sons to show a zeal for the law and a self-sacrifice for the covenant worthy of Israel's ancestors: "Remember the deeds of the fathers, which they did in their generations." Mattathias delivers a Ben Sira–like catalogue of resilient biblical heroes, from national origins to the Achaemenid monarchy—Abraham, Joseph, Phineas, Joshua, Caleb, David, Elijah, Hannaniah, Azariah, Mishael, and Daniel[142]—and then inveighs against the vain pretensions of Antiochus IV: "Do not fear the threats of a wicked man: remember that he will die and all his splendor will end with worms feeding on his decaying body. Today he may be highly honored, but tomorrow he will disappear."[143] The fullness, depth, and comfort found in Israel's pre-Hellenistic past is here counterposed to the Seleucid empire's transitory and empty present. In this way, 1 Maccabees, which had opened with the conquests of Alexander the Great and the foundation of the Seleucid line, establishes two discrete temporal landscapes—that of the protagonists' historical action (the Hellenistic age), and that of their own historical reflection (the pre-Hellenistic age).[144]

Similarly, 2 Maccabees reports, albeit as a condensed *oratio obliqua*, Judas Maccabee's speech to his army of believers before the great battle with the Seleucid general Nicanor: "He exhorted his men not to fear the attack of the Gentiles, but to keep in mind the former times when help had come to them from heaven. . . . Encouraging them from the Law and the Prophets (παραμυθούμενος αὐτοὺς ἐκ τοῦ νόμου καὶ τῶν προφητῶν), and reminding them also of the struggles they had won (προσυπομνήσας δὲ αὐτοὺς καὶ τοὺς ἀγῶνας, οὓς ἦσαν ἐκτετελεκότες), he made them the more eager."[145] 2 Maccabees' Judas here establishes an antithetical contrast between a sealed, textualized past (Torah and *Nevi'im*) and the contemporary experience of endured suffering and triumphant victory.

The most intriguing of our historical reviews is found in the book of Judith. This novella or historical fable, composed in the Maccabean-Hasmonean context of Jewish self-emancipation from Seleucid overlords,[146] tells the tale of the heroic resistance of its eponymous heroine to imperial aggrandizement. The book archly and anarchically jumbles historical elements from all of Israelite-Judean pre-Hellenistic history, like a mix-and-match game in which famous names, places, and adjectives are shuffled in a hat to be joined at random.[147]

The story begins in the form of a royal chronicle—"In the twelfth year of the reign of Nebuchadnezzar, who ruled over the Assyrians in the great city of Nineveh"[148]—that conflates the greatest of the Babylonian kings with the ancestral enemy of Babylon overthrown by his father. As Philip Esler aptly observes, this is akin to making Napoleon emperor of Russia.[149] The Assyrian Nebuchadnezzar, in an echo of Cyrus the Great, seeks to make war against "Arphaxad, king of the Medes," and summons troops for this purpose from all lands between Ethiopia and the river Hydaspes in the Punjab.[150] After capturing Ecbatana, killing his rival, and celebrating a victory banquet at Nineveh for 120 days,[151] this fictional Nebuchadnezzar then determines to punish all the peoples of the west who had refused to join his army. He dispatches for this purpose one general Holophernes, as had the historical Artaxerxes III Ochus,[152] to whom he assigns a certain Bagoas, a name that recalls the famed Persian court eunuch.[153] Like Darius I and Xerxes, he demands the earth and water of submission.[154] The novel sets this Levantine expedition in the Assyrian Nebuchadnezzar's eighteenth regnal year—that is, the precise date in which the historical Neo-Babylonian Nebuchadnezzar II had campaigned against Jerusalem and razed her Temple (587 BCE).[155] Holophernes's vast army of 120,000 infantry, 12,000 cavalry, and uncounted chariots, camels, and so forth ravage and sack land after land, city after city, until only Judea stands unconquered.

The fictional town of Bethulia ("Virginity"[156]), something between a Thermopylae and Asterix's indomitable Gaulic village, guards the path to Jerusalem. With its wells in the hands of Holophernes's army and its thirsty inhabitants reduced to thoughts of capitulation, the beautiful, noble, and wealthy widow Judith (simply and archetypally "the Jewess") steps forward to save her people. Departing in makeup, tiara, and anklets for the Assyrian-Babylonian-Persian camp, she seduces Holophernes, gets him as drunk as the Cyclops, and like a latter-day Jael[157] hacks his head off with his own scimitar (the distinctly Achaemenid *akinakēs*[158]). Judith then steals back to Bethulia to display her gruesome trophy from the battlements. The next morning, on discovering her butchery, the imperial army panics and the Jews are saved.[159]

Inserted into this congeries of historical incongruities is a straightforward historical review. When Nebuchadnezzar's army approaches the Israelites, barricaded in their hill forts, Holophernes interrogates their trans-Jordanian neighbors about the background and nature of this resisting population: "Tell me, now, you sons of Canaan, who is this people which lives in the hill country . . . and why have they, alone of all who dwell in the west, scorned to meet me?"[160] Achior, chief of the Ammonites, responds by delivering a historical review that gives the idealized *Heilsgeschichte* from the patriarchal discovery of the One God through Nebuchadnezzar II's capture of Jerusalem, burning of its Temple, and exile of its inhabitants, right up to the Achaemenid-sponsored restoration and rebuilding.[161] On the basis of this history lesson, Achior exhorts Holophernes to pass them by, "lest their Lord and their God cover them with His shield and we be shamed in the eyes of all the earth."[162] The review makes a mockery of the historical mash-up that is the rest of the book, incorporating and passing beyond the events of the wider, fictional frame: in a supreme anachronism, Nebuchadnezzar II's destruction of the Temple and the Jews' subsequent return from Babylonian exile is reported to the fictional king's general in the very regnal year of historical Jerusalem's capture. As has long been recognized, Achior the Ammonite here resembles the literary type of the wise adviser, foolishly ignored by hubristic aggressors. In particular, the episode is closely modeled on the report of the exiled Spartan king Demaratus, in Herodotus's account of the battle of Thermopylae, when asked by king Xerxes about the kind of men he would face.[163] But where Herodotus's Demaratus had emphasized to the Persian despot the Spartans' unbreakable commitment to their contemporary law, Achior's Jews are made invulnerable by history.

The book of Judith amounts to a compensatory, Tarantino-esque revenge-gratification fantasy,[164] unwriting not only Nebuchadnezzar II's 587 BCE conquest but also, more urgently, the surpassingly cruel persecution of Antiochus IV. As with the book of Daniel, the tale engages with and undermines the Seleucid empire without directly representing it. The horizon of total history is doubly reinforced, first as a playful conceit and then as a serious review. For the author(s) of Judith, all books are simultaneously open, all Israel's pre-Hellenistic past an equally absent "Before," its Near Eastern imperial foes interchangeable because they are transitory. The novel's ironic and knowing attitude to history plays time like an accordion; it is as clear an expression of historical distantiation as may be found.

Too little survives from the other regions of the Seleucid empire for us to even hope to sketch the outlines of comparable responses.[165] Elsewhere in the Hellenistic world, beyond Seleucid space, we have a few hints of the possibilities of a similar totalization and distantiation of pre-Alexander history. Roughly contemporary with Berossus and possibly under his influence, the

Egyptian Manetho composed a king-list-type history of the pharaohs that stopped before Alexander at the Nectanebid, or Thirtieth, Dynasty; like twelve in Israel, thirty was a number of perfected completion in the Egyptian tradition.[166] Similarly, in the second half of the third century BCE, the Cyrenean polymath Eratosthenes systemized in tabular form the historical chronology of Greece as a series of countable intervals between milestone events, such as the return of the Heraclids, the first Olympic games, the end of the Peloponnesian War, and the battle of Leuctra.[167] The outer frames for this "diastematic" matrix were the fall of Troy at the start and the death of Alexander at the end: Eratosthenes's computations made chronographically comprehensible a pre-Hellenistic Greek world that was complete and passed.[168] More subtly, the so-called Lindian Chronicle of 99 BCE, a long stone inscription from the sanctuary of Athena Lindia on the island of Rhodes, listed in (mostly) chronological order an inventory of sacred dedications to the cult. According to this epigraphic catalogue, many of these objects were inscribed by their donors with dedicatory inscriptions: the Chronicle's first dedication, for example, is a *phiale* given by Lindos, the community's mythical eponymous founder, "on which had been inscribed (ἐπεγέγραπτο), 'Lindos to Athena Polias and Zeus Polieus.'"[169] With the appearance of Alexander and his successors, the verbal tense changes: the Chronicle had employed the pluperfect tense (ἐπεγέγραπτο, "it had been inscribed") for all inscribed objects predating the great Macedonian conquests, but it has used the perfect tense (ἐπιγέγραπται, "it has been inscribed") for all subsequent dedicatory labels.[170] Thereby, the Lindian Chronicle distinguishes two temporal fields, a closed and lost world from earliest mythic origins to the period of Alexander and, thenceforth, a contemporary and accessible one still inhabited by the inscription's authors and readers.[171] But these are the exceptions. For the ancient Greek *poleis* and the other Hellenistic kingdoms, historiography, chronographic lists, and other forms of deep-time thinking tended to efface a horizon, not to seal one, incorporating the new world into the old—the Parian Marble, the local history of Nymphis of Heraclea Pontica, the Pride of Halicarnassus inscription, the Atthidographers, Apollonius of Rhodes's *Ktiseis*, the *politeiai* of Heraclides Lembus, and so forth. In short, the emphatic closure of the pre-Macedonian world appears unusually prominent within the Seleucid empire, gaining a charge and reinforced outline from the Seleucid temporal regime.

Conclusion

Most treatments of ancient historiography, much like the preceding pages, adopt modes of redescription, attempting to clarify for modern readers the substance of ancient historical narratives. But these histories, lists, chronicles, and novels cannot be explained from within, nor can the ancient authors be trusted to recognize the conditions of their own possibility. It is necessary to step back from the locutionary content (what the authors are saying about the past) to the illocutionary act (what they are doing when they say this).[172]

Every statement about the past presupposes the present, in that the past assumes opposition to some present and that knowledge of the past can only be a present knowing. But the "total histories" examined in this chapter do not bleed into the contemporary, Seleucid world of their authors. Their pasts are fenced off and at rest, over and done with, antecedent—like a race that has run its course, even if its glories or humiliations may be recalled and its results processed into lessons.[173] These writings by Seleucid subjects, all situated somewhere between an inward-directed consolation and an outward-facing apology, as between auto-ethnography and local historiography, reproduced their past in a manner in which they could take full possession of it. It is as if their contemporary, Seleucid situation were a point of Archimedes in time: the break with the past was not only the structural limit of these works but also, in some sense, generative of them. This perspective accords with certain modern reflections on the writing or contemplation of history that have emphasized its basis in temporal interruption, in the differentiation of past from present. It has been argued that, since the physical time that we perceive spontaneously does not have a historical character, its unrolling must be halted for historicity to be possible: history must treat the past as unitary, factual, and finished.[174] Accordingly, the objectification, even incarceration, of the past that we see in these Hellenistic total histories inscribes as the fundamental characteristic of this past time its absence.[175] To adopt the framing metaphor of the apocalyptic histories, to be discussed in Chapter 5, the past was sealed.

Theorists in a number of disciplines have debated the workings of distantiation, from the Brechtian *Verfremdungseffekt* (distancing effect) to Viktor Shklovsky's aesthetics of *ostranenie* (defamiliarization), Georg Simmel's "stranger," and Martin Buber's *Ich-Es* (I-it) relationship. But historiographical and historical-philosophical discussions have tended to be interested more in the problem of origins and causation than that of endings and closure, as if assuming the past naturally breaks off from the present, like an icicle under its own accumulated weight.[176] I would tentatively suggest that the temporal distantiation we witness among the Seleucid empire's subject communities made possible three stances—or "doors"—to the past. First, the pre-Hellenistic world could become a closed space of retrospection, a turning away from the present and all that this permitted: a political retreat, a nationalist fantasy, a cultural bereavement, a repertoire of exempla, a body for learned exegesis, and so forth.[177] This is Hegel's owl of Minerva taking wing at dusk. Second, the historicized past could become an existential consolation, a shelter from the traumatic experience of time as transitory, and a compensation for the condition of ephemerality. While such a transposition of pathological categories to the historiographical plane was a distinctly modern discovery, associated with Friedrich Nietzsche, Mircea Eliade, and their heirs, I would suggest that the "terror of history" was made unusually and precociously acute by the employment of the Seleucid Era.[178] For the Era's numericalization of time (reframing the eternal as the innumerable)[179] and its open futurity (eliminating possibilities of completion or return) may have produced a profound, if not directly visible, crisis of meaning regarding temporal duration. The total histories' retrospective illusion of fatality suggested to a newly opened world that it was directed and full. We shall see in Chapter 5 that the apocalyptic imagination more comprehensively answered this need. Third, putting the past at a distance could be liberating, giving a freedom from precedent and custom. Constantin Fasolt has argued that the feeling of uplift from reading about the past is the satisfaction of a temporal self-affirmation: "I am here and now."[180] Its chief attraction, he argues, consists of its promise of enhanced control over self and society. From this perspective, of historiography as a ground-clearing operation or entombment,[181] historical distantiation permits a range of consciously adopted cultural changes with less self-alienation than we might otherwise expect.

"*Was* is not *is*." The Seleucid temporal regime, with its absolute distinction from the past, generated a system of two ages—the imperial present and the pre-imperial past. Indigenous total histories consolidated this horizon from the other side. These periods were correlative: where one began, the other stopped. In this sundering of the present from a totalized, pre-Macedonian past we see the first emergence of two perspectives that underlie scholarship to this day: the textbook "end" of the Near East as an ancient and ongoing historical culture, and the invention of the Hellenistic period as a discrete and differentiated age. While this was by no means the only mode of indigenous thinking about durational time and local pasts within the Seleucid empire, as we shall see, it was sufficiently extensive and impactful to order imperial ambitions, local self-assertions, and cultural interactions into the temporalized categories of past and present, traditional and modern. The consequences: the periodization of time, the distinction between discrete historical ages, emerges as a central concern of historical and religious thought in the Hellenistic east (see Chapter 5); and it became a politically aggressive act to suggest that indigenous history either had not ended or could be re-opened (see Chapter 6).

Total History 2:
Periodization and Apocalypse

Chapter 4 argued that the Seleucid temporal regime made possible and significant for its subject populations a new kind of historiography, characterized by the objectification, totalization, and distantiation of the pre-Hellenistic past. Here, we will examine a further development in total histories. Like Berossus's *Babyloniaca*, the Torah-*Nevi'im* (Pentateuch-Prophets) pair, the Uruk List of Kings and Sages, or the numerous historical reviews from second-century BCE Judea, the total histories examined in this chapter ran through a full and extended timeline, from the deep reaches of the past and through the successive Near Eastern empires. But in contrast to those works that excluded the Hellenistic present from their accounts, these eschatological total histories continued into and extended beyond the Seleucid empire: the horizon of historical record was pulled forward and over the imperial monarchy to the very threshold of the end-time; the empire was prospectively historicized and so made preliminary; and history writing was transformed from a point of observation into a tribunal of justice, from a location of vantage for stabilizing the past to a frontier for redeeming it.

The ancient works examined below are typically classified as "apocalyptic" literature—that is, Jewish, Babylonian, and Iranian texts held to share features with or have family resemblances to the Christian Bible's Apocalypse of St. John.[1] I have selected from this disparate and somewhat arbitrary collection those third- and second-century BCE works that function as total histories—that is, accounts of world events in chronological succession through the sequence of Near Eastern empires and up to the predicted end of the Seleucids. These are, at a minimum, the biblical book of Daniel, the Apocalypse of Animals and the Apocalypse of Weeks from 1 Enoch, the Babylonian Dynastic Prophecy, and whatever lies behind the Iranian *Zand ī Wahman Yasn*. All these works have deep local genealogies and evident theological concerns. Indeed, decades if not centuries of scholarship have been dominated by the exploration of their intellectual and generic roots, including prophecy and scribal wisdom in Judea, omen texts, royal inscriptions, and astronomical research in Babylonia, Zoroastrian and Achaemenid traditions in Iran, and these regional traditions' mutual influence. They have formed the heart of fierce polemics on Second Temple Judaism, the Essenes and the Dead Sea Scrolls, and messianism and early Christianity.[2] Yet such analyses, however important and necessary, have not paid sufficient attention either to these works' interest in the past or to their Seleucid imperial context.

These writings, although mostly treated as prognostic, future-oriented compositions,[3] are works of historiography. Their lengthy historical accounts do not merely authenticate the authors' much briefer predictions. Rather, they are attempts to narrate, order, and find meaning in the outplaying of centuries' worth of historical events, with a focus on the central concerns of most ancient historiography—political change, military conquests, imperial rule, and the injustices of despotic kingship. Of course, this is not to suggest that those total histories that continue into the future function just as, say, a Herodotus or Berossus. Indeed, the titles of this and the last chapter, "Total History 1" and "Total History 2," are designed to do more than link arms between a long argument split in two. The terms are modeled on Dipesh Chakrabarty's influential counterpoint of "History 1" and "History 2," as outlined in his *Provincializing Europe: Postcolonial Thought and Historical Difference*. Chakrabarty teased out of a close reading of Marx an important distinction between two histories of capital: the first, Chakrabarty's History 1, consists of the past retrospectively posited by capital itself, "the historical process in and through which the logical presuppositions of capital's 'being' are realized"; History 2, by contrast, is a remainder, the incommensurable past, which cannot be caught within capital's own life-process and interrupts, punctuates, and extends into, through, and beyond capital.[4] Similarly, the historiographic texts explored in Chapter 4 ("Total History 1") present their indigenous pasts as the completed precondition of the Seleucid empire, which operates not so much as their historical telos as "a perspective point from which to read the archives."[5] Conversely, the works of apocalyptic eschatology discussed in this chapter ("Total History 2") constitute a historiography of irreducible excess and affective immediacy, a dialogical history, preceding, incorporating, and surpassing the Seleucid empire and exploding its logics of time.

Second, the basic, central fact of this body of evidence, even if mostly bypassed by the historians of the Hellenistic world, is that it emerged first and only within the developed Seleucid kingdom. Indeed, while sophisticated recent interpretations have drawn from the postcolonial toolbox to characterize these writings as resistance literature, as "apocalypses against empire," they have underplayed the specificity and force of the Seleucid thought-world at the same time as they overplay the empire's disciplinary capacities.[6] Accordingly, this chapter will suggest that the determining context of these new ways of thinking about history, periodization, agency, justice, and the future was the common experience of the Seleucid temporal regime (as outlined in Chapters 1–4)—the ever-present Era count, the dynastic temporality, the Seleucid horizon, and the distantiation of the pre-Hellenistic past. As I hope to demonstrate, it is, above all, the shared inhabiting of the Seleucid world, and not a particular social location, theological perspective, literary genre, or chain of influence, that undergirds the common concerns and structures of these works.

This chapter explores the origins, character, and function of this particular, if highly influential, counterdiscourse about historical time; alternative viewpoints, other movements, and the entangled complications of anti-Seleucid action will be addressed in Chapter 6. The analysis here will proceed by discussing in turn the major, extant eschatological total histories before treating their common technology of periodization. I begin with a close reading of the book of Daniel, for this is the longest, most sophisticated, and best-known example of an anti-Seleucid eschatological total history, in relation to which the other works can most helpfully be understood.

Daniel

The book of Daniel, the only canonized work under discussion, is an interrogation of political change, Hellenistic imperial rule, and the nature of meaning and justice in history. The book, in its final form, can be dated to the last year or two of the reign of Antiochus IV Epiphanes, circa 165 BCE: by the standard method of fixing the composition of *vaticinia ex eventu* (pseudonymous works that present historical events as prediction), its closing vision's fine-grained account of Ptolemaic-Seleucid conflict is accurate up to but not beyond that date.[7] The book is arranged into distinct episodes, for which the chapter divisions provide mostly satisfactory boundaries. Formally, these fall into two halves. Chapters 1–6 consist of third-person tales of the Judean exiles Daniel, Hananiah, Mishael, and Azariah in the courts first of Nebuchadnezzar II, the Neo-Babylonian conqueror of Jerusalem, then of Belshazzar, his supposed son (in fact, the son of king Nabonidus), and finally of Darius the Mede, an invented Median conqueror of Babylonia.[8] In a series of trials, contests, and ultimately triumphs—including the famous lions' den and fiery furnace—the Judean heroes are challenged to reconcile the earthly authority of the gentile king with their faithful service to the Jewish God.[9] By contrast, chapters 7–12 are presented as Daniel's own first-person record of visionary angelophanies that predict, in bizarre figuration, the future history of the Seleucid kingdom. Date formulae set these revelations in the successive reigns of the Babylonian king Belshazzar, the Median king Darius, and the Persian king Cyrus the Great. Linguistically, chapters 2:4b to the end of chapter 7 are composed in Aramaic, and chapters 1:1 to 2:4a and all of chapters 8–12 in a late biblical Hebrew. These formal and linguistic differences do not precisely overlap, and this is crucial: it allows us to nuance the standard, diachronic model, first proposed by Baruch Spinoza, of the book's gradual accretion as a composite of Aramaic court tales of late Achaemenid or third-century BCE date and Seleucid-period Hebrew apocalyptic visions.[10] In fact, the book as we have it displays a carefully constructed, programmatic unity: the sequence of languages, Hebrew-Aramaic-Hebrew, traces the shift from covenantal independence to imperial world empire and then to the eschatological reclaiming of that national autonomy;[11] the not-quite-perfect linguistic distinction between the court tales and the visions simultaneously melds the Seleucid with the pre-Hellenistic kingdoms and marks it as separate; and the generic distinction between court tales for the Babylonian, Median, and Persian kingdoms and apocalyptic visions for the Seleucid proposes a real contrast of political character. The following analysis, progressing in order through the book of Daniel, seeks to demonstrate its overriding concern with the periodization of history and the temporal location within it of the Seleucid kingdom. Daniel offers, in some respects, a deeper insight than Polybius or the ever-increasing library of inscriptions into the workings and experiences of the Seleucid regime. Moreover, the book rewards close attention. Not only does it unite a sharp opposition to empire with a repeated interrogation of the structuring of temporal duration. It also had an impact that cannot be overestimated: for almost two millennia its periodization of history was employed as the central structuring device of deep historical thinking and future prognosis in Christian, Jewish, and even Enlightenment thought.[12]

The book of Daniel immediately establishes for itself an exilic political and geographic landscape. Its first chapter opens with a dating formula, in the manner of traditional biblical historiography: "In the third year of Jehoiakim, king of Judah, Nebuchadnezzar, king of Babylon, came to Jerusalem and laid siege to it. Into his hand the Lord gave Jehoiakim, king of Judah, and some of the vessels of the Temple of God, and he brought them to the land of Shinar."[13] We are placed at the historical watershed of Jewish history: the Neo-Babylonian conquest of Jerusalem, the end of the Davidic state, and the exile of the king, Judeans, and sacred loot to Nebuchadnezzar's capital.[14] The chapter's chronological marker assumes all earlier biblical history, from creation to conquest, as described in the Law and Prophets; but it fixes as the book's relevant historical field only the events after Israel had fallen out of the holy land. For the author(s) of Daniel, the exile opened a special period of history, when Jerusalem and the Davidic line were replaced by gentile empires and kings, and in consequence, Israel's sacred history became coterminous with the fate of these universal monarchies.[15] All the book's tales and visions belong to this postconquest world of foreign, despotic kings and their unstable, competitive courts. Importantly, the continuation of the book's fictive chronology from the first trauma of exile (1:1) to the reign of Cyrus the Great (1:21)—indeed, into his third regnal year (10:1)—refuses to recognize or countenance the Persian-sponsored return to Judea.[16] In other words, according to the book of Daniel, the period of history that had begun with the fall of Jerusalem will continue, without earthly redemption, until the *eschaton*. Accordingly, the book establishes three distinct epochs: the richly detailed gentile monarchies of its courtly present (Babylonian, Median, Persian) and visionary future (Seleucid); the undescribed but entirely assumed past from creation to exile; and the faintly outlined kingdom of God beyond time's end.[17]

Following this opening dating formula, the first chapter immediately reveals a concern with the delimitation of discrete periods of time; this characteristic, appearing first in the narrow, folktale setting of the royal palace, will later emerge at the cosmic scale of providential history. The heroes of the book's court tales (Daniel, Hananiah, Mishael, and Azariah—handsome, unblemished, wise youths of the exiled Judean nobility in Babylonia)[18] are handed over to the chief eunuch for three years. For this specified duration

Daniel and his companions are compelled to undergo a thorough assimilation into the intellectual, linguistic, dietary, and bodily culture of their imperial master: educated in the language and learning of the Chaldeans ("*ûlᵃlammᵉḏām se)er ûlᵉšôn kaśdim*"), fed from the king's table ("*mippaṯ-bap*," an Old Persian word[19]), redesignated with the Babylonian or Persian theophoric names Belteshazzar, Shadrach, Meshach, and Abednego, and, at least by implication, reduced to eunuchs.[20] At the end of this fixed time they were to be brought to serve before the king. Needless to say, as we shall see, such a reconfiguring of personal identity and bodily behaviors to that of the imperial ethno-elite was a central dynamic of Hellenistic Judea in general and of the Maccabean moment more acutely. Furthermore, the three-year period prescribed for this transformation recalls or, from the book's pseudonymous perspective, anticipates Antiochus IV's three-year profanation of the Jerusalem Temple.[21]

The youthful Daniel responds to this timed cultural effacement by determining not to defile himself with the royal food and drink. To assuage the fears of their caretaker, the chief eunuch, that he would be liable for the Judeans' starved features, Daniel proposed a trial: "Please test (*nas-nā'*) your servants for ten days" by feeding them vegetables and water and then, at the end of this delimited period, comparing their appearance to those who had stuffed themselves with the king's *pat-bag*. Needless to say, after ten days of this diet, the four Judeans look healthier and chubbier than all their fellow courtiers.[22] Obviously, the mini-victory valorizes the retention of dietary purity, a key concern in the context of the Antiochid persecution, but it also establishes the idea of time as trial, of the testing of faith for a numericalized duration, with a retrospective justification for pious suffering. We shall see that the numeral ten, the period of days established by Daniel, recurs in several eschatological periodizations. Thus, already in this brief Hebrew introduction to the Aramaic court tales, set at the very beginning of the exile, with a conciliatory dynamic and apparently low stakes, the book of Daniel establishes the control of time as the major site of conflict, opposing gentile-imperial and Jewish-divine authority over temporal duration.

The book of Daniel's second chapter develops this theme into a deep historical perspective. It is an altogether remarkable piece of symbolic historiography. The episode opens "in the second year of king Nebuchadnezzar" (2:1), preferring the regnal year of the gentile ruler to the exilic date of the Davidic king employed in other biblical texts.[23] The story tells how king Nebuchadnezzar, in bed at night, with his spirit troubled by disturbing dreams, summons the full department of eastern mantic experts—magicians (*ḥarṭummim*, an Egyptian loanword), astrologers (*'aššā)im*, an Akkadian term), sorcerers (*mᵉkaššᵉ)im*, Hebrew), and Chaldeans (*kaśdim*)—setting the stage for a competition of nations and their wisdom traditions. The king demands of this faculty of scholars that they produce not only the interpretation of the dream that had disturbed his sleep but also its content ("the dream and its interpretation [*ḥelmā' û)išreh*]"). When the wise men of Babylon, addressing the king in Aramaic, protest the unfeasibility of such a challenge, Nebuchadnezzar accuses them of trying to "buy time"—a new idiom of temporal commodification, discussed in Chapter 2—and condemns all of them to death, including Daniel and his companions.²⁴ But on the eve of the mass execution, "the mystery" (*rāzâ*, a Persian loanword) of the king's dream is revealed to Daniel, who in response addresses a thanksgiving poem to God, to which we shall return below. The next day, Daniel was brought before Nebuchadnezzar to reveal the dream and its interpretation:

(31) You, O king, were watching, and—behold!—a single great statue (*ṣᵉlem ḥad śaggi'*); this statue, mighty and exceedingly dazzling, stood before you (*qā'em lᵉqoblāk*), and its appearance was dreadful. (32) The head of this statue was of pure gold, its breasts and arms of silver, its belly and thighs of bronze. (33) Its legs were of iron, its feet partly of iron and partly of clay (*raplôhi minnᵉhewn di)arzel ûminnᵉhewn di ḥᵃsa)*²⁵). (34) You were watching until a stone was cut out, not by (human) hands, and it struck the statue on its feet of iron and clay and broke them into pieces. (35) Then the iron, the clay, the bronze, the silver, and the gold, were crushed all as one (*dāqû kaḥᵃdâ*) and became like the chaff on the summer threshing-floors; and the wind carried them away, and not a trace of them was found; and the stone, which struck the statue, became a great mountain and it filled the whole earth.

Daniel then interprets. Addressing king Nebuchadnezzar, Daniel fixes the dream's symbolic code: "You are the head of gold." Then, continuing into an ever more distant future: "After you will arise another kingdom, inferior to you (*'ᵃra'*) and then a third kingdom, of bronze, which will rule over all the earth. Then there will be a fourth kingdom, strong as iron." As Daniel explains, just as iron shatters everything, so this kingdom "will break and crush all these" former kingdoms. Yet in as much as Nebuchadnezzar saw the feet and toes partly of clay and partly of iron, this fourth kingdom will be divided ("*malkû*)*ᵉlipâ*"), partly strong and partly brittle. The *pesher* concludes by elucidating the function and identity of the stone: "In the days of those kings (*ûbyômehôn di malkayyā' 'innûn*)"—note the dynastic, not monarchic, temporality—"the God of heaven will set up a kingdom that will never be destroyed, nor will it be left to another people . . . but it will itself endure forever (*wᵉhi' tᵉqûm lᵉᶜolmayyā'*)."[26] Nebuchadnezzar, recognizing his dream and Daniel's inspired interpretation, falls in cultic homage before the Jewish exile,[27] recognizes his God, and promotes him to the charge of Babylon and all its wise men.

This self-standing episode is framed as a court contest, betraying the typical "ruled ethnic perspective" that takes vicarious pride in the success and elevation of a compatriot over his court competition—Mordecai before Ahasuerus, Ahiqar at the Neo-Assyrian court, and Joseph in Egypt.[28] Indeed, it has long been recognized that the chapter is related to Genesis 41, where the Israelite ancestor Joseph was promoted from prison to the viziership for correctly interpreting pharaoh's troubling dreams: seven fat and seven lean cows, seven plump and seven thin ears of wheat, signifying seven years of agricultural plenty to be followed by seven years of famine.[29] Yet Nebuchadnezzar's dream and Daniel's interpretation are different in scale and type. The chapter is above all concerned to establish a double periodization: first, it orders into discrete periods the whole postexilic history, exposing the hidden but already determined structure of successive world empires; and second, it predicts the destruction of this half-millennium of gentile monarchy, presented as a coherent if segmented unit, and its replacement by God's everlasting kingdom.

The dream aligns three different systems for its postexilic history. First, Daniel orders a succession of numbered empires: after the Babylonian present, "another kingdom," "a third kingdom," and "a fourth kingdom." Although names are not assigned to these successive rules, both the book's date of composition and its eighth chapter, where the identifications are made explicit, fix the ordering as Babylon, Media, Persia, and Alexander's Macedonia and its Seleucid and Ptolemaic successors. Second, the dream sequences metals of declining worth from gold to silver, bronze, iron, and an iron-clay mix. This is an ancient, obvious, and much-paralleled index.[30] As we shall see below, it also appears in apocalyptic traditions from Iran. Third, and too often overlooked in exegesis, Nebuchadnezzar's gaze moves down through the parts of a human body, from head to toe.[31] Not only is this a progressive move from the most to the least noble limbs (*ᵃraᶜ* in Aramaic, meaning "earthward" or "inferior"),[32] but it also demonstrates the ever-increasing fragmentation and splintering of "the kingdom," from head and chest to belly and thighs and then to legs, feet, and (in Daniel's interpretation) toes. Indeed, the Seleucid-Ptolemaic iron-clay feet or toes neatly capture both the fracturing of Seleucid authority across the early second-century Near East and the Hellenistic breakup of universal empire into a self-consciously peer-kingdom system.[33] The homology of these periodizing schemes—dynastic, material, and bodily— inscribes a clearly staged historical sequence that is also a trajectory of decline: the Seleucid kingdom is fourth, least precious, most lowly, and broken.

At the same time, Nebuchadnezzar's dream presents a completed, total history. The statue is a body—unitary ("a single image [*ṣᵉlem ḥaḏ]*"), if divided into limbs, and necessarily eschatological, for there can be nothing beyond the feet. Moreover, the number four was a widespread Near Eastern figure for totality.[34] Focalized in the Babylonian king, the dream projects a viewing gaze that is exterior and out of time, as much a distantiation and objectification of the author(s)' Seleucid present as of Nebuchadnezzar's own reign.[35] This is a kind of synchronic knowing that can look from head to toe, along time's arrow, and also from toe to head, as the stone destroys the iron-clay, bronze, silver, and gold body parts. Indeed, the unity of this half-millennium of gentile monarchy is confirmed by its simultaneous destruction: the four metals "were crushed as one (*dāqû kaḥᵃḏâ*)." The history is absorbed by its effect. The kingdom of God is opposed to gentile empire as a whole—just as an unworked, natural stone, not cut by human hands, eternal, and unchanging, is aesthetically, materially, and temporally antithetical to the artificial, processed metals and burned clay of this transitory, destructible, unstable, idol-like image.[36]

The statue's destruction has long been situated within a distinctly Jewish prophetic discourse and anti-idol polemic. Certainly, the wind blasts that scatter the shattered statue fragments as mere chaff from the threshing floor are built up from common metaphors of God's judgment.[37] The earth-filling glory of God is an eschatological trope.[38] But the central image, which is without close parallel in the Hebrew Bible, should rather be understood within the developed Hellenistic political culture of statues, their unmaking, and the self-periodizing of communal histories. Statues of monarchs and political notables were among the most visible features of the Hellenistic cityscape, in both old *poleis* and newly founded colonies. Public squares, sanctuaries, gymnasia, council houses, city gates, and a host of other prominent sites were populated by images of rulers, standing on podia or seated on horseback. These statues were often the end result of a euergetical exchange, of a transaction between—according to their public transcript—a well-governed populace and an accepted, acceptable political authority.[39] While few of these statues survive, many publicly inscribed decrees in Greek record political assemblies voting to honor someone with a statue, often specifying the metal to be employed. A decree from Ilium for king Antiochus I, for instance, prescribed that the city "set up a gold statue of him on horseback in the most prominent place in the temple of Athena ([στῆσαι δὲ αὐτοῦ εἰ]|κόνα χρυσῆν ἐφ᾽ ἵππου ἐν τῷ ἱερῷ τῆς Ἀθηνᾶς ἐν τῷ ἐπιφα[νεστάτῳ τόπῳ])."[40] The materialization of royal honor and acknowledged legitimacy in publicly placed statues paradoxically made these images the very foci of resistance to the intrusive political regime they manifested. In other words, as with Nebuchadnezzar's dream, the same statues that at first view expressed the fixity and permanence of political power were among the chief media for exposing its transitoriness.

Such manipulation of statues is known across the Mediterranean and west Asian world: we have already seen, in Plutarch's anecdote at the opening of Chapter 4, that the Macedonian takeover of the Persian empire was encapsulated in Alexander's gazing upon the fallen statue of Xerxes.[41] Hellenistic Athens provides two especially illuminating episodes of statue destruction. The first has to do with Demetrius of Phalerum, the Peripatetic philosopher and later the founding father of the Library of Alexandria, who governed Athens from 317 to 307 BCE on behalf of Alexander's successor Cassander. Demetrius had been honored by the city with an unprecedented number of statues. But when democracy was restored to Athens in 307 BCE, Demetrius was expelled and his statues targeted by a number of outrages: all were pulled down, some thrown into the sea, others melted into coin, and—the basest of humiliations—still others turned into chamber pots.[42] It is, perhaps, no accident that this same Demetrius would observe, in his treatise on *Tychē* (introduced in Chapter 4), the fickleness and unpredictable collapse of imperial fortunes.[43]

The second episode took place a little over a century later, in 200 BCE, when Athens was allied with Rome against the Antigonid king Philip V, in what is known as the Second Macedonian War. The city's assembly formally decreed "that the statues and portraits of Philip and all their inscriptions, as well as those of his ancestors, both male and female, be removed and destroyed (*ut Philippi statuae imagines omnes nominaque earum, item maiorum eius virile ac muliebre secus omnium tollerentur delerenturque).*"[44] Several decades ago archaeologists uncovered from one of the wells in the Athenian agora the deliberately shattered fragments of an equestrian statue of Demetrius Poliorcetes, Philip V's great-grandfather (and the father-in-law of Seleucus I); these *membra disiecta* were found in an archaeological context dating to around 200 BCE, confirming their connection to this event.[45] Just as Nebuchadnezzar's dream toppled all four metal empires, not just the Seleucid feet of iron and clay, so the Athenians went back through time to destroy the statues of Philip V's once-honored ancestors.

These two cases, which can be much paralleled, indicate that statue toppling went beyond a simple erasure: as in Daniel 2, we witness first the powerful symbolization of personified imperial power and then that symbol's inversion, desacralization, fragmentation, and ultimately elimination.[46] The very statues that had substituted the absent king's body for a subordinate community now permitted fantasies of vicarious violence against it. We are not far from the logics of "voodoo dolls" and sympathetic magic. Such topplings of rulers' statues were physically enacted declarations of the end of subservience. In ideology, if not necessarily in effect, they inaugurated the arrival or restoration of autonomy. Thus, within the broad civic culture of the Hellenistic world, destroying the king's statue was a widespread, public, and self-conscious idiom for periodizing a community's history. This is key to understanding the temporal texture and historical shaping of Nebuchadnezzar's dream.

To sum up: the second chapter of the book of Daniel schematically segments a half-millennium of political history into four gentile empires and then sets these face-to-face with the eternal, eschatological kingship of God. The lesson of the episode is given programmatic, theological formulation in Daniel's doxology, or thanksgiving proclamation, delivered after the mystery of Nebuchadnezzar's dream was revealed to him:

> Let the name of God be blessed from eternity and for eternity (*min-ʿolmā' wᵉʿaḏ-ʿolmā'*),
>
> For the wisdom and the power are His.
> He is the one who changes the times and the seasons (*wᵉhû' mᵉhašne' ʿiddānayyā' wᵉzimnayyā'*),

Who removes kings and establishes kings *(mᵉhaʿdê malkin ûmᵉhāqem malkin).*[47]

In the typical manner of biblical poetry, the prayer is constructed as a series of semantic parallels or "thought-rhymes," where the first verset is strengthened, heightened, advanced, or specified by the second:[48] accordingly, as God controls the passage of time, so he determines the fate of empires. The word ʿiddānāʾ ("the time"), which Nebuchadnezzar had accused Babylon's wise men of buying when they could not tell him his dream (Daniel 2:8; see Chapter 3), is here the plaything of God alone, "the one who changes the times *(mᵉhašneʾ ʿiddānayyāʾ).*" The key point is that temporal control and earthly kingship are identified as linked, synonymous phenomena. The Jewish God—emphatically identified by the Aramaic coordinated pronoun *wᵉhûʾ* ("it is He who")—is the architect of both, mastering time and ruling kings.[49]

The book's next four chapters consist of court tales that demonstrate, in folkloric idiom, the exclusive and uniquely salvific sovereignty of the Jewish God. Thus, in Daniel 3's martyr-type legend, Daniel's companions Shadrach, Meshach, and Abednego refuse to join the rest of the Babylonian court in worshipping a new golden idol erected by Nebuchadnezzar; condemned by the king to an agonizing death in a superheated fiery furnace, the pious Judeans remain unsinged.

The following episode, Daniel 4, takes the form of an encyclical edict of Nebuchadnezzar to all his subject peoples. Like Daniel 1, it concerns the delimitation of an appointed temporal duration. The king reports how he was disturbed by troubling dreams of which, once again, only Daniel could make sense: a great, mighty, and flourishing cosmic tree, suddenly cut down by the decree of a heavenly Watcher, is interpreted as the humbling of Nebuchadnezzar with a seven-year loss of kingship and a bestial sojourn in the wilderness.[50] And so it came to pass: twelve months later, the prideful *roi bâtisseur,* promenading on his palace rooftop, boasts of his capital: "Is not this Babylon the great, which I have built as a royal seat, by the might of my power, for the sake of my glory?"[51] No other ancient source, it should be noted, more neatly captures the ideology of the Seleucids' urban foundations.[52] At the very instant of his self-satisfied bragging, Nebuchadnezzar is driven from society to eat grass like the cattle, his hair growing long like an eagle's feathers, his nails curling into talons. Only his recognition of God's eternal dominion "at the end of the period *(wᵉliqṣāṭ yômayyâ)*" restores the monarch to civilization and his throne.

The next tale, Daniel 5, puts us in the reign of Belshazzar and on the eve of Babylon's fall to Darius the Mede. This last Babylonian king's debauched banquet, for which the Jerusalem Temple's vessels were profaned, is interrupted by a spectral hand inscribing on the palace's plaster wall words that Daniel alone can understand: "*mᵉneʾ mᵉneʾ tᵉqel û)arsin*" (that is, the weight units mina, mina, sheqel, and half-minas). As we have seen in Chapter 2, the graffito riddle employs these four devices of market trade to connote the divinely determined succession of imperial rule.[53]

In the following, sixth chapter, Daniel now serves Belshazzar's conqueror, Darius the Mede. In a contest tale that shares theme and structure with Daniel 3 (the fiery furnace), the book's eponymous hero is thrown into a den of lions for continuing to worship his God. As the piety of Shadrach, Meshach, and Abednego kept them cool amid the flames, so Daniel's steadfast fidelity maintains him unscathed. Then, in an inversion typical of such tales, Daniel's competitors at court, together with their families, are condemned to the starving beasts, which no longer restrain their hunger, and Darius the Mede issues a panimperial edict recognizing Daniel's God.

These four chapters have been treated cursorily, but nonetheless it should be apparent that, even if built up from a pre-second-century BCE kernel, they work hard against the Seleucid imperial power: neutralizing the humiliations of Graeco-Macedonian rule, satirizing the royal forms and heralded achievements of Hellenistic kingship, and urging religious fidelity and perseverance amid the traumas of imperial persecution.

A dominating and ever more urgent concern with the periodization of history returns in the four apocalyptic visions that make up the second half of the book. Daniel 7 occupies a pivotal and unifying place—composed in Aramaic, like the court tales that precede it (Daniel 2–6), yet inaugurating the first-person Hebrew visions that complete the book (Daniel 8–12). Although the chapter is dated to the first year of Belshazzar, it is Daniel, not the Neo-Babylonian king, who is now the visionary witness, anticipating by focalization the eschatological relocating of political sovereignty. The chapter is presented as Daniel's scribal record of his own experience:

(2) I was watching in my vision during the night and—behold!—the four winds of heaven were stirring up the great sea (*ʾarbaʿ rûḥe šᵉmayyāʾ mᵉpiḥān lᵉyammāʾ rabbāʾ*). (3) And four great beasts (*wᵉʾarbaʿ ḥewān raḇrḇān*) came forth from the sea, each different from the other.

These beasts are numbered and described in turn. The first appeared as a lion with the wings of an eagle; Daniel saw it being somewhat humanized, its wings plucked, lifted onto two feet, and granted a human heart. The second was as a bear, raised up on one side, with three ribs between its teeth; it was told, "Arise, devour much flesh!" "After this *(bāʾtar dᵉnâ)*" appeared the third, as a leopard, with four wings and four heads, "and dominion was given to it *(wᵉšolṭān yᵉhib lah)*."

> (7) And after this *(bāʾtar dᵉnâ)* I was watching in visions of the night and— behold!—the fourth beast, dreadful and terrible and exceedingly strong; it had great teeth of iron *(wᵉšinnayîn di-)arzel lah rᵉbiʿāyâ)*; eating and smashing and trampling the remnant with its feet; it was different from all the other beasts that were before it *(mᵉšannᵉyâ min-kol-ḥewāṭāʾ di qāḏāmayh)*; it had ten horns *(wᵉqarnayîn ʿᵃšar lah)*. (8) I was contemplating the horns and—behold!—another horn, a little one, came up in their midst; and three of the former horns were uprooted before it, and—behold!—there were eyes like human eyes in that horn and a mouth speaking great things.

Daniel, not comprehending this disturbing night vision, seeks elucidation from an angel. These four beasts, Daniel is informed, are four kingdoms. The fourth beast's ten horns are ten kings, and the eleventh is different from its predecessors. This eleventh horn-king, the angel continues, will bring about the downfall of three kings and "will speak words against the Most High and will afflict the holy ones of the Most High." Then, in a passage of the utmost significance, Daniel is told that this eleventh king "will think to change times and law *(wᵉyîsbar lᵉhašnāyâ zimnin wᵉḏāt)*, and they will be given into his hand for a time, and times, and half a time *(ʿaḏ-ʿiddān wᵉ ʿiddānin û)ᵉlap ʿiddān)*" (7:25).

Spliced into this first-person report of the four beasts is a vision in poetry of the heavenly court of God: the "Ancient of Days *(ʿattiq yômin)*," served by myriads upon myriads, takes His seat on a flashing throne of flowing fire;[54] then, a trial of judgment *(dināʾ)* begins, and books are opened *(wᵉsi)rin pᵉṭiḥû)* (7:9–10). In consequence, Daniel sees the fourth beast put to death and its body committed to the burning flames. The first three beasts, although they have their dominions taken away *(heʿdiw šolṭonhôn)*, are granted an extension of life "for a season and a time *(ʿaḏ-zᵉman wᵉ ʿiddān)*" (7:11–12). Then, in a moment of great importance for later messianic and Christological thinking, Daniel sees "one like a son of man (k^{e}ḇar ^{e}nāš)" approach before the Ancient of Days, riding with the thunderclouds; to him is given "dominion and glory and kingdom (šolṭān wiqār ûmalḵû)" over all peoples: "His dominion is an everlasting dominion (šolṭāneh šolṭān ʿālam), which will not pass away, and his kingdom is indestructible" (7:13–14).

Daniel's vision constructs two periodizing schemes that precisely map onto the multimetal statue and stone mountain of the book's second chapter: the sequence of four kingdoms; then, following the heavenly trial and punishment, the commissioning of "one like a son of man" to eternal rule. Although unnamed, we can again be certain that the beast-kingdoms are the Babylonian, Median, Persian, and Macedonian-Seleucid empires. The precise significance of each kingdom's bodily attributes mostly escapes confident reconstruction,[55] but the bestiary as a whole is a mode of representation that, among other things, doubly engages the Seleucid imperial thought-world.

First, the vision draws directly from the empire's own self-representation. Alone among predecessor, peer, or successor states, the Seleucid dynasty had established a particular animal—the Indian elephant, than which there was no more fearsome or exotic beast—as a blazon of its identity and symbol of its might.[56] The vision's central symbolic opposition (beastly, hybrid monsters and humanlike heavenly or Jewish figures) takes this imperial self-symbolization as a disclosure of its true obscenity: a grotesque, violent foulness slouching out of the chaos-waters. Indeed, certain attributes of the fourth beast (its iron teeth, probably tusks, and the trampling underfoot) precisely recall the characteristics of the war elephant.[57] We shall see below that contemporary Enochic texts characterized the war elephant as a primordial evil, the monstrous offspring of the fallen angels.

Second, the chapter adopts the underlying imagery and common structure of Canaanite and Mesopotamian *Chaoskampf* myths: the great cosmic sea as the spawning ground of monsters, which are defeated by a youthful warrior-creator god.[58] In this context it may be significant that, as I discussed in Chapter 1, the Seleucid kingdom's formalized origin in the Seleucid Era epoch of 311 BCE was coordinated to the Babylonian New Year festival, in which the reciting of the *Enūma eliš* myth and the procession out of and back into Babylon reenacted Marduk's victory over the sea demon Tiamat. It was also suggested that this commemoration had acquired an imperial Seleucid character, and that Seleucus's foundational defeat of Antigonus Monophthalmus may have been assimilated to its paradigm. Accordingly, if this association was known in Judea, as I think likely, then Daniel's vision turns the origin myth of imperial time against itself.[59] Psalm 74, if dated to the Maccabean period, is a helpful comparison, marshaling similar ideas against the empire: the enemy's destructive outrages against the Temple; the absence of prophecy (see Chapter 4); the imposed religious change ("Your foes roar inside Your meeting-place; they take their signs for true signs"); the divine, not royal, control of time and calendar ("The day is Yours, the night also is Yours. . . . Summer and winter—You made them"); and God's defeat of the chaotic sea demons ("It was You who drove back the sea with Your might, You who smashed the heads of the monsters in the waters").[60]

The fourth beast and its last horn, the vision's central focus, represent the culmination of worldly history. Daniel's vision is impelled toward this absolute, ultimate monstrosity. The Seleucid empire is repeatedly and explicitly marked off as something qualitatively unprecedented: "it was different from all the beasts that came before it (*meᵉšanneᵉyâ min-kol-ḥewātāʾ di qāḏāmayh*)" (7:7); its body, without named faunal identifications, is characterized only by its surpassing violence—it cannot be named; and it is uniquely condemned to the hell-flames. Moreover, unlike the other beast-empires, the fourth has an anatomy that is refined into a dynastic king-list of individual monarch-horns, once more adopting and recoding official imperial imagery. The eleventh horn (Antiochus IV Epiphanes) is, like the fourth beast itself, without precedent ("and it was different from the earlier ones [*weᵉhûʾ yîšneʾ min-qaḏmāyêʾ*]" (7:24), a phrase with lexical roots identical to those of the previously quoted phrase). Antiochus IV's knocking out the three other horns likely refers to his displacement of Seleucus IV's legitimate sons, his nephews Demetrius (I), abandoned as a hostage in Rome, and a child Antiochus, soon done away with, and, perhaps, the vizier Heliodorus or, retrospectively, Seleucus IV.[61] In other words, the vision seeks to delegitimize the persecutor king according to the empire's own temporal regime, as I laid out in Chapter 3: determined vertical succession and continual reference to the precedent of the dynastic *progonoi*.

The terminal horror of earthly kingship concerns temporality. The eleventh horn, who utters words against God and wears down the holy ones of God, "will think to change times and law (*weᵉyisbar leᵉhašnāyâ zimnin weᵉḏāt*)" (7:25). It has been plausibly suggested that this refers not just to Antiochus IV's abolishing of the divinely ordained Jewish cult and sacrifices but also, more pointedly, to an enforced replacement of the Temple's solar, 364-day, sabbatical

calendar with the empire's luni-solar, 360-day, Babylonian intercalatial calendar. Certainly, the calendrical polemics in 1 Enoch, the book of Jubilees, and
the early Qumranic literature suggest that the abandonment of the God-given
solar calendar contributed to the sect's subsequent break from the Jerusalem
community.[62] Whatever Antiochus IV's precise actions, now irrecoverable in
detail, this final horn of the last beast is characterized as a usurper of the divine prerogative of temporal control. As we have seen above, the seer's doxology in Daniel 2:21 had hymned God as "the one who changes the times and
the seasons *(weˁhû' meˁhašne' ˁiddānayyā' weˁzimnayyā')*" and, in the parallel
verset, "the one who removes kings and establishes kings *(meˁhaˁdê malḵin
ûmeˁhāqem malḵin)*." Here, Antiochus Epiphanes brings about the downfall of
three kings and attempts to change the times and law. This goes far beyond a
lexical echo: it is a direct confrontation of the new Seleucid king (an erupting
horn) and the eternal God (the Ancient of Days) over the shaping of time and
the control of history.[63] Indeed, the king's pretense to such power is exposed
by God's determining of his profanation's length: "a time, times, and half a
time *(ˁad-ˁiddān weˁiddānin û)eˁlap ˁiddān)*"; "times *(weˁiddānin)*" is pointed as
a plural but has usually and rightly been understood as a dual ("two times"),
amounting to, as elsewhere in the book of Daniel, three and a half years.

Daniel 7 presents a periodized total history. It extends from the wind-
spumed cosmic waters—a clear allusion to the world's first emergence in the
book of Genesis,[64] and so to God's creative mastery of historical process—right
up to the burning fires of judgment. The four beast-empires appear in a historical sequence; note their ordinal numbering and the repeated temporal
marker *bā'ṭar deˁnâ* ("after this"). Yet at the same time, Daniel's external, distanced spectating presents a grotesque menagerie[65] that is closed and simultaneous. Indeed, as we have already seen in Nebuchadnezzar's dream statue
(Daniel 2) and the writing on the wall at Belshazzar's feast (Daniel 5), the
number four, thematized throughout this vision, was a figure of totality.

The climax of this history takes the form of a heavenly trial that gazes
back upon and evaluates all this past: the seating of the court, the reading
of accusations or proofs from the opened books, and the execution of
sentence.[66] Indeed, much as with the tale of Judith, which I discussed in
Chapter 4, all books are simultaneously open. The grant of "a season and a
time" to the first three beasts, even after the fourth's condemnation, undoes
every Seleucid attempt to seal itself off from preceding empires. Perhaps (as we
shall see in my Chapter 6), such an extension points to Judean knowledge of

and identification with contemporary anti-Seleucid activity in Babylonia, Armenia, and western Iran.[67] Finally, as with the stone that smashes the statue and grows to a mountain in Daniel 2, so this vision discloses an absolute spiritual, bodily, and temporal difference between the common monstrosity of gentile empire and the eschatological kingdom that will follow, however faintly outlined. Whoever the "one like a son of man" may be—the collective symbol of Israel or the righteous Jews, the archangel Michael or Gabriel, Judas Maccabeus, or the messiah[68]—the sempiternity of his kingdom, after or outside of history, cannot be more strongly emphasized: the holy ones will possess this kingdom "forever and forever and forever (*'ad-'ālmā' w^e'ad 'ālam 'olmayyā'*)" (7:18).

The next vision, the book's eighth chapter, takes place in the third regnal year of Belshazzar, and thus two years after Daniel 7, beside the Ulai canal of the Elamite capital Susa, in southwestern Iran. For the first time we are given explicit, political identifications of the symbolic imagery. The chapter recounts, again in Daniel's voice, first a distressing vision and then archangel Gabriel's interpretation. Daniel sees a ram with two horns, the second of which grows up later and greater; this ram thrusts in the cardinal directions, overcoming all opposition. Gabriel explains that this ram is the "kings of Media and Persia" (8:3-4, 20). Suddenly, a he-goat, with a single horn between its eyes, charges from the west at full speed and in fury against the ram, breaking its horns and trampling it underfoot. This he-goat grows mighty until, at the height of its strength, its single great horn is shattered and "four conspicuous ones grew in its place (*watta^lenâ ḥāzût 'arba' taḥtehā*)," in the directions of the four winds of heaven. As we might have guessed, Gabriel identifies the he-goat as the "king of Greece (*melek yāwān*)" and the single horn as its "first king (*hammelek hāri'šôn*)," that is, Alexander (8:5-8, 21); indeed, the horn's adjective, "the great (*hagg^e dôlâ*)," may be an early allusion to the employment of this epithet for the conqueror.[69] The subsequent four horns are the successor kingdoms. Daniel watches as another horn sprouts from one of these four successors, at first small, but "growing exceedingly great (*wattipdal-yêter*)" toward the south, the east, and the holy land of Israel. Gabriel identifies him as a "fierce-faced king (*melek 'az-pānim*)," adept at duplicity, mighty in power, devising wondrous things, arising in "the latter time of their kingdom (*ûb'aḥ^rit malkûtām*)" (8:9, 23-25); evidently, this is the persecutor king Antiochus IV Epiphanes. Daniel sees the Temple's daily offering taken away ("*huraym hattāmid*"), the sanctuary cast down ("*w^e hušlak m^e kôn miqdāšô*"), and truth thrown to the ground *("w^etašlek ^{ʾe}met ʾarṣâ")* (8:10–12). Then, Daniel overhears a holy being asking another "for how long *(ʿaḏ-māṯay)*" this abomination would continue, and the response "for two thousand three hundred evenings and mornings," at which point, this horn "will be broken without human hand *(ûḇ^{eʾ}e)es yāḏ yiššāḇer)*" (8:13–14, 25). Gabriel, having explained all this to Daniel, instructs him, in the language of archival bureaucracy, to "seal up the vision *(w^{eʾ}attâ s^eṯōm heḥāzôn)*" as it pertains to the distant future. And so Daniel, traumatized by sights and words he still cannot fathom, returns to his duties at Belshazzar's court (8:26–27).

This vision at Susa, though sharing with Daniel 7 a sequence of beasts and plurality of horns, simplifies its historical scheme and restrains its imagery. The Babylonian kingdom goes unmentioned, the Median and Persian are two horns of the same ram, and similarly the Hellenistic kingdoms are grouped as a single he-goat. Nothing is said of what will follow Antiochus IV's fall. The vision's key concern and central contribution, asked for the first time in the book, is the question "for how long *(ʿaḏ-māṯay)*?" Offered in response is a divinely predetermined, numericalized count of a predictable, regular, future time: 2,300 evenings and mornings, or 1,150 days, a little over three years.[70]

The following episode, Daniel 9, is set in the first year of Darius the Mede, "who was made king over the kingdom of the Chaldeans" (9:1); the use here of the Hofal *homlak* ("he was made king"), a passive, causative verbal form, indicates that the imperial baton passing is under God's control. I have already discussed Daniel 9 in Chapter 4, so we can be brief here. As we have seen, it forms an exegetical rather than a symbolic vision: to Daniel, pondering on the texts of Jeremiah's seventy-year prophecy, the archangel Gabriel reveals that these seventy years are, in fact, seventy weeks of years—that is, 490 years, or ten jubilees.[71] This is the chronological backbone for the full extension of Daniel's biography and visions, from Nebuchadnezzar's sack of Jerusalem to the end-times: the quadrimetallic statue (Daniel 2), the parade of beasts (Daniel 7), and the ram and the he-goat (Daniel 8) are here numericalized as a determined count of years.

Gabriel divides these seventy weeks of years into three unequal periods. The first seven weeks (49 years) run from the destruction of the First Temple "until there is an anointed prince *(ʿaḏ-māšiaḥ nāpiḏ)*"—that is, until the Second Temple's reconstruction under Zerubbabel, the Davidic governor of the early restoration, Joshua, the high priest, or Cyrus, the liberating Persian conqueror (9:25). The next sixty-two weeks of years (434 years), "times of distress (ûḇᵉṣôq hā'ittim)," are used up in the rebuilding of the city of Jeru-salem (9:25). The final week (7 years) begins in the reign of Antiochus IV; the Hebrew is difficult, ungrammatical, and, in some places almost impenetrable. This gentile king will cut off an anointed one (a likely reference to the murder of the high priest Onias III)[72] and ruin the city and sanctuary. For half of this final week (that is, for three and a half years), Antiochus Epiphanes will sup-press the sacrifice and offerings. But "his end will be with a flood (wᵉqiṣṣô ḇaššeṭe))," with war and horrors, right up "until the pre-determined de-struction of the desolator ('aḏ-kālâ wᵉneḥᵉrāṣâ tittaḵ 'al-šōmem)" (9:26–27). In this way, Daniel 9 is concerned to establish a deep enumeration of contin-uous and irreversible years. Yet these are not schematized, as before, to the successive waves of imperial conquest. Here, it is Jewish time that structures history: the fate of the Jerusalem Temple, with its three periods of destruc-tion, rebuilding, and abomination, and the cultic rhythm of the sabbatical and jubilee calendar.

The book of Daniel culminates in a final, lengthy, and detailed vision that extends over chapters 10–12. This is the most precisely dated of all the epi-sodes, to 24th Nisannu in the third regnal year of Cyrus the Great (10:1). It finds the seer Daniel on the bank of the river Tigris at the end of a three-week fast of mourning and bodily self-affliction. Suddenly, an angel appears (per-haps Gabriel), his face like lightning and voice as the rumble of multitudes, to explain to Daniel what will befall the people of Israel "in the end of days (bᵉʾaḥᵃriṭ hayyāmim)," according to the heavenly "Book of Truth (biḵṯāḇ ᵉʾmeṭ)." Before beginning his report, which will make up all of Daniel 11 and part of Daniel 12, the angel describes a conflict between the respective patron angels of Israel and the imperial conquerors: he had been opposed for twenty-one days by "the prince of the kingdom of Persia (śar malḵûṯ pāras)," until the arch-angel Michael had come to give succor. And when he will return, the angel reports, "the prince of Greece will come (śar-yāwān bāʾ)" (10:2–21). This mys-terious, otherworldly introduction characterizes the earthly geopolitics of im-perial succession and subjugation as a parallel or reflex or reflection of a timed combat between national patron angels.[73]

After a brief prologue that narrates the arrival of Alexander and the Mace-donians, the bulk of the angel's account is a detailed narrative of the still undecided Syrian Wars between the Seleucid and Ptolemaic empires for control of the southern Levant. In fact, Daniel 11 offers the single extant, uni-tary, and continuous historiographic narrative of this multigenerational conflict, and one deriving from the region in dispute. We should reject modern scholarly surprise at the finely granulated detail of the chapter's historical knowledge.[74] Rather, this episode illustrates the degree of political and dynastic information required of indigenous elites throughout the Hellenistic kingdoms' borderlands.[75] The rulers of the Seleucid and Ptolemaic empires appear under the polar geographic titles of the "king of the north (melek̲ haṣṣā)ôn)" and the "king of the south (melek̲ hannepeb̲)," generic identifications that, from the Levantine perspective, reactivate long-standing prophetic oracles against Assyria-Babylonia and Egypt and blacken the Seleucid dynasty with the ancient enemy-from-the-north chaos tradition.[76] The vision, still in need of close study from the perspective of political and military history, at first glance seems like an unschematized and flat account of interdynastic rivalry, one thing after another, a turning into narrative of the split iron-clay feet of Nebuchadnezzar II's dream statue in Daniel 2. In fact, the angel's report works out a model of original rebellion and successive failure that locates the end of history as much in Seleucid-Ptolemaic imperial relations as Antiochus IV's desolating sacrilege in Jerusalem.

The angel's recital immediately undoes the Seleucid empire's official, chronographically formalized accounting of its own origin. This ultimate period of earthly empire, as in the book of Daniel's previous episodes, opens with Alexander, not Seleucus I. The archangel foretells how the fourth and last of the Persian kings will, in his vast wealth, provoke the kingdom of Greece. "And a great king will arise (w°ʿāmad̲ melek̲ gibbôr)" (a formulation that recurs in the Dynastic Prophecy from Babylon, discussed below) and will possess great dominion; but his kingdom, broken and divided to the four winds of heaven, "will not be for his posterity" (11:1–4). In its place, the angel continues, "the king of the south (Ptolemy I) will be strong; but one of his princes (Seleucus I) will be strong over him (ûmin-śārāyw w°yêḥ°zaq ʿālāyw) and will rule and his rule will be a great rule (ûmāšāl mimšāl rab̲ memšaltô[77])" (11:5). The angel has narrowed down on the precise origins of the Seleucid empire and its time system—Seleucus I's 311 BCE restoration to Babylon from his exile at the Alexandrian court—and transformed this epochal moment of heroic, sanctified return into a rebellion of a subordinate (śar, or "prince") against his overlord (melek̲, or "king").[78] Recall that, as I discussed in Chapter 1, the Seleucid empire's foundational origin was similarly acknowledged and reframed by the Parian Marble, the Ptolemaic-aligned epigraphic chronicle from the Aegean island of Paros, as the deliberate sending of a subordinate on a mission.[79]

The angel's report transforms the Seleucid empire's moment of creation, its *akītu*-aligned Big Bang of 1 SE, into a kind of original sin. From it flow the subsequent decades of rivalry, murderous marriage, false peace, and finally history's close. The Syrian Wars are presented as a tale of repeated Seleucid attempts to conquer and incorporate the territories of their original Ptolemaic masters. The narrative runs from Antiochus II's marriage to the Ptolemaic princess Berenice and the consequent Laodicean, or Third Syrian, War through the Fourth Syrian War to Antiochus III's conquest of Judea and Jerusalem in the Fifth Syrian War: "And he (Antiochus III) will take his stand in the beautiful land and it will all be in his hand *(kālā beyādô)*" (11:16). But even this monarchic success, the winning of the long-claimed province of Coele Syria and Phoenicia, is undone. For Antiochus III's invasion of Asia Minor and Greece ("he will set his face to the coastlands") leads to his defeat by Rome, and the king's third and final *anabasis* ends with his dishonorable death in Elymaïs ("and he will set his face to the strongholds of his own land *[lemāʿûzze ʾarṣô]* and he will stumble and fall and not be found") (11:17–19). Seleucus IV is, as ever, swiftly dealt with: "In a few days he shall be broken, but neither in anger nor in battle" (11:20). Thus, from the birth-moment of the Seleucid empire until the arrival of Antiochus IV, we are given a history of ongoing, worsening conflict and successive royal stalemate:[80] the armies get bigger, the suffering increases, but the dynamic of conflict does not change—"and he shall cast down tens of thousands; but he shall not prevail" (11:12).

The mutual containment of the Seleucid and Ptolemaic kingdoms is broken by the arrival of Antiochus IV. As in Daniel 7, Antiochus Epiphanes does things never seen before; he is a rupture in his own imperial tradition, acting repeatedly in ways unsanctioned by or at variance with his dynastic *progonoi*: "There will arise in his (Seleucus IV's) place a despicable man *(weʿāmad ʿal-kannô nibzê)*"; "the royal majesty *(hôd malkût)* was not given to him"; "he will arrive in stealth and steal the kingdom by deceit"; "he will do what his fathers and his fathers' fathers did not do"; "he will honor the god of strongholds *(weleʾelōha māʿuzzim)*, a god his ancestors did not know" (11:21–24, 37–38).[81]

The angel's account is shaped around Antiochus IV's three invasions of Egypt, the first two historical (the Sixth Syrian War) and the third an unfulfilled prediction that, by its inaccuracies, provides the dating for the book's final composition. The first invasion (170–169 BCE) ends with a short-lived peace between Antiochus IV and Ptolemy VI ("they shall speak lies at one table") and the Seleucid monarch's return home with great booty (11:25–28).[82] The second invasion (168 BCE) is interrupted at the so-called Day of Eleusis, when the Roman legate C. Popillius Laenas drew a circle around the Seleucid monarch and the Republic's protection over Egypt: "Ships of Kittim will come against him and he will be cowed and he will return and he will rage against the holy covenant (wᵉzāʿam ʿal-bᵉrit-qôdeš)" (11:29–30). Antiochus's desecration of the Jerusalem Temple, persecution of Jewish practices, and distribution of Judean lands to settlers are interpreted as a furious response to this Roman compulsion (11:31). Finally, shifting from *ex eventu* prophecy to actual prediction, in a third invasion, "at the time of the end (ûbᵉʿet qeṣ)," Antiochus IV will muster a vast army of cavalry, chariots, and ships and, like a whirlwind or flood, conquer the entire Ptolemaic empire in northeastern Africa: "He will gain control of the hidden stores of gold and silver, and all the precious things of Egypt; the Libyans and the Ethiopians shall be in his train" (11:43). At this point, rumors from the east and north will draw him into the holy land, where, pitching his palatial tent between the Mediterranean Sea and Jerusalem, "he shall come to his end (ûbāʾ ʿad-qiṣṣô)"—and with the king, history (11:45).[83]

The inaugural rebellion of Seleucus I against "the king of the south" reaches its consummation in Antiochus IV's conquest of Egypt:[84] the Ptolemaic kingdom no longer exists, and Antiochus IV directs his aggression heavenward. The geopolitical logic of the vision resembles, in outline, that of the Roman Republic: a world order sustained by the counterbalancing of territorially bounded Hellenistic kingdoms and undone by their unification. Daniel 12 concludes the vision's narrative with the angel's report that Michael, patron protector of Israel, will arise. There will be "a time of distress (ʿet ṣārâ)" such as had never before come to pass. The dead will be resurrected, some to eternal life and others to everlasting disgrace (12:1–3). Then, on finishing his history, the angel instructs Daniel, in the language of the archive, to "stop up the words and seal the book until the time of the end (sᵉtōm haddᵉbārim waḥᵃtōm hasse)er ʿad-ʿet qeṣ)" (12:4, 9). With this recitation from the Book of Truth complete, Daniel overhears angels on either side of the river Tigris asking the now familiar question "how long (ʿad-mātay)" until this end? The responses— "for a time, times, and a half (lᵉmôʿed môʿᵃdim wāḥeṣi)," "1,290 days," and finally "1,335 days"—are a series of successive revisions (12:5–12).[85] The book's final verse ends with the promise of Daniel's personal resurrection: "Now go! You shall rest and rise to your destiny at the end of days (lᵉqeṣ hayyāmin)" (12:13). The dragon of years is almost done.

I have been exploring the periodization of history episodically, as it plays out in the individual court tales and visions. Turning to the book of Daniel as a whole, we can identify this same succession of empires as the work's determining and unifying structure. This is achieved in two modes. First, the book's pseudonymous conceit allows this scheme to take on biographic form. The exilic visionary serves as a wise adviser in the successive courts of the Babylonian, Median, and Persian empires;[86] the book's very last verse promises Daniel a future resurrection in the eschatological kingdom.[87] In other words, the book's narrative constructs a yoke-team of Jewish seer and imperial ruler that runs through the gentile monarchies in order and then leapfrogs over the entire Seleucid dynasty to the kingdom of God that will follow its destruction.[88] As in the Uruk List of Kings and Sages or the biblical book of the Twelve minor prophets, both discussed in Chapter 4, the book of Daniel establishes a normative pattern of Jewish mantic wisdom that crosses conquests and regimes only to be broken in the Seleucid world. Put differently, the first three kingdoms are lived and knowable, while the fourth is only horrifyingly and hazily envisioned. Many of the formal and theological differences between the book's two halves—Aramaic court tales and Hebrew apocalyptic visions— in fact work to construct this distinction between the Seleucid empire and its predecessors. Furthermore, the scope of the book's visions progressively narrows, as the baseline of historical focus is pulled ever closer to the Seleucid horizon: from four empires in Daniel 2 and 7 to three (or perhaps two) in Daniel 8, and to one divided against itself in Daniel 10-12.[89] There is a perceptible increase of urgency and detail as we read through the book, from the deferred eschatology in the reign of Nebuchadnezzar,[90] gazing ahead half a millennium, to the final countdown at the end of chapter 12.

Second, the book of Daniel spatializes this succession of empires by the seer's location. Daniel, exiled with king Jehoiakim from Judea, is educated in the wisdom of Babylon and serves Nebuchadnezzar, Belshazzar, and Darius the Mede in the city's palace (chapters 1-7). Then, in chapter 8, in the first year of Belshazzar, Daniel finds himself "in the fortified city of Susa in the province of Elam, by the canal Ulai *(beʾšûšan habbirâ ʾăšer beʿelām hammeʿdinâ . . . ʿal-ʾûbal ʾûlāy)"* (8:2),[91] where he hears a man's voice "in the midst of the Ulai" *(ben ʾûlāy)* (8:16). It is unclear whether Daniel is physically present in the Elamite capital or, like Ezekiel, had been transported in spirit. The final vision, in the third year of Cyrus the Great (chapters 10-12), takes place "on the bank of the great river, that is, the Tigris *(ʿal yaḏ hannahār haggāḏôl hûʾ ḥiddāqel)"* (10:4);

Daniel sees an angel on each side of this waterway *("liś)aṭ hay'ōr")* (12:5). (Chapter 9, the inspired reading of Jeremiah's seventy-year prophecy, is unlocated.) To an attentive Hellenistic reader these three locations in succession—Babylon, the Ulai river by Susa, and the Tigris—signified the Mesopotamian or western Iranian capitals of the Neo-Babylonian, Persian, and Seleucid empires, respectively: Susa, the old Elamite city, was to become the Achaemenid administrative center from at least the reign of Darius I, and the eastern terminus of the Royal Road from Sardis; and Seleucus I Nicator would densely colonize the middle Tigris, establishing Seleucia-on-the-Tigris as his first and largest city foundation, regional capital, and successor to Babylon as the *āl šarrūti* (city of kingship).

Moreover, each individual vision locale is crafted into a palimpsest of kingdoms. Thus, the inconcinnity of time and place foregrounds the passing on of the imperial baton: Daniel finds himself by the future Achaemenid capital during the reign of the Neo-Babylonian Belshazzar, and beside the river of the still-to-be-built Seleucia in the third year of Persian Cyrus. The waterway locations are especially important for achieving this sense of succession. Achaemenid Susa was refounded under the Seleucids as Seleucia-on-the-Eulaeus (that is, on the Ulai); it is likely that this potamonym supplanted the Achaemenid-period name Choaspes.[92] Pliny reports that the Ulai (or Eulaeus) river originated in Media, flowed underground, and then reappeared to encircle Susa's citadel.[93] If known to the author(s) of Daniel, such a river-run expressed the chronological ordering and ultimate convergence of the Medes and the Persians. Chapter 8's vision—in the reign of Belshazzar, at Susa, on the Ulai—perfectly catches the book's periodization of Neo-Babylonian, Median-Persian, and Macedonian empires. Similarly, the Tigris, on whose banks Seleucid history is revealed to Daniel, is introduced as "the great river *(hannahār haggāḏôl)*." This, the standard biblical epithet for the Euphrates,[94] here has been transferred to the Tigris, just as the early Seleucid monarchs had relocated the region's center of political gravity from the old Euphratene core to the formerly marginal Tigris river. The character Daniel moves in order through the stations of history.

Flowing fresh water had numinous associations in biblical and parabiblical texts: Ezekiel, son of Buzi, beheld the throne-chariot of God at the Chebar canal, near the exilic town of Tel Aviv;[95] Enoch was sent winging by dreams beside the waters of Dan.[96] But the use of rivers to figure historical connection, debt, and process was more highly developed in contemporary Greek literary and artistic culture.[97] Rivers were identified as focal points of landscapes, connecting distinct zones and lending their character to the populations inhabiting their banks. Like the Median Eulaeus, for instance, the river Alpheus was thought to flow underground from Elis in the Peloponnese, beneath the Ionian Sea, to the Dorian colony of Syracuse in Sicily,[98] an assertion of both the fact and direction of kinship. The famous Nile mosaic from Palestrina, likely a version of a Ptolemaic original, coordinated the Nile's progress with a civilizational and historical development, from hunter-gatherers at the Nubian source to a Greek symposium at the Mediterranean mouth.[99] Additionally, and perhaps more germane, the Seleucid monarchs prominently deployed rivers in their naming of colonies to distinguish between homonymous dynastic foundations; indeed, the two most important Seleucid settlements in the Mesopotamia–western Iran sphere were Seleucia-on-the-Tigris and Seleucia-on-the-Eulaeus (or Ulai). To an insurgent indigenous population, recent witnesses to the Seleucid colonization of the Levant and the attempted foundation of an Antioch in Jerusalem, it can be no accident that Daniel's visions occur on the Eulaeus and Tigris. The apocalypse turns imperial hydronomy against the kingdom. These rivers do not appear as the sites of an imperial urbanism that gives material form to the dynasty's glorious longevity—"built," to quote the book's Nebuchadnezzar, "as a royal seat, by the might of my power, for the sake of my glory." Instead, they are the locations for soul-shattering epiphanies that strip away every Seleucid pretense to temporal agency and imperial continuity.

Let me conclude this exposition. Daniel is a book about history and its segmentation, repeatedly slotting its author(s)' contemporary Seleucid world into the final niche of different, overlapping temporal schemata. Its visions are steeped in the experience of temporal duration opened by the Seleucid kings even as they work hard to undo it. They display an entirely unprecedented and hugely influential interest in the numericalization and predictability of time, in the chronography of the future as well as of the past. Indeed, Flavius Josephus identified this as Daniel's key and unique characteristic: "He did not only prophesy future events, as did the other prophets, but even fixed the time of their accomplishment (ἀλλὰ καὶ καιρὸν ὥριζεν εἰς ὃν ταῦτα ἀποβήσεται)."[100] For the author(s) of Daniel, the end was not only near, it could be calculated[101]—this is apocalypse as science.[102] The seer's dreams and visions, though in no way exhausted by rationalizing decipherment, nonetheless "testify to an irresistible facticity of the fictive":[103] conjured from the present, directed to the future, sensitive to the forms and idioms of imperial power, yet disclosing their essential monstrosity. At the same time, the book's episodes take possession of, reenunciate, misshape, relocate, invert, and thus thoroughly undermine the imperial worldview, both politically and temporally. The book's determinism, while neither mechanistic nor empty of personal responsibility,[104] undoes the most fundamental assumption of empire: that the king, as the limit-subject of action, makes history. In its place, it opens a representation of providential time, of history as revelation, "decided by an inaccessible Subject who can be deciphered only in the signs that he gives of his wishes."[105] If the Seleucid empire had pared time down to transcendent number, the book of Daniel offered instead a violently imagistic, richly symbolic representation of its historical duration: what could be more unlike the Seleucid Era count than the fourth beast of Daniel 7?

1 Enoch

Two other eschatological total histories survive from Seleucid Judea, known by the titles "Apocalypse of Animals" and "Apocalypse of Weeks." These second-century BCE compositions are pseudonymously attributed to the antediluvian patriarch Enoch. The man Enoch (*Ḥᵃnôk*), seventh in line from Adam and great-grandfather of Noah, makes only a brief, if intriguing, appearance in the Hebrew Bible, in the priestly genealogy of Genesis 5:18–24:

> When Jared had lived 162 years, he fathered Enoch. . . . And all the days of Jared were 962 years, and then he died. And Enoch lived 65 years and fathered Methuselah. And after he fathered Methuselah, Enoch walked with God (*wayyithallek Ḥᵃnôk 'et-hā ᵃᵉlōhim*) for 300 years; and he fathered sons and daughters. All the days of Enoch were 365 years. Enoch walked with God, then he was no more, for God took him (*ki-lāqaḥ 'ōtô ᵃᵉlōhim*).[106]

On the basis of this unique and enduring access to heaven (walking with God and rapture by God) and genealogical position and age (seventh generation and 365 years, of evident sabbatical and annual significance), Enoch was transformed in the Hellenistic period into a primeval sage and the first transcriber of divinely revealed cosmic, astronomical, and chronographic knowledge. Starting in the third century BCE, an entire corpus of Judean writings in Aramaic was pseudonymously composed in his name. These Enochic works are characterized by a scientific interest in geography and astronomy, a heavenly accounting of cosmic evil (the myth of the Watchers; see below), a down-playing of both Moses and the Sinaitic covenant, veiled or explicit criticism of the Second Temple and its priesthood, and a commitment to the coming eschatological judgment.[107]

This remarkable library of Hellenistic texts, spurned by both Jewish and most Christian biblical canons, was entirely lost to western scholarship and exegesis until 1773, when James Bruce, a Scottish explorer of the Blue Nile, brought back to Europe from Ethiopia three manuscripts of the *Maṣḥafa Henok Nabiy*, the "Book of the Prophet Enoch." This Ethiopic Book of Enoch, now better known as 1 Enoch, was a late antique translation into Geʿez of the Greek version of an original Aramaic compilation. The book, along with many other works rejected by the rabbis and the northern Christians, was accepted into the Ethiopian orthodox church and played a significant role in the nation's religious and literary life.[108] The discovery of several Aramaic Enoch fragments from among the Dead Sea Scrolls at Qumran confirms that 1 Enoch is made up of five major documents of independent authorship and circulation:[109] the *Book of the Watchers* (1 Enoch 1–36), composed in the third century BCE, in Ptolemaic Judea; the *Similitudes of Enoch*, also known as the *Book of Parables* (1 Enoch 37–71), dating to the first century BCE or after; the third-century BCE *Astronomical Book* (1 Enoch 72–82); and the *Book of Dreams* (1 Enoch 83–90) and the *Epistle of Enoch* (1 Enoch 91–108), both composed in Seleucid Judea in the first half of the second century BCE. The two second-century BCE creations discussed below, the Apocalypse of Animals from the *Book of Dreams* and the Apocalypse of Weeks from the *Epistle of Enoch*, adapt the mythic and calendrical ideas of Ptolemaic-period Enochic thought to the writing of total history. For in the Enochic tradition, just as in the Judean wisdom writings discussed in Chapter 4, the Seleucid conquest of the southern Levant encouraged, provoked, or catalyzed a profound shift from cosmic to historical concerns.[110]

Before examining each of these eschatological total histories, I will permit myself a little proselytizing. 1 Enoch is among the most significant textual corpora from the ancient Mediterranean, yet it remains almost entirely un-exploited by Hellenistic historians. The book offers accounts of the Judean experience of Ptolemaic and Seleucid rule, a narrative of the Maccabean revolt, and close and critical engagement with the scientific, geographic, and religious thought of the contemporary Greek and Mesopotamian worlds. Even without learning Geʿez (a language with remarkably consistent morphology and simple syntax), translations, commentaries, and other resources are now easily sufficient to serve the needs of interested readers.[111]

The Apocalypse of Animals (1 Enoch 85–90) presents a full accounting of world events, along the entire axis of historical time, as an allegory of beasts and birds. In origin an independent composition, it is now incorporated as the second of the two dream-visions that form the *Book of Dreams*, the fourth part of 1 Enoch (83–90).[112] There is a general scholarly consensus that it dates to some point during the Maccabean revolt in the later 160s BCE, as it describes the ascendancy of the Jewish rebel leader Judas but does not appear to know of his death.[113]

The Apocalypse of Animals is presented as Enoch's first-person report of his dream-vision (*ḥelma*), witnessed before his marriage to Edna and later delivered to his son Methuselah. In contrast to the book of Daniel and other apocalyptic works, the symbolic vision requires no interpretation by an angelic figure, for its narrative constitutes a fairly transparent, if selective, abbreviation of the biblical history given in the books of the Pentateuch, the Former Prophets (Joshua to 2 Kings), Ezra, and perhaps also Nehemiah and Chronicles.[114] The Apocalypse of Animals builds up its total history of Israel and the world by activating and combining four harshly evaluative allegorical systems: first, the kind of being—animals, representing the peoples of the earth, and men, representing celestial beings (good angels are "white men," wicked angels "seventy shepherds," and God "the Lord of the sheep"; Noah and Moses, alone of the animals, are turned into men in the course of the narrative);[115] second, the animals' dietary and sacrificial (that is, their kosher) status—clean and domesticated cattle and sheep for primordial humanity, the people of Israel, and eschatological humanity, unclean or wild predators and scavengers for all the gentile nations of the postdiluvian world;[116] third, color—white (elect), black (nonelect or wicked), and red (unclear, perhaps neutral);[117] and fourth, sight—the sheep (Israel) have their eyes opened (revelation), closed or darkened (apostasy), and pecked out (external oppression or "Hellenization").[118] Along these four axes, Enoch maps out his full historical narrative of election, political oppression, moral deterioration, and eschatological justice.

Enoch's vision account opens at the birth of humanity, with God's creation of a pure-souled Adam: "Behold! A bull came forth from the earth, and that bull was white" (85:3). Enoch swiftly moves on to report the proliferation of violence and oppression in the antediluvian world, with Cain's murder of his brother Abel and the fall of the so-called Watchers to earth. This latter myth was central to the Enochic understanding of earthly evil: the *Book of the Watchers*, the opening section of 1 Enoch, tells of a primordial breach of the supernatural sphere, when wicked angel-stars fell from heaven, mated with the daughters of men, and educated humanity in illicit cultural techniques, including metalwork, sorcery, and astrology.[119] In the Apocalypse of Animals, Enoch watched as these fallen Watchers "let out their [sexual] organs like horses, and they began to mount upon the cows of the bulls; and they all conceived and bore elephants and camels and asses (*wawaladā nagayāta wa'agmāla wa'a'duga*)" (86:4). These monsters bit, gored, and devoured the cattle (86:5–6), were given a sword to fight one another (88:2), and then were killed off in the great flood: "And the elephants and the camels and the asses sank to the earth as well as all the animals. And I could not see them and they were unable to come out, and they perished and sank in the depths (*watahag^wlu watasaṭmu westa qalāy*)" (89:6). The elephant, alone of the Watchers' offspring, was not a traditional biblical beast.[120] It has been introduced into the Apocalypse of Animals, just as into Daniel 7 (see above), to recharacterize the Seleucid imperial blazon as a primeval monstrosity.[121] Indeed, the image of a drowning elephant, flailing in vain to save itself from the punishing floodwaters, offers a grotesque prototype for the expected or longed-for destruction of the Seleucid empire in the author(s)' own day.[122]

In the generations after Noah are born the gentile nations that will oppress Israel for all of postdiluvian history: "And they began to beget wild beasts and birds and there came from them species of every sort: lions, leopards, hyenas, wolves, dogs, wild boars, foxes, conies, pigs, falcons, vultures, kites, eagles, and ravens" (89:10).[123] The people of Israel, beginning from the patriarch Jacob, are figured as a flock of sheep, from which the Apocalypse of Animals derives its basic metaphors of shepherding. Enoch's account moves swiftly through the patriarchal and exodus narratives, the arrival in the promised land, where the period of the judges is schematically presented as cycles of faith and apostasy—"And sometimes their eyes were opened, and sometimes they were blinded" (89:41)—and up to Solomon's construction of the First Temple. From Solomon's death onward, world history begins an ever-accelerating, irreversible moral decline, first with the apostasy of the divided kingdom, when the sheep "strayed from everything and their eyes became dark" (89:54), and subsequently with the succession of four fearsome gentile empires (89:59–90:19).

The final era of postdiluvian world history opens, like the book of Daniel, with the Babylonian conquest of Jerusalem in the reign of Jehoiakim and continues through the Persian restoration and up to the end-time.[124] In these closing centuries, Israel is cast off by God to the rule of wicked angels: Enoch reports that God, "the Lord of the sheep," delegated "seventy shepherds (sab'ā nolāwiyāna)" to tend and kill the flock (89:59).[125] Simultaneously, another angel was appointed to record in writing "how many [the seventy shepherds] destroy by My command and how many they destroy by themselves," for, Enoch continues, God wished "to know every deed of the shepherds (kama 'ā'mer kʷello gebromu lanolāwiyān)" (89:63). In other words, this age of vicious shepherds and gentile empires is also one of heavenly historiography.

The rule of the seventy negligent angels is divided into four periods, each closed by the recital and archiving of the angel's historical report. The first period, that of the Babylonian empire (lions, supported by leopards and wild boars), is that of the destruction of Jerusalem and her Temple and the devouring of the sheep. It concludes, "And the book was read before the Lord of the sheep (wamaṣhaf tanabba baqedma 'egzi'a 'abāgeʿ), and He took the book from his hand and He read (it) and sealed (it) and laid (it) down (wanaśʾa maṣhafa 'em'edu wa'anbaba waxatama wa'anbara)" (89:71). Then, in the second period, the imperial baton passes to the Persian empire and a new heavenly book is opened; notably, the rebuilt Second Temple is condemned from its origins as an impiety.[126] As before, the historical period is closed with the inscribing, recital, and sealing of the second angelic scroll (89:76–77). The third period (90:1–5) extends from Alexander's initial conquest until the expulsion of the Ptolemies from Coele Syria and Phoenicia by Antiochus III in the Fifth Syrian War. The coming of the Greeks and Macedonians is depicted as the descent of a mixed flock of sharp-beaked birds: "And then in my vision I saw all the birds of heaven come: eagles and vultures and kites and ravens. And the eagles were leading the birds, and they began to devour those sheep and to peck out their eyes (wayekreyu 'aʿyentihomu) and to devour their flesh" (90:2). The Apocalypse of Animals' shift from predatory mammals to birds of prey figures the Hellenistic age as qualitatively different from the previous millennia of Israel's history. Indeed, the birds' power to peck at the sheep's eyes and so blind them far exceeds the capacity of all earlier gentile violence.[127]

The fourth and final period, explicitly characterized as the worst of times, is that of the Seleucid empire, figured as black ravens (qʷāʿāt); recall the color symbolism outlined above. The beginning of this period coincides with the birth of lambs, which, for the first time since the reign of Manasseh, "began to open their eyes and to see *(wa'axazu 'a'yentihomu yekšetu wayer'ayu),"* indicating some kind of renewed revelation or religious revival (90:6). The Seleucid ravens "flew upon these lambs and seized those lambs and crushed the sheep and devoured them" (90:8). Even worse, the sheep of Israel for the first time were targeted by apostates among their own flock: the blind sheep "afflicted them *('asrexewwomu)"* (90:7), in what may be a reference to the Jewish allies of empire. Finally, one of the sheep sprouted a big horn *("baqʷala 'aḥadu qarn 'abiy la'aḥadu 'emenna zeku 'abāgeʿ")* and the flock's eyes were opened (90:9), which refers to Judas, the Maccabean revolt, and the widening popularity of militant religious resistance.[128] Enoch continued to watch until "the man who was writing down the names of the shepherds *(zeku be'si zayeṣehhef 'asmātihomu*[129] *lanolot)"* (90:14)—that is, the angelic historian—sought help from God. Reciting from his fourth and final scroll, the heavenly history of Seleucid imperial rule, "he showed the Lord of the sheep that they had destroyed much more than (those who were) before them" (90:17). At such provocation, God Himself descended for a final battle against the Graeco-Macedonians and their gentile allies: "All the eagles and vultures and ravens and kites assembled, and they brought with them all the wild <beasts>" (90:16). In a theophanic earthquake, the birds were sunk into the earth; and the sheep of Israel, granted a great sword *("sayf 'abiy"),* slaughtered the wild beasts (90:18-19).

Enoch's report closes with a graduated eschatological restoration: first, a final judgment at which the sealed books are opened, and the fallen stars (the Watchers), the seventy negligent shepherds, and the blinded sheep (the disobedient or Hellenizing Jews) are condemned to the pit of fire (90:20-27); then, the establishment of a heavenly Jerusalem, at which all beasts and birds bow down before the open-eyed sheep (90:28-30); and finally, the birth of a white bull and the transformation of all species of the earth into white cattle (90:37-38). This eschatological conclusion restores humanity to the primordial, Adamic condition of Eden.

This extended faunal allegory is concerned with the segmentations of its historiographical narrative. World history is organized into three great epochs, antediluvian (1 Enoch 85:3-89:8), postdiluvian (1 Enoch 89:9-90:27), and eschatological (1 Enoch 90:28-38)—each marked by the birth of a white bull. The postdiluvian, historical age is itself fully periodized: first, when the flock of Israel are under direct divine control and second, when the sheep suffer under the delegated rule of the seventy shepherds. This final half-millennium of successive Near Eastern empires unites key temporal schemata. We have seen already, in Daniel 10, the notion that a patron angel had been assigned to each of the seventy nations of the earth. But in the Apocalypse of Animals this spatially distributive model is temporalized into a succession of seventy chronological periods of gentile domination. Indeed, the words "shepherds" (*nolot* or *noläwəyān*) and "times" (*saʿāt*) are interchangeable in the vision.[130] These seventy shepherd-times are, evidently enough, yet another exegetical reworking of Jeremiah's famous seventy-year prophecy for Israel's desolation.[131] We have already seen its importance to Daniel 9 and, perhaps also, to Simon Maccabee's introduction of his Year 1 in 170 SE (discussed in Chapter 3).

Additionally, these seventy units are arranged into four numbered groupings of twelve, twenty-three, twenty-three, and twelve shepherd-times, respectively, each closed by the angelic completion of a heavenly scroll of historical record.[132] These establish a sequence of four gentile empires—here, appropriately enough for the history of the southern Levant, the Babylonian, Persian, Alexandrian-Ptolemaic, and Seleucid kingdoms. Such a symmetrical division of the seventy periods is in no sense chronologically accurate; the Apocalypse of Animals, importantly, does not count years. The scheme corresponds to historical reality only in the sense that there are two short rules at the beginning and end (Neo-Babylonian and Seleucid) and two longer periods of dominance in the middle (Achaemenid and Alexandrian-Ptolemaic).[133] Finally, the allegory develops a dynastic rather than a monarchic mode of representation: no individual gentile king, not even the conqueror Alexander or the persecutor Antiochus IV Epiphanes, is identified; despite their common origin, the Hellenistic empires appear as discrete species of birds, of which the darkest and most fearsome is the Seleucid.

1 Enoch incorporates another eschatological total history, the so-called Apocalypse of Weeks. Composed around the time of the Maccabean revolt, likely a little earlier than the Apocalypse of Animals,[134] this brief text runs through the entire history of the world in a mere seventeen verses (1 Enoch 93:1–10 and 91:11–17). It is known in full only from the Ethiopic translation, but a mid-first-century BCE manuscript from Qumran, 4QEn^g (4Q212), preserves parts of the Aramaic original in its proper order.[135] The Apocalypse of Weeks is concerned, above all, to establish a schematic periodization of history, from creation to *eschaton*, sequenced as ten numbered weeks (Aramaic *šabuʿin*;

Ge'ez *sanbatāt*).[136] Given the seven divisions to each week, this amounts, once more, to a closed and completed system of seventy chronological units.[137]

The Apocalypse of Weeks presents itself as Enoch's repetition to his posterity of what had been revealed to him by angels and heavenly tablets. Enoch opens the account by introducing his own birth, seventh in line from Adam: "I was born (in) the seventh (part) in the first week, until which point justice and righteousness lasted" (93:3). Week 1 is characterized as a period of complete goodness, unblemished by wickedness. But in week 2, "great evil will arise" and "there will be the first end (*qadāmit feṣāmē*)," Noah's flood (93: 4); this "first end" anticipates and demands a second, the eschatological judgment that pressed upon the Enochic authors of the second century BCE. In week 3, we move forward to the election of Abraham, figured as "the plant of righteous judgment." Enoch reports that this righteous plant—a figure for the corporate, national Israel, a specific individual, or the circle from which the apocalypse derives—"will go forth forever and ever" (93:5). At the end of week 4, the Torah, "a law for every generation," is given at Sina (93:6), the only explicit reference to the Mosaic Torah in the entirety of 1 Enoch.[138] Week 5 concludes with the construction of Solomon's Temple: "And after this, in the fifth week, at the end, a house of glory and royalty (*bēta sebḥat wamangešt*) will be built unto eternity" (93:7). Week 6, that of the divided kingdom, is characterized by the spread of apostasy and closes with the Babylonian destruction of the First Temple (93:8). Week 7, extending from the Babylonian exile to the author(s)' own time in the Seleucid empire, receives the lengthiest description, although no mention is made of the Second (restored) Temple: "In the seventh week there will arise a wicked generation (*tewled 'elut*), and many will be its deeds, and all its deeds will be wicked," which is likely a reference to the so-called Hellenizing Jews (93:9). The week concludes with the formation of the very group to which the apocalypse's author(s) claimed to belong, characterized as the reemerging righteous branch of Abraham from week 3: "And at its end there will be chosen the chosen righteous ones from the eternal plant of righteousness, to whom will be given the sevenfold instruction" (93:10). Weeks 8–10 are a graduated eschatological age of ever-expanding scope. Week 8 is a time of retribution, in which the righteous will be armed with a sword and judgment exacted on "those who oppress (*'em-'ella yegaffe'u*)," who may be the Seleucid occupiers (91:12). At the end of this eighth week the eschatological Temple will be constructed: "And the house of the Great King (*bēta neguš ʿabiy*) will be built in glory forever" (91:13). Week 9 sees righteous judgment revealed to the world and the conversion of all humanity (91:14). In the seventh part of Week 10, there will be an eternal judgment, and "the first heaven shall disappear and pass away, and a new heaven shall appear, and every power of the heavens shall shine sevenfold forever" (91:15–16). The Apocalypse of Weeks concludes after the tenth week, with the opening of a true eternity that cannot be periodized, that is, with the total disappearance of counted time: "And after this there will be many weeks without number (*'albon xolqʷa*) unto eternity (*la-ʿālam*)" (91:17).

This is a spare and selective scheme, incorporating only a very few figures and turning points from Israel's sacred history, always with more of a religious or moral than a national significance. In contrast to the Apocalypse of Animals, which explores the sequence and termination of gentile violence against Israel, the Apocalypse of Weeks does not mention hostile, external forces.[139] Rather, it is oriented to the ethical opposition and historical outplaying of righteousness and wickedness, with a concern for the salvation of the elect and the fate of God's house. It presents a schematic rather than a chronologically accurate narrative:[140] historical meaning resides in the sabbatical chronography, built upon the numbers seven and ten.[141] The apocalypse's simple, quasi-tabular, numericalized schema allows its millennia-long duration to take shape, with patterns of repetition, typology, and mirroring. Thus, the fulcrum of world history is the construction of the First Temple at the end of the week 5, the precise middle of the narrative. In a kind of historical theology of chiasm, the weeks on either side of Solomon's sanctuary parallel and reverse each other:[142] the beginning of the eschatological process in week 8 with the election of Abraham in week 3; the punishment and conversion in week 9 with the deluge of week 2; the creation of the new, perfected heaven in the seventh part of week 10 with the untarnished righteousness of creation, up to Enoch's birth in the seventh part of week 1.

Seleucid-period Judea, as I hope to have demonstrated, offers richly sophisticated and well-preserved eschatological total histories, of long-lasting and widespread impact within Judaism and early Christianity. But it is crucial to the project and argument of this book that a contemporary, parallel, indeed

closely similar interest in the schematic periodization of history can be identified in the Seleucid empire's other core regions, Babylonia and Iran.

The Dynastic Prophecy

Babylonian scholars of the Hellenistic period took a lively interest in the fact
and dramatic texture of Near Eastern imperial succession. Several cuneiform
texts of Seleucid date probe the fall and sequencing of Mesopotamia's first-
millennium BCE royal dynasties. Recent years have witnessed a paradigm shift
in the understanding of these works: long treated as mere, photocopy-like
"late copies" of supposed pre-Hellenistic originals, they are now recognized
either as entirely new third- or second-century BCE historiographic compositions or as charged reapplications of older texts to the particular Seleucid situations with which they resonated.[143] For instance, the fall of the Neo-Assyrian
empire at the end of the seventh century BCE was dramatized as a letter exchange between Sîn-šarru-iškun, last king of Nineveh, and Nabopolassar,
founder of the Neo-Babylonian dynasty that would replace it. Nabopolassar's
epistle hurled charges of hostility, impiety, outrageous violence, and illegitimacy at Sîn-šarru-iškun, amounting to a multicausal explanation of the Assyrian empire's demise.[144] Fortunately, the very fragmentary tablet bearing
Sîn-šarru-iškun's letter response has its colophon preserved, which dates its
inscription to the reign of the Seleucid monarch Alexander I Balas (150–145
BCE). It is hard to doubt the potency of such an epistolographic fiction at this
historical moment, in the immediate approach to the Parthian conquest of
Mesopotamia and with the strife-riven Syrian dynasty beginning its breakup
into rival statelets.[145] Similarly, it is very likely that the so-called Nabonidus
Chronicle, the most detailed cuneiform narrative of the Neo-Babylonian empire's fall to Cyrus the Great of Persia, is a composition of Hellenistic scribes,
constructing out of sixth-century BCE chronographic and propagandistic texts
a coherent narrative of imperial replacement.[146] These transition moments
were brought together by the Babylonian priest-historian Berossus. The third
book of his *Babyloniaca* schematized the period after Nabonassar's biblioclasm
as a sequence of Asian empires: the extant fragments concern the establishment and fall of Neo-Assyrian rule in Babylonia, the rise of the successor Neo-
Babylonian empire under Nabopolassar and Nebuchadnezzar, and the Persian
defeat of Nabonidus; and, as we have seen in Chapter 4, the *Babyloniaca* takes

its limit point and authorial stance from the Macedonian conquest of the Achaemenids.[147]

A Hellenistic cuneiform text, known as the Dynastic Prophecy, absorbs and extends this imperial sequence into and over the Seleucid kingdom.[148] Composed at some point in the mid- or late third century BCE, it represents our earliest extant exemplar of what I have termed "total history 2." With only about a third of the original tablet extant, and with breaks in the worst places, we must exercise all necessary caution in analysis. Even so, it is sufficiently clear that the Dynastic Prophecy belongs in the well-established Mesopotamian textual tradition of *vaticinia ex eventu*—that is, pseudoprophetic narratives that depict known history as future revelation before passing on to genuine prediction.[149] In contrast to earlier cuneiform *ex eventu* prophecies, though, this tablet shapes history to a dynastic rather than a monarchic temporal rhythm (for the distinction, see Chapter 3).[150] This Dynastic Prophecy, while remaining distinctively Babylonian in outlook, form, and expressive range, also anticipates several characteristics of the second-century BCE Judean apocalyptic visions in Daniel and 1 Enoch discussed above. It frames itself as an unalterable divine vision, revealed by the great gods (1.1-6). The closing colophon ascribes the tablet's composition to one Munnabtum (6.16-18), a name that does not occur in any other published late Babylonian text but that was likely intended to suggest the authorship of the seventh-century BCE Babylonian Munnabtum, a famous court astrologer to the Neo-Assyrian kings Esarhaddon and Aššurbanipal.[151] Accordingly, much like "Daniel" in the court of Nebuchadnezzar, the pseudonymous "Munnabtum" would predict the future sequence of empires from the palace of Nineveh. Finally, although in itself not especially unusual,[152] the tablet prescribes the secrecy of its predictions: "you must not show it to the un[initiated] ([. . . . *la mu-du*]-*ú la tu-kal-lam*)" (6.14).

After this introduction, the tablet runs through the first-millennium BCE succession of Babylon's imperial rulers. The first column narrates the end of Neo-Assyrian rule in Babylon and the empire's fall to the Chaldean monarch Nabopolassar (1.7-25). The second column runs through the kings of the Neo-Babylonian dynasty to Nabonidus, his defeat by Cyrus the Great—"A king of Elam will set out (lugal ᵏᵘʳnim.maᵏⁱ *i-te-eb*) . . ."—and the establishment of the Persian empire (2.17-24). In all likelihood, a lost third column on the obverse and a lost fourth on the reverse ran through the reigns of the Achaemenid monarchs from Cambyses to Artaxerxes III.

Our fragmentary tablet picks up again on its reverse side in the fifth column. The Achaemenid imperial line comes to a close in a difficult passage, which has attracted much scholarly attention and disagreement. "Munnabtum" writes (5.8–23):

8   [He (Darius III) will exercise] king[ship] for five years.
       Hanaean troops (lúerín^meš kurḫa-ni-i) [
       will set out a[nd? . . .
       [his] troop[s . . .
12   They will take booty [and his spoils?] from him
       they will plunder (i-šal-la-lu). (But) later (ár-ka-nu) [his] tr[oops . . .
       he will assemble and his weapons he will ra[ise . . .
       Enlil, Šamaš, and [Marduk?]
16   will go at the side of his troops (da lúerín^meš-šú gin^meš) [. . .
       He will [bring about] the overthrow of the Hanaean troops (su-kup-tu
       lúerín^meš ḫa-ni-i ⌜i⌝-[šak-kan])
       He will car[ry off] extensive booty from him
       he [will bring it] into his palace
20   The people, who had [experienced] misfortune, (lúun^meš šá lum-nu
       i-[mu-ru])
       [will enjoy] well-being (dum-qa)
       The heart of the land [will be happy]
       Tax exemption [. . . [153]

The Dynastic Prophecy here "predicts" the five-year reign of Darius III and the arrival of the armies of Macedonia, identified as "Hanaeans." The name Hana, originally referring to a Bronze Age nomadic tribal confederacy on the middle Euphrates, was reapplied in subsequent centuries to barbaric populations that constituted a northern threat to Babylonia. It was used of Macedonians in hostile contexts only.[154] The Dynastic Prophecy, extremely fragmentary in this section, seems to narrate a complete and divinely ordained reversal of fortune: an initial Hanaean victory over an individual, whose name is lost to the lacuna, followed by the vanquished party's heroic return, together with the great gods, to bring about defeat for the Hanaeans and good fortune and tax exemption for the people of Babylonia.

The Dynastic Prophecy then continues into a mostly lost sixth column, which runs through the reigns of at least two more kings before predicting the establishment of yet another dynasty of conquerors:

9   [. . .] they will be purified *(ú-tal-la-lum)*
        [A king/prince will set] out and seize the land ([*zi-a]m-ma*
        kur dib-bat)
        [. . .]
12   [*After him his sons?*] will rule [the land] ([*kur*] ⸢*i-be-el-lu*⸣).[155]

The mysterious "purification" at line 9 of this final column parallels an identical event at the fall of the Neo-Assyrian empire and the succession of the
Neo-Babylonian dynasty, as described in column 1, lines 11–12: "they will be
purified (*kug.ga-ú*) | [A king . . .] will set out." Whatever the political or reli-
gious significance of such a process,[156] the equivalent context and phrasing
confirm that what is being predicted is the fall and replacement of this fourth,
post-Achaemenid dynasty. After a further line break, the Dynastic Prophecy
concludes with the colophon, its attribution of authorship to Munnabtum,
and the seal of secrecy.

There are two main interpretations of the Dynastic Prophecy's fragmentary
fifth and sixth columns. The original and still dominant opinion is that they
recount the arrival of Alexander the Great, his initial victory over Darius III
(whether at Issus or Gaugamela), and then a hoped-for return of Darius and
a fantasized Achaemenid defeat of the Macedonians.[157] While this solution
might at first seem attractive—for, from our disciplinary perspective, who
could be more important than Alexander?—it not only depends upon the historical untruth of an imaginary Persian victory but also fails to make sense of
the Dynastic Prophecy's continuation into at least two more reigns and a
further regime change.[158]

The alternative (and, to my eyes, correct) explanation is that the tablet's
fifth column offers an account of the fall of the Achaemenid and rise of the
Seleucid empires.[159] Seleucus Nicator was the only late-fourth-century BCE
political leader to be driven from Babylonia and to return with reunited forces
in triumph. According to this solution, "the overthrow of the Hanaean troops"
at 5.17 would describe Seleucus's defeat of Antigonus Monophthalmus; other
contemporary cuneiform texts characterize the Seleucids' Macedonian enemies as Hanaeans.[160] Moreover, Enlil, Šamaš, and probably Marduk return at
the side of the king (5.15–16), the only such episode in the extant portions of the
Dynastic Prophecy. It is crucial for understanding such explicit divine sanction
that Seleucus I assimilated his return to Babylon in Nisannu 311 BCE (the epoch
of the Seleucid Era count) to the religious paradigm of the *akītu* New Year

TABLE 5
The Dynastic Prophecy

Obverse	1 a	Introduction: Revealed knowledge
	1 b	Fall/rise Assyria → Babylonia
	2	Fall/rise Babylonia → Persia
	[3	Persian kings]
Reverse	[4	Persian kings]
	5	Fall/rise Persia → Syria
	6 a	Fall/rise Syria → ?
	6 b	Colophon: Revealed knowledge

festival; for, as we have seen in Chapter 1, the ritual climax of this celebration was the victorious king's march back into the city accompanied by the great gods. The idealized situation with which column 5 ends is supported by a number of sources that indicate the Babylonians' positive reception of Seleucus and their hostility to Antigonus.[161] Thus, according to this preferred interpretation, the Dynastic Prophecy's fifth column reproduced the empire's own myth of origins to represent the transition from Achaemenid to Seleucid rule. The subsequent reigns of column six can then be identified as, in all likelihood, those of Seleucus I's successors—Antiochus I, Antiochus II, and possibly Seleucus II—after which (in 6.10-12, quoted above) we find a genuine prediction of the Seleucid empire's conquest and the emergence of a new, unnamed dynasty. The identity of this new empire is unclear: if this tablet is from the mid- or late third century BCE, then the possibilities include the Ptolemies, who briefly conquered Babylonia in the Third Syrian War, a satrapal revolt, and any number of unhistorical and so irrecoverable political fantasies.

This second, Seleucid interpretation is supported by the physical shape and internal delineations of the Dynastic Prophecy. Although only parts of two columns on each side survive, it is sufficiently clear from the tablet's dimensions and historical content that it was originally composed in six columns.[162] This columnar layout fixed a historical rhythm of imperial succession, which can be most easily grasped in a tabular scheme (see Table 5).

The founders of the imperial dynasties are considered good and successful; the last kings of Assyria, Babylonia, and Persia are rebels who fall.[163] As in double-sided, multicolumnar cuneiform texts from this period, columns 1-3 proceed from left to right on the obverse side and columns 4–6 from right to left on the reverse, turning from obverse to reverse at the top edge. Accordingly, the Dynastic Prophecy establishes a chiastic, mirrored structure, with a set of back-to-back alignments: the revelatory introduction and pseudonymous colophon in columns 1 and 6; Babylonia's replacement by Persia and Persia's replacement by the Seleucids in columns 2 and 5; and the line of Achaemenid rulers in columns 3 and 4. Most provocatively, the fall of the Neo-Assyrian empire to Babylon in column 1 is answered by the predicted conquest of the Seleucid empire in column 6. Indeed, we have seen above that the Assyrian and Seleucid empires are each "purified" and conquered in precisely similar terms. Certainly, such identification of the Assyrian and Syrian kingdoms is common in anti-Seleucid writings from Judea and elsewhere,[164] and it is at the very least suggestive that the Assyrian kingdom had fallen to a Babylonian dynasty.

The Dynastic Prophecy is a pseudonymous, revealed, and secret total history that delivers an account of imperial succession and, in all likelihood, imagines the overthrow of the Seleucids. While it is without the explicit condemnation, imagistic violence, and eschatological punishments of the Judean and Iranian apocalyptic works (see above and below), it represents our earliest clearly systemized periodization of Near Eastern imperial succession extended beyond the Seleucids and into the future. Its lesson is that the Syrian house will pass, falling victim like its predecessors to dynastic decline, divine abandonment, and military conquest.

Iranian Traditions

Iran has endured two devastating foreign conquests, Macedonian and Arab. Though separated by a millennium, scholarship has held that the fall of the Achaemenid and Sasanid empires, and the humiliating dislocations that followed in each case, generated a common apocalyptic response; further, that the apocalyptic eschatological texts from the late first millennium CE were re-workings of original Hellenistic compositions.[165] The locus classicus for this bold claim is the first chapter of the *Zand ī Wahman Yasn* (also known as the *Bahman Yašt*), a ninth- or tenth-century CE Middle Persian compilation of earlier Zoroastrian eschatological traditions and exegesis:

As (is known) from the *Stūdgar* [*Nask*], Zarduxšt sought immortality from Ohrmazd. Then Ohrmazd showed the omniscient wisdom to Zarduxšt.

And therewith he saw the trunk of a tree on which were four branches, one of gold, one of silver, one of steel, and one on (which) iron had been mixed. Then he considered that he had seen it in a dream. Once woken from sleep, Zarduxšt said, "O Lord of beings in the world of thought and the Realm of Living Beings, it seems that I have seen the trunk of a tree on which were four branches." Ohrmazd said to Spitāmān Zarduxšt, "The tree trunk that you have seen, (that is the Realm of Living Beings which I, Ohrmazd, have created). Those four branches are four ages that will come. The one of gold is (the age) when I and you converse, and king Vištāspa accepts the Religion and breaks the bodies of the *dēws*, and (the *dēws*, from the state of being visible), take to flight and hiding. And the one of silver is the reign (of) Ardaxšīr, the Kayanid king. And the one of steel is the reign (of) Husraw of immortal soul, son of Kawād. And the one on which iron had been mixed is the evil rule (of) the parted-hair *dēws* of the seed of Xēšm (*dēwān ī wizārd-wars ī xēšm tōhmag*), when it will be the end of your tenth century, Spitāmān Zarduxšt.[166]

Zoroaster's dream-vision should seem immediately familiar from the texts discussed above in this chapter, in particular the dream of Nebuchadnezzar II in Daniel 2—its scenario as a revealed total history, requiring secondary explanation; its imagery figuring imperial power as a great tree, recalling both Daniel 4 and a fragmentary Dead Sea Scroll;[167] its periodization as a sequence of four ages, symbolized by metals of declining worth; its pseudonymous authorship, with Zoroaster at its top golden branch gazing forward through history; and the demonic degradation of its final age, figured as mixed iron.[168] As the text stands, Ahuramzda (or Ohrmazd) identifies the first branch of gold with king Hystaspes (or Vištāspa), the royal patron of the prophet Zoroaster; the second branch of silver with Artaxerxes (or Ardaxšīr) the Kayanid, apparently conflating a historical Achaemenid king with the Kayanian dynasty of Iranian myth;[169] the third branch of steel with the great Sasanian monarch Chosroes (or Husraw) I; and the fourth branch of mixed iron with the offspring of the wrath demon Xēšm. The fourfold periodization is combined with a chronographic, millenary scheme, in which the age opened by Zoroaster comes to a close at this final branch of mixed iron.

It is an abundantly precarious endeavor to propose a Hellenistic origin for this post-Sasanid text or to consider the *Zand ī Wahman Yasn* a reliable source for Iranian theological-historical ideas from a millennium earlier. Indeed, the very antiquity of Iranian apocalyptic ideas has been called into question by several scholars,[170] and it cannot be doubted that any pre-Sasanid version of the *Zand ī Wahman Yasn* is, as Jacques Duchesne-Guillemin cautions, "tout à fait conjecturale."[171] Yet three historical controls make the existence of some kind of anti-Seleucid original all but certain.

First, eschatological concepts were already central to Bronze and Iron Age Mazdean theology. Avestan texts, including songs that were recited during the Zoroastrian *haoma* sacrifice,[172] describe a dualistic battle for dominance between the two independent principles of Ahura Mazda and Anra Mainyu (approximately, order and chaos, or truth and lie) that would culminate in the coming of a savior figure (the *Saošyant*), the resurrection and judgment of the dead, and a glorious and wholly good *eschaton* (known as the *Frašō.kərəti* or *Frašegird*).[173] Even if no revelatory relevance seems to be ascribed to the specific events of political or "national" history, the directional passage of time is the medium for this cosmic drama, advancing through the dualist world to the ultimate and absolute rule of Ahura Mazda. We see the politicization of this (or a closely related) eschatology in the imperial rhetoric of the Achaemenids. As Bruce Lincoln has demonstrated, Darius I and his descendants constructed a legitimating political theology based on Ahuramazda's creation of the world, his divine election of the monarchs, their enacting of his will in history, and their ultimate restoration of creation, from the grip of liars and rebels, to its original, pristine, and perfect state.[174]

Second, a variety of Iranian millenarian and eschatological traditions are attested in Greek and Latin sources. Already in the fifth century BCE Xanthus of Lydia dated the founder prophet Zoroaster six thousand years before Xerxes's bridging of the Hellespont, perhaps relaying eschatological propaganda from the Achaemenid invasion of Greece.[175] A century or so later, Eudoxus and possibly also Aristotle fixed Zoroaster six millennia before the death of Plato, a reperiodization that may have implied a typological identity between the Iranian prophet and the Greek philosopher or, possibly, the closure of a coherent world period.[176] More explicit is the fourth-century BCE historian Theopompus's representation of the Mazdean tripartite millenary eschatology, as paraphrased in Plutarch's late treatise on the cult of Isis and Osiris: the good god Oromazes and the evil demon Areimanius, having dominated each other for three thousand years in turn, will engage in conflict for a further three thousand years. Then, "at the destined time (χρόνος εἱμαρμένος)," Areimanius will be utterly vanquished, the mountains will be flattened, and all the peoples of the earth will be brought under one mode of life, one government, and one language ("ἔνα βίον καὶ μίαν πολιτείαν ἀνθρώπων μακαρίων καὶ ὁμογλώσσων ἀπάντων γενέσθαι").[177] Theopompus's account closely parallels, down to specific details, the Zoroastrian eschatology laid out in the Middle Persian cosmological text *Greater Bundahišn*.[178]

Importantly, these traditions could be turned against imperial domination. The so-called *Oracles of Hystaspes* (a lost work paraphrased by Justin Martyr, Clement of Alexandria, and Lactantius) predicted the fall of the Roman empire, a great battle of kings, an ordeal by fire in which the godless would be exterminated, and the descent from heaven of a royal savior figure.[179] Even as reported by these Christian apologists, the *Oracles of Hystaspes* contain several indubitably authentic and specifically Zoroastrian eschatological tropes, which closely parallel the *Zand ī Wahman Yasn*.[180] Additionally, its narrative frame—"a wonderful dream interpreted by a prophesying boy *(admirabile somnium sub interpretatione vaticinantis pueri)*"[181]—recalls the scenario and structure of the *Zand ī Wahman Yasn*, quoted above, as well as the Middle Persian *Žāmāsp-Nāmak* and, of course, the dream of Nebuchadnezzar II in Daniel 2. In all likelihood, we can identify in the text's eponymous Hystaspes (or Vištāspa) the royal patron and first follower of Zoroaster and in the *vaticinans puer* the youthful prophet himself.[182] Scholars have persuasively located the text in a Zoroastrian or Zoroastrian-influenced Graeco-Iranian context, most likely Hellenistic or early imperial Asia Minor; several have suggested its development from an earlier, anti-Seleucid oracle, although there is no direct evidence for this.[183]

Third, the *Zand ī Wahman Yasn* itself contains internal evidence of progressive textual development. Its compiler cites four earlier sources—the *S(t)ūdgar Nask* and the *zands* (or Middle Persian commentaries) on the *Wahman Yasn*, *Hordād Yasn*, and *Aštād Yasn* liturgies—of which all but the *Wahman Yasn* are attested or extant in some form.[184] Moreover, the key image of the multimetallic branched tree is revised and updated in the course of the collection: its four branches of gold, silver, steel, and mixed iron are increased, in the book's third chapter, to seven branches of gold, silver, copper, brass, lead, steel, and mixed iron. Furthermore, we find explicit mention of the post-Achaemenid Macedonian conquerors. The fourth branch of the third chapter's seven-branch tree represents "Alexander the Ecclesiastic *(Aleksandar-ī Kilīsāyīg)*," who will be "cancelled from the religion" by the Arsacid kings, likely reproducing the primary identification of the fourth and final branch of mixed iron of the original four-branch tree of the first chapter.[185] More certain is the prediction that, in the final days, an army of demons will reach to "the Greeks dwelling in *Asūrestān (yōnān ī asūrestān-*manišn),*" meaning the Graeco-Macedonian colonists in Babylonia, even "up to their residence," glossed by one exegete as "the dwelling of the *dēws (gilistag ī dēwān),*"[186] which may refer to Seleucia-on-the-Tigris (against which the hostility of the Magi is otherwise attested).[187] Carlo Cereti, one of the text's editors and more cautious than most about the apocalypse's antiquity, has noted that the ethnic label used here, *yōnān,* "Greeks" (cf. Old Persian *Yauna*), is practically unknown from Parthian and Middle Persian texts and indicates that at least this passage goes back to a Hellenistic original.[188]

The circulation of normative Zoroastrian (or, more loosely, Mazdean) eschatological and millenarian ideas, their politicization against foreign empire, and traces of early strata within the *Zand ī Wahman Yasn*—in the historical context of imperial collapse, devastating conquest, and the pains of provincialization, these suggest an anti-Macedonian periodization that configured the Seleucid world as the final, ultimate stage of history. If we retain the template of the four-branch tree, with metals of declining worth, the fourth and final branch of mixed iron necessarily would represent the Macedonian conquerors of Persia. Significantly, in the extant version of the *Zand ī Wahman Yasn,* this mixed-iron branch represents not an individual ruler but an entire wicked population; and it is noteworthy that other Zoroastrian sources identify the wrath demon Xēšm, progenitor of this horde of the last age, with Alexander the Great.[189] The other identifications remain uncertain: Hystaspes (or Vištāspa) should be retained in gold place, as the apocalypse's pseudonymous conceit requires; Artaxerxes (or Ardaxšīr) is, perhaps, the Achaemenid Artaxerxes II; and he may have been preceded in silver by a Median monarch or followed in steel by Darius III. Whatever the precise shape and sequence, we can be reasonably confident that certain Iranians of the Seleucid period shared with their fellow subjugated Judeans and Babylonians a historical scheme of imperial succession in which time ran out on the kings of Syria.

The Politics of Periodization

The eschatological total histories discussed in this chapter—from Seleucid-governed Judea, Babylonia, and Iran—share a common technology of historical periodization. Their full narratives are cut up into sequences of ages, sharply delineating coherent periods of time. As we have seen, these take shape as discrete or numbered entities in a series, whether figurative and allegorical (monstrous beasts, animal species, metals, body parts, market weights, or tree branches) or scribal (tablet columns or parchment scrolls). We must not dismiss these periods as mere narrative punctuation, rhythms in the forward flow of history; rather, this is a new schematics of time that constitutes the primary concern of our apocalyptic texts and an implicit historical philosophy. For those works that establish a succession of Near Eastern empires—among them Daniel 2, 7, 8, and 10-12, 1 Enoch's Apocalypse of Animals, the Dynastic Prophecy, and, probably, the *Zand ī Wahman Yasn*—there is an absolute and complete identification between the rule of an ethnically defined imperial state and a segment of historical time. The transition from one of these empire-periods to the next is immediate and total, a principle neatly formulated in the early rabbinic chronographic text *Seder 'Olam:* "The rule of one people does not overlap with that of another people, nor the rule of one government with that of another government; but a government whose time has expired during the day will fall during the day and that whose time has expired during the night will fall during the night."[190] Each discrete age follows directly upon its predecessor, and we find the same pattern for the sabbatical schemata of Daniel 9 and the Enochic Apocalypse of Weeks. Such segmentation of history is simultaneously a pressing forward of time, with each age, like a caryatid, supporting the floor that the next will stand upon; a temporal dislocation, as each owes little or nothing to its predecessor; and a drawing forth of time, with each pulled toward the goal rather than impelled by antecedent events. Indeed, the eschatological redemption, whether conceived as a final period of eternal rule or as the transcendence of time itself, has no clear causal relationship to previous history: it is a breaking in upon the Seleucid world, a leap out of history, "an intrusion in which history itself perishes."[191]

There has been protracted scholarly debate over the source of this apocalyptic periodization, particularly the sequence of four empires. Lines of debt and descent have been traced between the historiographic and religious traditions of Greece, Iran, Judea, and to a lesser extent Mesopotamia, reaching back into the Bronze Age.[192] The disciplinary claims mirror, perhaps a little too closely, Hellenistic apologetics over national priority. Every possible path of derivation has been explored, as if an origin, like the source of the Nile, can be found (or would suffice as explanation).[193] I would suggest, instead, that what must be accounted for is not the various, widespread, and necessary anticipations of the phenomenon, but the concurrent emergence in the third and second centuries BCE and within the core regions of the Seleucid empire of a historiographical and theological obsession with such totalizing, highly schematized periodizations. These segmentations of history are the central concern of the eschatological total histories, as I hope this chapter has demonstrated; they are not outlined with such clarity or centrality in earlier texts; and they do not appear elsewhere in the Hellenistic world (including Ptolemaic Egypt).[194] We are dealing with a phenomenon specific to the Seleucid empire's subject communities.

The periodizing schemes discussed in this chapter respond to, reproduce, and reduplicate the Seleucid temporal regime. The first part of this book argued that the Seleucid court worked hard to understand and project itself as a discrete and unitary period of history. Its imperial rule was ubiquitously equated with an ongoing dynastic duration. Seleucid historiography, appeals to precedent, the ruler cult, Apollonine ancestry, the naming of colonies, and, perhaps above all, the Seleucid Era epoch generated for this historical period of empire a crisp beginning, marking it off from a past to which it claimed to owe nothing. This is precisely the texture of historical segmentation we find in our eschatological total histories.

Moreover, we can detect a staging of imperial succession in the Seleucid kingdom's inhabited and spectated landscapes, a space-making that corresponds to the chronological shaping of our texts. Take, for instance, the main palace of Babylon, exilic residence of Daniel and his companions and still in use under the Seleucid monarchs. Excavation has disclosed that onto the complex's Neo-Babylonian core the Achaemenids had added a distinctly Persian hypostyle pavilion, and subsequently the Macedonians a stuccoed peristyle courtyard of Hellenic form.[195] Working in or walking through this site of governance and ceremony would have spatialized, by culturally marked aesthetic distinctions, the succession of imperial masters.

Elsewhere we find a deliberate ruination of the pre-Seleucid past. For example, in distant Bactria, the Achaemenid-period fort of Kohna Qala was deserted in favor of the new Hellenistic colony at Aï Khanoum, a mere two kilometers southwest; the spatial relationship translated a historical one, periodizing the transition from Persian to Seleucid rule.[196] Similarly, at Seleucia-on-the-Eulaeus, the refounded Susa, Darius I's famed royal palace was left outside the new Hellenistic settlement, abandoned to decay, employed for burial, and partially occupied by squatters:[197] a demotion from Achaemenid courtly center to Seleucid slum, a ruin that gave physical form to the historical sequence. Such periodization by destruction is even apparent at the minor administrative center of Kedesh, on the edge of the Hula basin in northern Israel, discussed in Chapter 3. The original Achaemenid complex was repurposed, first under the Ptolemies, who disarticulated the column drums of its monumental pi-shaped entrance into the walls of new storage rooms,[198] and then, after the Seleucid conquest, with the insertion of an archive and reception suite (see Chapter 2), the floors of which were lined with smashed Ptolemaic pottery.[199] Presumably, in one way or another, such periodized landscapes were common throughout the empire, from Sardis to Samarkhand.[200] We may suppose that the overlaying and ruination of earlier built infrastructure produced contrastive landscapes, generating a differential sense of historical time and a comparatistic perception.[201] These ruins, by merging history with setting, naturalized the Seleucid historical succession. The dilapidating palace at Susa, wind-gutted Kohna Qala, and the column drums and sherd floor of Kedesh were witnesses to political cataclysm, testifying to a sharply delineated past that would not return. "Ruins are not elegiac but rather dialectical":[202] sensitivity to imperial succession was an effect of the Seleucid empire's landscape as well as its explicit temporal regime.

In sum, schematic historical periodization, while drawing on deep and important roots, took on a new prominence, systemized clarity, urgency, plausibility, and explanatory efficiency within the core regions of the Seleucid empire and on account of this Seleucid imperial context. Put another way, while the fourth beast of Daniel 7 may be the last monster to emerge from the cosmic waters, or the black ravens of the Apocalypse of Animals the final predators to descend on the sheep, they were the first in order of conceptualization. According to the Arabic maxim, approvingly quoted by the great French historian Marc Bloch, people resemble their times more than they do their parents;[203] so, I am convinced, do ideas.

It is a curious thing to oppose an empire by segmenting history. What did this apocalyptic periodization achieve? Two answers can be given—broadly speaking, political and cognitive. By incorporating this fourth empire into their narratives, the authors prospectively historicized it, and the historicization of political authority undermines its legitimacy and necessity. The Seleucid kingdom, despite all of its efforts to establish a limit-horizon of historical reference, was dragged into the same space of experience as the empires that had preceded it. Apocalyptic periodization implied a similitude of fate and thereby encouraged typological thinking: the patterning of history guaranteed the fall of the Seleucids and the salvation of the righteous. Moreover, by promising a dramatic end to Seleucid rule, these texts made it possible to imagine a liberated world beyond empire. Indeed, the consummation of the Judean apocalypses in the trial and punishment of the Seleucid beast transformed a royal prerogative into a reckoning of the dynasty itself. The most basic fact about the imagined futures of these works is that the Seleucids are nowhere to be found; in the Judean and, probably, Iranian works, they have been condemned to the hell-flames. In a world coordinated to the Seleucid timeline and pervaded with assertions of monarchic agency, the apocalyptic total histories revealed the ultimate and underlying sovereignty of God. They staged a battle over the control of time and the architecture of history, exposing the claims of empire as illusory and relocating the fate of nations to heaven: in its way, this was as radical a decentering as the Copernican revolution, which obliged us to renounce a sense of the immobility of our world and recognize a movement we do not feel. The emancipatory potential of such periodization is even, on occasion, made explicit: in the book of Daniel and the Apocalypse of Animals, the prototypical resister is the historian—the *maśkilim* of Daniel educate the Judean masses in the book's theology of history,[204] and 1 Enoch's angelic scribe brings about the divine intervention by reciting his historiographical record before God.

Yet the hermeneutic of "resistance," whether political or epistemic, is inadequate to these works. In addition to their targeting of the Seleucid worldview, they also answered to what F. A. Hayek termed "the craving for intelligibility."[205] All history writing, to a greater or lesser degree, is an attempt to disclose meaning in or attribute meaning to the formless infinity of past data; but the eschatological total histories, composed by the Seleucids' subjects, mark the extreme of articulation in this regard. They should be understood as a particular kind of response to a crisis of historical meaning generated by the imperial regime, a temporal pathology already outlined in brief in the conclusion to Chapter 4: the disinterestedness of transcendent time; the monotonization and disenchantment of the future; the depersonalization, commodification, and portability of temporal texture; the absence of rhythm, striation, or restart; the alienation from one's own past; and a vertiginous endlessness that, by implication, overwhelms eternity. Seleucid time was a mere passing, and so a loss.

In such a world, "without an echo to its cry," apocalyptic historiography filled a lack. According to these eschatological total histories, everything, including the future, was determined; all that happened to you happened for you; history was shaped, directed, and reaching toward a conclusion; all events, however dislocated, were part of a single story; and your own place in it could be located. Periodizing schemata fixed the stations of history, replacing a temporal parataxis with the enchainment of a temporal syntax. By this morphogenesis,[206] events were rendered intelligible according to their ordering within the total form. Like spatial boundaries, periodizing distinctions allowed the conceptualization of identity and difference.[207] Like the miniature or miniaturized, they offered a totality intended for cognition, an exterior and simultaneous view of a world of parts, outside-in, whose unity, containment, and completion was absolute.[208] Like liturgy or ritual, periodization imposed formalized discontinuities on seamless, continuous, and indistinct processes;[209] by cutting up historical time, periodization unambiguously distinguished succeeding from preceding, serialized ordered stages, invested both those periods and the segmented whole with significance, and *in fine* produced for the outplaying of centuries meaning or (what may be the same thing) the illusion of meaning. Furthermore, describing or calling forth the *eschaton* did not only fantasize a world without the Seleucids, imagining the empire as already over. It also brought an ever-unfurling, homogeneous time to a close, like the backing a mirror needs if we are to see anything. Just as the conceit of pseudonymy allowed the actual authors of the apocalyptic histories to unify in prospect past, present, and future, so the end-times achieved in retrospect a temporal integration, converting simple chronicity into a plot.[210] Finally, if the Seleucid time regime was unjust in its essence, both in its evacuation of social therapies and the ever greater disproportion between crimes and impunity, the culmination of these total histories in a redemptive trial and retributory punishment proposed—when all evidence held against it—the identity of history and justice: *die Weltgeschichte ist das Weltgericht*.

To close, this long chapter has suggested not only that the Seleucid empire provides the political context for the emergence of eschatological total histories, but also, more significantly, that its temporal regime—institutionalized, ubiquitous, present-focused, past-effacing, and future-opened—was the inspiration and provocation to schematic periodizations and fantasies of finitude. In some most important respects, apocalyptic eschatology was Seleucid by derivation and anti-Seleucid by consequence.

CHAPTER 6

Altneuland
Resistance and the Resurrected State

The fall of the Seleucid empire was not the end of history, despite the perfervid hopes of our apocalypses. Rather, the kingdom's collapse was the multiplication and equalization of operative political agency in the Hellenistic east. For in place of God and His angels—or, at least, more amenable to historical investigation—three sets of actors bear responsibility for the empire's destabilization, contraction, and disintegration. First, the outside players: primarily Ptolemaic Egypt, the perennial adversary, whose blows could always be parried, and the Parthians and Romans, whose westward and eastward expansions proved as irreversible as tectonic plates. Next, internal subordinates: rebel satraps and princes, who snapped off whole regions or struck for the throne. And, finally, the subject of this chapter: local elites, who framed their insurrections against empire or bids for increased power as revivals of a long defunct indigenous rule. While the first two sets of actors—external conquerors and internal pretenders—have been well and recently studied and can be understood on their own terms,[1] the third set acted, in significant ways, in response to the Seleucid temporal regime. As we shall see, certain priests, aristocrats, and vassals weaponized history against the empire. These indigenous elites fashioned a "future-past" that found its political form in, to borrow Theodor Herzl's programmatic title, an *Altneuland* (an old-new land).

Up to this point I have deliberately eschewed a regional or ethnic focus, in favor of panimperial phenomena and interregional developments. But here, at last, the Seleucid breakup demands that we also disaggregate our account. Accordingly, I will treat four regions in the order in which they were incorporated into the empire—Babylonia, Persia, Armenia, and Judea. Needless to say, other populations may well have pressed for greater autonomy at around the same time and in similarly atavistic modes—perhaps in Commagene and Sophene in eastern Anatolia, in Osrhoene and Adiabene in northern

Mesopotamia, in Media Atropatene in the upper Zagros, in the Nabatean desert, and along the Phoenician coastland[2]—but the surviving evidence is too exiguous to have purchase.

Babylonia

The Seleucid empire had been born in Babylon, but the city was more a midwife than a mother. Alexander's favor and Seleucus I's seeking for legitimacy had given a heightened significance to the old imperial capital, the Sumero-Akkadian temple elites, and the region at large that has been abundantly enthused over by modern scholars.[3] But we must recognize that these inaugural years of empire were exceptional. Already in the early third century BCE, Babylon, its cultic centers, and its ancient neighbors were being irreversibly marginalized. Seleucus I's foundation of Seleucia-on-the-Tigris absorbed much of the governmental functions, residents, trade wealth, and primate status of the old city; the new colony's foundation narrative, reported in Appian's *Syriaca*, recounts the determined opposition of the local priesthood.[4] The old Babylonian elite was further demoted by Antiochus III's or Antiochus IV's establishment within the city of a *polis* or *politeuma*, which took over the temples' administrative and judicial functions, political prerogatives, and perhaps also agricultural estates; another *polis* may have been introduced into the southern city of Uruk.[5] It was probably not meant kindly when a Babylonian Chronicle characterized the citizens of its new, internal colony as "the Greeks, their name being '*politai*,' who entered into Babylon at the order of king Antiochus and who cover themselves with oil like the *politai* who live in Seleucia, the royal city, which is on the Tigris and the Royal Canal"[6]—a bodily alterity, foreign vocabulary, and imperial imposition as jarring to Babylonians as to their Jewish contemporaries. Furthermore, cuneiform was not employed by the Seleucid administration; Antiochus I's Cylinder from the Ezida of Borsippa seems to have been a one-off and in any case deeply impressed with Seleucid ideology.[7] Sacrifices "in the Greek style" were marked aberrations.[8] The Babylonian sanctuaries' fiscal independence was increasingly curtailed.[9]

Nonetheless, cuneiform and classical sources attest no massed, militarized uprising against the Seleucid kings of the kind that Babylon and its immediate neighbors had launched against Darius I[10] and Xerxes;[11] it is unlikely that one has slipped through the cracks. Until the late 140s BCE, Babylonia seems to have remained a stable and central part of the Seleucid empire, only ephemerally lost to Ptolemy III Euergetes in the Third Syrian War and to the rebel Seleucid satraps Molon and Timarchus in the early reigns of Antiochus III and Demetrius I, respectively. The Babylonian Chronicles and Astronomical Diaries do mention episodic violence: street battles in Babylon and Seleucia-on-the-Tigris, conflict between different Seleucid officials, an army mutiny in the Babylonian palace, and several instances of theft.[12] Yet these do not add up to anything approaching a determined, on-the-ground opposition to the kings of Syria. Samuel Eddy saw here a lack, the failed spirit of an exhausted and essentially passive "Babylonian man," brooding over his vanished past in an atmosphere of impotent regret, discouraged from action by the pusillanimity of his gods, "bewitched by a sense of his own worthlessness and sin and bemused by the conviction of his own weakness and frailty."[13] We might prefer instead to emphasize demographic and structural factors. The population of Babylonia was more mixed, and self-consciously so,[14] than anywhere else in the empire: several centuries of imperial expansions, deportations, migrations, and colonizations had fragmented regional hierarchies, atomized urban communities, and deposited populations of Arameans, Jews, Carians, Iranians, Arabs, Greeks, and others,[15] hindering the easy formation of the anti-Seleucid solidarities and coordination that we witness elsewhere and channeling political resentments toward intercommunal rivalries.[16] Moreover, it appears that the local populations were effectively demilitarized: no Babylonian forces are attested in any Seleucid army; the region's garrisons appear to have been Greek, Macedonian, and Jewish;[17] and even from the mid-second century BCE, amid the chaos of Seleucid collapse, the Babylonians did not join the Parthians, Elymeans, and Characenians in the contest for regional spoils.[18] Finally, the first Seleucid kings may have successfully neutralized the region's heroic history, co-opting the figure of Nebuchadnezzar II in particular, around whom ambitions would otherwise have coalesced.[19]

In the absence of mass insurgency we can identify a more ambiguous situation. It may be helpful to distinguish between two main periods: a first, the high Seleucid empire from the third to the middle of the second century BCE, characterized by the general disinterest and occasional intervention of a mostly unchallenged monarchy; and a second, from the 140s to the late 120s BCE, when the region eroded into a zone disputed by powers centered elsewhere, a frontier situation for which Babylonia had neither experience nor cultural and diplomatic strategy.[20] The first period witnesses cuneiform scholars insisting on the fathomless antiquity of their scribal and religious culture, relocating monarchic capacities to temple notables, and carving out a world independent of empire; the second, a re-enlivening of this lost past in feverish dreams and reconstructive hopes.

Local interest in Mesopotamia's deep past was, of course, no new thing: earlier in the first millennium BCE Aššurbanipal of Assyria had claimed knowledge of antediluvian texts; and Nabonidus, last of Babylon's indigenous kings, had excavated ruins, studied old tablets, and revived defunct religious practices.[21] But where this pre-Hellenistic research had been concerned to align the region's deep history with centralized imperial power and drew much of its élan from this, the antiquarianism of the Seleucid period moved in precisely the opposite direction, crevassing between past and power and exposing their sharp disjuncture. Such a discrepant pragmatics had begun under the Achaemenids, who—unlike earlier Amorite, Kassite, and Chaldean conquerors—did not assimilate themselves to the Babylonian royal ideology or culture.[22] But even they had adopted Akkadian as a royal language, maintained Babylon as the region's capital, and could easily be mapped onto the (perhaps not flattering) historical prototype of Elam.[23] The Seleucid empire, by contrast, with its new imperial centers to the west and east, its distinctive time regime, and its own highly developed and self-satisfied intellectual world, was altogether less familiar, more radical, and harder to fit into the region's past.

Our data set from Hellenistic Babylonia is more skewed and partial than we often allow ourselves to recognize. Undoubtedly, the major medium of writing in the region in this period was alphabetic Greek or Aramaic on parchment and papyrus. But this vast and dominant "paper" archive has been entirely lost to us. Instead, we build up our interpretative edifices from the thousands of clay tablets that survive. These represent a self-consciously antiquarian, systematically conservative mode of record and composition: Hellenistic writing in cuneiform on clay was a deliberate choice to affiliate with a local and non-Greek tradition, an ancient, sanctified inheritance, and a visibly, undeniably sinking world.[24] Indeed, over the course of the Hellenistic period, and increasingly from the second century BCE on, legal, administrative, and literary texts jumped ship from clay to ink. At Uruk the last cuneiform slave sale contract dates to 274 BCE, the last letter to 171 BCE, the last literary composition to 162 BCE, the last prebend sale to 140 BCE, the last house plot sale to 138 BCE, and the last slave dedication contract to 127 BCE.[25] Ritual and historiographical texts held on longer in Babylon, but the pattern of decline was the same. New discoveries may change the details,[26] but not the overall picture. It had been unusual in the millennia of pre-Hellenistic Mesopotamian antiquity for inscribed clay tablets to be baked—typically, this has been effected by excavators and museums[27]—yet a great proportion of Seleucid-period tablets attest to such treatment; we can follow Lauren Ristvet in identifying here a heightened anxiety over preservation and survival.[28] Indeed, several carefully shaped and beautifully inscribed tablets from Uruk copied fiscal transactions also or already recorded on parchment, which was likely a legal redundancy but certainly a performance of cultural identity.[29] Even seal rings could appeal to this deep tradition: several members of the Urukean Ekur-zākir family revived as their personal signature types the fish-man sage from the dawn of Babylonian civilization, the *apkallū* celebrated in Berossus's *Babyloniaca* and the Uruk List of Kings and Sages.[30] Of course, as Near Eastern scholars have rightly demurred, we must recognize cuneiform innovation, most notably in the remarkable flourishing of observational and mathematical astronomy in this period.[31] But this was a narrow furrow, both as an intellectual niche captured by the tiny circle of cuneiform scholars and as a typecast role (the horoscoping Chaldean) to perform to the modern world.

If the very form of writing asserted a fathomless indigenous antiquity, reaching back to Adapa-Oannes, certain cuneiform works underlined the political and religious inadequacy of the Seleucid present. We have already seen, in Chapter 4, that the Uruk List of Kings and Sages depicted a contemporary Uruk with neither a king nor a sage. Ritual scripts, long an established genre in Mesopotamia and the wider Near East, offer another good case study. These tablets, surviving primarily from Seleucid Babylon and Uruk, textualized the sequences of proper cultic actions for various liturgies or ritual scenarios, a process that "demanded the selection of religious ideas, institutions, and practices and their articulation into a legible object," in the helpful terms of Duncan MacRae's recent study of comparable Roman efforts.[32] For instance, a tablet from Uruk in the late third or early second century BCE, perhaps composed by the same Anu-bēlšunu of the Uruk List,[33] prescribes a building ritual to be performed in the event that "the wall of the temple of Anu buckles." Among the sacrifices, libations, and lamentations we find this picturesque rite: "You will make the king recite (*tu-šad-bab*) the *eršemšaḫungû*-prayers 'O lord, what turned about in my mind? It is Anu' before Anu, 'I want to offer my life to my lord,' to Enlil (and) Ea, (and) 'I want to speak a prayer to Šamaš' to Šamaš. He will then prostrate himself. He submits to treatment by the barber. You will gather his body hair into a porous *laḫannu* bottle and send it off to the enemy's border."[34] If this is performed appropriately, "evil will not approach the king."[35] In another example, the Seleucid-period ritual text for the Babylonian New Year festival, the great *akītu* described in Chapter 1, outlines for 5th Nisannu a spectacular humiliation of the king: "When (the king) has arrived [before] Bel, the high priest will go out (of the cella) and lift up the scepter, the loop, the mace [of the king]. He will (also) lift up the Crown of Kingship. . . . He will strike the cheek of the king. He will place [the king] behind him. He will make him enter before Bel. He will pull his ears, make him kneel to the ground."[36]

These ritual prescriptions have been investigated as archaic religious expressions of purification, protection of the land, and renewal of kingship.[37] But, taken as the Seleucid-period documents they are, we must ask if it is really plausible that Antiochus II permitted the dispatching of his depilated back hairs to Pelusium or the Parthian frontier? Would Antiochus III have been dragged by his earlobes beneath Marduk? Did Seleucus II recite a prayer in Sumerian? Even if such performances did take place on rare occasion, by representation, or in some approximation, the normative ideal established by these tablets was one that could never be achieved. There was an inescapable, irreparable incongruity between the reality of the Seleucid world and these ritual texts' paradigm of politics and religion, of god and king. Why prescribe an impossibility, and one that could only emphasize the absence, foreignness, and indifference of the reigning dynasty? Were these texts merely "for the file," virtual rituals, substitute kings in clay?[38] Did the tablets constitute, at least in part, scribal fantasies of control—note the causative verbs and priestly agency—that gleefully imagined manipulating or humiliating the body of the reigning monarch?[39] Did these tablets, baked for preservation, represent hope for the restoration of a properly engaged Babylonian monarchy, preparing the ground like the flood heroes of Berossus—*rex quondam, rexque futurus?*

Another important ritual text from Uruk, an illegally excavated tablet from the first half of the second century BCE, gives instructions for the daily offerings of beer, bread, and meat at the Bīt Rēš (the temple to Anu, Uruk's tutelary deity) and at the city's other sanctuaries.[40] The colophon, at the tablet's close, gives a remarkable account of the text's derivation:

(Copied from) a writing-board (containing) the rites for the Anu-cult, the sacred purification rites, the rites of kingship, including the divine purification rites of the Bīt Rēš, the Ešgal, the Eanna, and all the (other) temples of Uruk, the ritual activities of the *āšipus* (exorcists), of the *kalûs* (lamentation priests), of the musicians, and of the craftsmen . . . in accordance with the tablets that Nabopolassar, king of the Sealand, plundered from Uruk. Then at that time, Kidin-Anu, the Urukean, the *āšipu* of Anu and Antu, descendant of Ekur-zākir, the high priest of the Bīt Rēš, saw these tablets in the land of Elam. During the reign of Seleucus and Antiochus, he copied them and brought them back to Uruk.[41]

The cult of Anu, the Mesopotamian sky god, about which this tablet is chiefly concerned, had risen to prominence in Uruk only in the mid- to late-Achaemenid period, as an effect of the decommissioning of the city's main temple, the Eanna sanctuary of Ištar, and the sudden disappearance of its high cultic personnel after the Babylonian revolt against Xerxes in 484 BCE.[42] The worship of Anu, though framed here as an antiquarian recovery, did not exist in a major form in Nabopolassar's Uruk.[43] In fact, this tablet gives the first undisputed reference to Anu's vast temple complex, the Bīt Rēš ("First Temple" or "House of the Beginning").[44] In short, there is scholarly consensus that the colophon's narrative is a pious fraud, its "restored" ritual instructions in fact a compilation and reworking of sources originally directed at other deities.[45]

The colophon's narrative fiction adapts the well-established Near Eastern historiographical trope of the seizure and return of divine statues or other cultic objects. For instance, the Assyrian king Aššurbanipal, on conquering Elam, returned to Uruk the Nanaya statue that had been seized by the Elamites;[46] Cyrus restored the vessels of the First Temple looted by Nebuchadnezzar II from Jerusalem;[47] Ptolemy III claimed to have brought back to Egypt from the Seleucid kingdom temple statues originally taken by the Persians;[48] and Seleucus I and Antiochus I restored the Apollo of Miletus, and perhaps also the Athenian Tyrannicides, abducted by the Achaemenids.[49] These traveling statues gave a material form to the kind of interaction between a particular community and its monarchic overlord, to claims of the authentic revival of an interrupted cult, and to historical periodization, with one empire committing a sacrilege and a successor rectifying the sin. Thus, this tablet's colophon constructs a Daniel-like sequence of empires: stolen by Nabopolassar, the founder of the Neo-Babylonian dynasty; deposited in Elam, likely at Susa, the regional capital of the Achaemenids, and not returned;[50] copied back onto clay and restored to Uruk in the early Seleucid empire. Yet in this tablet the trope has been doubly displaced. On the one hand, what has been seized and restored is not the object of cult but knowledge of the cult—that is, the encyclopedic, textualized prescriptions for all parts of the temple service, including, we should note, a whole section dealing with the appropriate rites of kingship ("*sak-ke-e* lugal-ú-tú"). On the other hand, while these Urukean ritual texts had been stolen by a king (the Babylonian Nabopolassar), as the trope demanded, they were restored by a priest (Kidin-Anu of the Bīt Rēš). Seleucus I and Antiochus I, who, as we have seen, were known for their own statue returns, function here merely as chronographic markers of the new imperial situation. As in the Uruk List of Kings and Sages, the Seleucid monarchs no longer follow the script.

All in all, we see cuneiform scribes and priests coming to terms with a world that had marginalized them: an apprehension of externality as to their own cultural traditions, an antiquarianizing nostalgia, a recognition of the inefficacy of the ruling kings, and so an absorption of their prerogatives and responsibilities. No doubt, this could enhance the prestige of local notables—for instance, the Bīt Rēš of Uruk was rebuilt not by Seleucus II or Antiochus III but by the Urukeans Anu-uballiṭ-Nicarchus and Anu-uballiṭ-Cephalon in 244 and 202 BCE, respectively; and Anu-uballiṭ-Cephalon's building inscription even credited his conveying of cedar logs down from the distant mountains, a standardly monarchic conquest of nature since the adventures of Uruk's heroic scion, Gilgamesh.[51] In an unparalleled self-assertion, this Anu-uballiṭ-Cephalon installed behind the cult image of Ištar, in the holy of holies of the Ešgal temple, a band of glazed bricks that repeated in Aramaic his own double name and nothing more, without mentioning either the king or the goddess.[52] But there remained an undeniable imbalance. A worldview of geopolitical centrality and monarchic participation, justified by millennia of Mesopotamian history, was broken by a century of Seleucid rule. It is toward the end of this first, extended period of Hellenistic dominance that the temples of Uruk and Babylon were being transformed into fortified refuges, their outer walls supported by battlements and defensive towers,[53] perhaps providing in addition to a defense of bodies a form appropriate to their interior life.

The privileging of the deep past, modeling of ideal practice, and fantasizing of restoration—all latent, implicit, and virtual during the first period of strong Seleucid rule—broke to the surface in the later second century BCE.

The evidence is slim, as ever, but two episodes suggest, in this troubled twilight of Seleucid Babylonia, a reanimation of the past.

Between the Seleucid dynasty's first retreat from the region in 141 BCE and the final consolidation of Parthian dominance under Mithridates II in the later 120s BCE, the Mesopotamian alluvium must have been one of the most ghastly regions in all of Eurasia: a universal prey, assaulted from west, north, south, and east by the shrunken sovereigns of Syria, the new horse-kings from Central Asia, aggrandizing, self-emancipated satraps, and marauders out of the mountains or from the desert. The Seleucid brothers Demetrius II Nicator and Antiochus VII Sidetes each launched massive expeditions of reconquest, in 140–139 and 131–129 BCE, respectively, briefly capturing this birthplace of empire only to provoke more lasting Parthian countercampaigns. Elymeans from the southern Zagros marched between cities, burning to the ground the Seleucid colony of Apamea-on-the-Selea.[54] Hyspaosines of Characene, the former Seleucid governor and now the independent king of the Mesene Sealand, attacked Seleucia-on-the-Tigris and for several months occupied Babylon.[55] Arab raiders tunneled through the city's walls—once a wonder of the world!—to demand and receive protection money.[56] The Astronomical Diaries of the 130s and 120s BCE read as an unrelenting, pitiless chronicle of famine, fear, locust attack, rumor, conflagration, the collapse of glazed-brick buildings, and the killing or seizure of noncombatants.[57] It is highly significant that these Diaries, for the first time in almost two centuries, record the arrival of monstrous births—a baby goat without a right thigh, conjoined twins, a pig that resembled a dog—an objective correlative of these baleful times and also a searching for their rhyme or reason.[58]

At this midnight in the century an unlikely figure stepped forward. Two overlapping Astronomical Diaries for September and October 133 BCE, unusually extended and dramatic, give the following account:[59] "One boatman (1-en ^{lú}dumu má.laḫ$_4$) . . . his mind was altered. He set up a dais (1-en bár)[60] between the Temple of Sîn and the Gate of Marduk" in Babylon. "He placed an offering upon it and spoke a good instruction to the people: 'Bel has entered Babylon (^{d}en ana e^{ki} ku$_4$-ub)!' People, men and women alike, came and placed offerings on that dais; in front of that dais they ate and drank and hummed and rejoiced (kú-ú nag-ú i-ḫa-am-mu-ú i-ru-uš-šu-ú)." People of high rank approached the boatman, who now proclaimed: "Nanaya has entered Borsippa and Ezida (^{d}Na-na-a-a ana Bar-⸢sìp⸣^{ki} ana é.zi.da i-te-ru-ub)!" Immediately, the boatman and his followers proceeded to Nanaya's city, all gates opening before them. Something or someone was paraded in a chariot—the passage is fragmentary—and "in the streets and the squares people listened to [the boatman's] proclamation." This was the word of the boatman: "I am the messenger of Nanaya! I have been sent (šap-ra-ku) concerning/against the strong god, the striker of your gods (⌈ana⌉ muḫ-ḫi dingir kalag ma-ḫi-ṣu dingir^{meš}-ku-nu)."[61] There then follows a trial-type scene, in which the boatman and his followers appear before the kiništu (the assembly of notables of a Babylonian temple, here probably the Ezida of Borsippa). The exchange is narrated, uniquely for the Astronomical Diaries, as a dialogue in direct speech:

Temple kiništu to the crowd: "Return to your cities! Do not let the city be plundered! Do not hand over (la tu-še-ṣa-a) the gods, like the city, as booty!"
The boatman, as their spokesman: "I am a messenger of Nanaya! I shall not hand over the city to robbery and plunder. Like the hand of the strong, hitting god (ki$^{?}$-ma šu^{II} dingir dan-nu ma-ḫi-ṣu) on Ezida . . . <broken>"
Temple kiništu to the crowd: "Do not listen to the words of the hothead (šá-bi-ba-an-nu)![62] [Save] your lives! Preserve yourselves!"

But the people did not obey the kiništu. Another Diary fragment, providing an overlapping if briefer account, delivers the violent denouement: "they killed in their midst . . . (ina lìb-bi-šú-nu gaz^{meš}) . . . in Babylon and Borsippa."[63] Presumably, as we hear nothing further of them, the followers of the boatman, and perhaps also the boatman himself, were among the dead.

This episode is known solely from the report of the social group it seems most to have threatened, the priests and scribes of Babylon's Esagil temple; it is as if our only source on the Jesus movement were the record of Caiaphas. The authors of these cuneiform tablets, contemporary witnesses from within their walled compound, characterize the boatman as a charismatic leader of an unsanctioned popular movement. He is not from the elite priestly families and is left unnamed throughout the narrative, attracts a mass following that includes women, fixes a new topography of worship, with daises between rather than at preestablished ritual locations, and turns the streets and city squares of Babylon and Borsippa into sites of feasting, carousing, and reverent expectation.

Yet at the same time, the boatman preaches restoration. First, he confidently hails the entry or return of Bel-Marduk and Nanaya to their home cities.[64] Such arrival implies, of course, their earlier absence, perhaps a divine abandonment long associated in Mesopotamia with a fallen state of impiety and defeat. Now, according to the boatman, the proper and original order is being recovered. The boatman goes on to present himself as the apostle of the goddess Nanaya, sent to speak about or to speak against—the prepositional phrase *ana muḫḫi* can mean either—a new, nameless, and violent deity. This is "the strong, striking god *(ilu dannu māḫiṣu)*," who "strikes your gods *(māḫiṣu ilīkunu)*."[65] Without more evidence it is impossible to identify this hostile divinity.[66] All that can be said is that, according to the boatman's word, a confrontation was taking place between the ancient patron deities of Babylon and Borsippa (Bel-Marduk and Nanaya), who were returning to their sanctuaries, and a hostile and aggressive god.

The popularity of the boatman's movement, which even the authors of the Astronomical Diaries acknowledge, suggests some kind of crisis consciousness or widespread religious hysteria. In this, the original Seleucid province, recently lost to the dynasty, soon to be reclaimed by Antiochus VII Sidetes, and then to slip away once more and forever, we appear to have an anxiety over cultural and theological traditionalism. While the boatman's teaching does not appear to be eschatological, at least in its depiction in the Astronomical Diaries, this sense of gods doing battle over a religious and political community comes remarkably close to the final apocalyptic visions of Daniel, discussed in Chapter 5.

For a second case study, we travel two hundred kilometers southeast of Babylon and Borsippa to the mud mounds of Girsu, the modern Tell Tello. This Sumerian city, well placed at the Tigris delta between the Iraqi marshes and the plains of Elam, had been the capital of the Second Dynasty of Lagash, a prosperous state in the late third millennium BCE that had regained its independence after the fall of the Agade empire.[67] Its most famous ruler, the ensí ("city governor") Gudea, was an extravagant builder, constructing or repairing numerous temples, dedicating statues of himself to the Lagashite pantheon, and depositing an enormous corpus of inscribed and sculpted artifacts— possibly the largest material legacy of any Mesopotamian monarch. After his death Gudea was honored in the Lagash state as a deity, his name employed as a theophoric element in personal nomenclature, and his statues receiving offerings at the "place of libations" and paraded through the fields on feast days.[68] But not much later, certainly by the early second millennium BCE and perhaps on account of the shifting of the Tigris river-run, Girsu sunk into provincial insignificance.

The city reappeared to history two millennia later, in the mid-second-century BCE chaos that followed the collapse of Seleucid authority in southern Babylonia. Archaeological excavation by Ernest de Sarzec, the French vice-consul at Basra, uncovered on Girsu's highest mound a Hellenistic palatial residence or administrative center (Figure 16).[69] This had been the location of extensive Neo-Sumerian construction, perhaps of the Eninnu temple complex, rebuilt by Gudea for the city's chief deity Ningirsu.[70] The Hellenistic building was a rectangular construction (53×31m), with all interior divisions at right angles. Its thirty-six rooms were arrayed around three courtyards of decreasing size and ease of access.[71] The walls of this palace were built of Gudea's two-thousand-year-old baked mud bricks (each 30×30×7.5cm), both standing elements and reused deposits, interlayered with and added to by the newly manufactured yet almost identical bricks (each 31×31×7.5cm) of the Hellenistic dynast. Certain palace walls seem to have extended or reconstructed those of Gudea; others emulated their recessed exterior decoration. Just as Gudea's bricks had been stamped with the ensí's titulature and building formula in Sumerian, so the bricks of this second-century BCE local ruler were impressed with a bilingual inscription, consisting of two lines of Aramaic followed by two lines of Greek: "hddn|dn'ḥ, Ἀδαδν|αδιναχης" (Figure 17).[72] The stamp used to impress these bricks may have been buried as a foundation deposit deep beneath Gudea's temple ruins.[73] It is from these bricks that we learn the identity of the new ruler of Hellenistic Girsu: Adad-nādin-aḫḫē, a good, traditional Babylonian theophoric name. His stamped bricks, also employed for still unidentified buildings at the nearby site of Medaïn,[74] were a hoary gesture of sovereign rule, but he remains otherwise unattested.[75]

In addition to the small object finds appropriate to a Hellenistic context—coins, ceramics, horseman figurines, and so forth[76]—de Sarzec and his team uncovered around ten diorite statues of the third-millennium BCE ruler Gudea.[77] One of these, a colossal sculpture, was placed in a deep niche beside the main entrance to the palace. All the others were set up in the large public courtyard (Courtyard A, 17×21m), gathered into a seated group and a standing group. According to their original Sumerian inscriptions, Gudea had dedicated these statues to various Lagashite deities at different cultic locales. Thus, it seems likely that Adad-nādin-aḫḫē or his agents had excavated these statues from across the ruined templescape of Girsu, gathered them together at the palace, and reerected them for secondary display in its main courtyard. One of these ancient statues depicted Gudea as an architect, holding on his

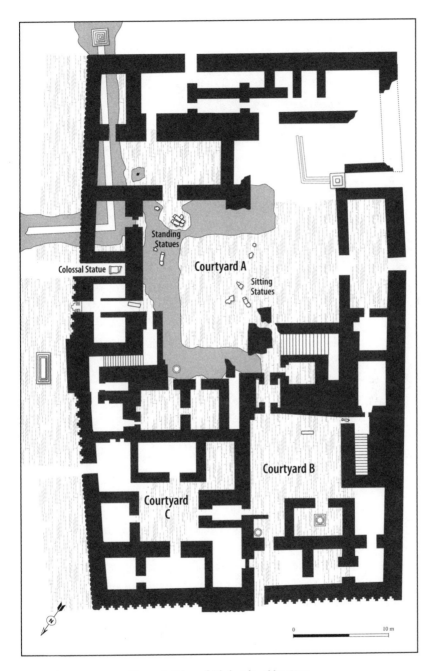

Figure 16. Palace of Adad-nādin-aḫḫē, Girsu

Figure 17. Stamped brick of Adad-nādin-aḫḫē

lap a measuring rule, a stylus, and a writing board, etched with the ground plan of a rectangular buttressed enclosure with six entrances (Figures 18 and 19). Each reader must decide for herself, but the alignment between the architectural sketch on Gudea's lap and the external plan of Adad-nādin-aḫḫē's Hellenistic palace—note, in particular, the relative position of the entrance ways—is almost too close to be accidental. If so, whatever the architectural plan's original referent—Girsu's walls, a city quarter, or the Eninnu sanctuary complex[78]—its depiction may have served as an inspiration for Adad-nādin-aḫḫē's second-century BCE building.[79] At the very least, this remained a possible interpretation of Adad-nādin-aḫḫē's contemporaries.

Figure 18. Statue B of Gudea

The brick stamps and palace suggest the new assertion of a centralized local power. Much as the fall of the Agade empire to the Gutians of the Zagros had made possible a resurgence of nonhegemonic states in Elam and southern Babylonia in the late third millennium BCE, so the Parthian conquest of Seleucia-on-the-Tigris and Babylon opened spaces for the emancipation

Figure 19. Statue B of Gudea, detail with architectural plan

and rivalry of Characene, Elymaïs, and, as we see here, Hellenistic Lagash. While Adad-nādin-aḫḫē's actions—reanimating the past underfoot, staging its union with the present—suggest political metaphors of revival or resurrection, nothing but the building and its objects can speak to his intentions. Unfortunately, de Sarzec excavated these with all the dash and dispatch of a first discoverer, recording and preserving little, more thrilled to the Neo-Sumerians than the post-Seleucids:[80] the Hellenistic palace has been excavated away. We cannot know whether the memory and antiquity of Gudea—almost as distant from Adad-nādin-aḫḫē as Adad-nādin-aḫḫē is from us—had passed unbroken down to Girsu's second-century BCE inhabitants.[81] We cannot know whether his statues' Sumerian inscriptions could be understood in Girsu.[82] Perhaps Adad-nādin-aḫḫē engaged cuneiform scribes from Uruk, only sixty kilometers to the west, restoring (if only briefly) the otherwise sundered yoke-team of scholar and ruler. Yet the surviving data from this Hellenistic palace—its location above and among the remains of the Eninnu temple, the touching and joining of ancient and new materials,[83] the excavation and reverent display of Gudea's statues, and perhaps even the re-realization of the Tell Tello lap map—all suggest some kind of deliberated, conscious evocation of the deep past. It may be that we are dealing only with a museumizing spectacle of local antiquities;[84] but the careful selection and curated presentation of the Gudea statues instead points to something like a cult of a local ancestor[85] or a legitimizing claim of descent.[86]

We would be guilty of an icy positivism, an undue narrowness of interpretation if we did not permit ourselves to wonder how this remarkable old-new palace moved the imagination of its builders, residents, and visitors. I would suggest that the concept of "sublime historical experience," proposed by Frank Ankersmit, may capture some of its antidistantiating force:[87] the complementary movements of the discovery ("loss") and recovery ("love") of a past that had "broken off"; the physical protrusion of past artifacts and, thereby, of the past itself into the present; the cohabitation and sensory encounter, rather than observed separation, of the two; the collapsing of two temporally distant moments into a single metahistorical experience; and so the blasting open of the directional continuum of history.

West Iran

Persis, the heartland of the Achaemenids, was integrated into the Seleucid empire shortly after Seleucus Nicator's return to Babylon in 311 BCE. Colonies were established on its borders, at Antioch-in-Persis, at or near Bushir, at Laodicea-in-Persis, still unlocated, and perhaps also at a site named Stasis.[88] A Seleucid garrison occupied the Tall-e Takht fortress at Pasargadae,[89] and Thracians were settled somewhere in a military colony.[90] No textual source suggests a third-century BCE loss or reconquest of Persis by the Seleucids, and we know of a Seleucid governor in place from the very start of Antiochus III's reign.[91] But starting in the early second century BCE, signs of instability begin to accumulate. Whereas Persians had fought in the Seleucid army against Ptolemy IV at Raphia in 217 BCE, less than three decades later, at the battle of Magnesia, none are found taking the field for their king against Rome and the Attalids.[92] Pliny reports an amphibious victory of a certain Numenius, a general of Antiochus IV, *contra Persas* at the Straits of Hormuz in the Persian Gulf.[93] A passage in 2 Maccabees recounts king Antiochus IV's battle or skirmish with the inhabitants of Persepolis and his panicked flight from the city.[94] Polyaenus, the second-century CE tactical writer, supplies a couplet of stratagems that suggest the decaying into suspicion, hostility, and mass murder of a precarious cooperation between Seleucid military settlers and the local Persian population. In the first episode, set during the reign of an unspecified Seleucus, three thousand "rebellious Persians (Περσῶν νεωτερίζοντας)" are annihilated by Thracian and Macedonian soldiers under the command of a Seiles or Cheiles;[95] in the second, a Persian named Oborzos, learning that the military settlers in Persis ("τῶν ἐν τῇ Περσίδι κατοίκων") were plotting against him, got them drunk, had them killed, and did away with the bodies.[96] Neither episode can be securely dated.

Since our written sources on Hellenistic Persis are few, equivocal, and Mediterranean, historians have turned instead to the region's own numismatic evidence. A rare silver coinage attests a sequence of Persian rulers, known, according to the coins' inscribed legends, by their title *frataraka* ("commander" or "governor"; plural, *fratarakā*). While there is something approaching consensus as to the relative chronology of these *fratarakā* dynasts,[97] scholars are fiercely divided about their absolute dating: either the *fratarakā* first emerged briefly in the third century BCE, whether in the chaos that followed the assassination of Seleucus I or after the Third Syrian War, and then returned to power in the early second century BCE; or they appeared in the late third or early second century BCE, during the twilight reign of Antiochus III or the rule of his successor Seleucus IV, and continued as independent dynasts until the Parthian conquest.[98] The numismatic evidence permits both scenarios, but the historical data from Persis and the wider region better fits the late dating, which I follow. In either case, however, we can be certain that in Persis, at various moments in the first half of the second century BCE, and possibly also before, the Seleucids were militarily challenged by local Persian dynasts.

The coins of these *fratarakā* appealed to the forms, sites, and memories of the defunct Achaemenid empire. The identification can be illustrated by the numismatic types of the first three *frataraka* rulers, Ardaxšīr I, Vahbarz, and Baydād (Figure 20a, c, d). Each of these dynasts appears on the coins' obverse sides in profile portrait, facing right, moustachioed or bearded, sometimes earringed, and wearing the so-called satrapal *kyrbasia* headdress, over which a Hellenistic diadem has been knotted.[99] Three reverse types are known. The earliest and most enduring (Figure 20a) depicts, on the left, a standing figure, resembling the *fratarakā*, with hand raised in a cultic adoration pose derived directly from the figure of the Persian king on the sculpted Achaemenid tombs

Figure 20. Coins of (a) Ardaxšīr I, (b) Vadfradad I, (c) Vahbarz, (d) Baydād, and (e) Seleucus I

Figure 21. Tomb of Darius I, Naqš-i Rustam

at Naqš-i Rustam (Figure 21); on the coins of Vadfradad I this *frataraka* holds a double bow in his left hand (Figure 20b).[100] A building, with large double doors, a dentilated cornice, and three double-horned parapets on the roof stands in the center of the visual plane. Typically, this has been identified as the Achaemenid-period stone tower at either Naqš-i Rustam or Pasargadae, now known as the Kaba-i Zardusht ("the Cube of Zoroaster"), and the Zandan-i Suleiman ("the Prison of Solomon"), respectively. The purpose of these buildings is not established and may have been unknown to the *fratarakā* themselves; suggestions include a fire temple, fire altar, burial tower, and coronation site.[101] Alternatively, the coin type's building may represent the Persepolis palace terrace, the southwestern edge of which was lined with precisely the kind of horned merlons that run along the coins' parapet (see below). For our purposes, the crucial point is that the *frataraka* dynast stands in ritual reverence before an Achaemenid-associated structure. From the reign of Vadfradad I on, the winged figure of Ahuramazda (sometimes identified as the personified royal glory), also familiar from Achaemenid imperial art, hovers above the tower (Figure 20b). On the right-hand side of this

206

building stands a tasseled standard, recalling the Achaemenid royal or religious banner depicted on the Persepolis reliefs.[102]

The second coin type (Figure 20c), of which only two examples are so far known, depicts on the obverse the portrait head of the *frataraka* Vahbarz and on the reverse a Greek or Macedonian soldier, in standard hoplite armor, sprawled on the ground, without his helmet, his left knee bent, his right leg stretched out; a figure in the costume of an Achaemenid monarch or high official looms over this soldier, seizes his hair, leans his full weight on him, and thrusts forward a dagger—the characteristically Achaemenid *akinakēs*—to do him in. The visual formula precisely recalls Achaemenid military iconography, from engraved gemstones and painted sarcophagi.[103] The *frataraka* Vahbarz, whose coin issue this is, can plausibly be identified with the Persian Oborzos who, according to the Polyaenus stratagem discussed above, slaughtered three thousand Seleucid military settlers on a single night. If the coin commemorates this massacre,[104] then it reframes the episode in the anachronistic register of the Achaemenid empire's anti-Greek victory art.

The third reverse type (Figure 20d), associated with the *frataraka* Baydād, shows the dynast seated on a throne, before the tasseled standard, holding a scepter in his right hand and a flower in his left. This closely emulates the throne and pose of the Achaemenid Great Kings from the Persepolitan audience bas-reliefs (Figure 22).[105]

These close visual links to Achaemenid monarchic traditions are reinforced by the coins' legends. Their script and language—a cursive Aramaic, likely rendering an early Middle Persian—inscribe themselves in the Iranian imperial tradition. Note, by contrast, that the great majority of coins circulating within the wider region—not only Seleucid, but also Graeco-Bactrian, Characenian, Elymean, and most Parthian—employed only Greek. Similarly, the dynasts' names—Ardaxšīr (Greek, Artaxerxes), Vahbarz (Greek, Oborzos), Baydād (Greek, Bagadates), and Vadfradad (Greek, Autophradates)—associated them with the Achaemenid empire and Mazdean religious tradition; none took or displayed a Greek or Seleucid name. The whole series is devoid of any kind of numerical dating.[106] Instead, the dynasts are identified as *"prtrk' zy 'lhy'"*—a title that normalizes to the Aramaic *frataraka zī elāhayyā* and Persian *fratarak ī bagān*, meaning "the commander of the gods" or "the prefect of the gods." In Achaemenid-period papyri from southern Egypt, the *frataraka* appears as a subsatrapal, nonreligious imperial commander,[107] though we cannot be certain that its meaning and status remained stable following the

Figure 22. Throne bas-relief, Persepolis

disintegration of the Persian imperial bureaucracy. The compound title *"frataraka* of the gods" is altogether new; it suggests that these dynasts claimed not kingship or royal descent as such,[108] but an authorization or delegation by the pantheon of Persis,[109] which accords with the adoration iconography of the first reverse type. The coins of Ardaxšīr I, Vahbarz, and perhaps also Vadfradad I carry the additional title *"br prs"* ("son of a Persian" or "son of Persis"),[110] an explicit assertion of an epichoric, rooted, territorial origin and ethnic affiliation and perhaps also an echo of Darius I's self-fashioning as "a Persian, son of a Persian" (DNa 13–14: *"Pārsa Pārsahạyā puça"*). Finally, Vahbarz, on his rare hoplite-slaying coin type, takes the title *"krny,"* identified as the Achaemenid military office of *karanos*; as before, we cannot know the force of this title in the post-Achaemenid world. The same title was adopted in the mid-third century BCE by Arsaces I, founder of the Parthian kingdom.[111]

Such Achaemenid revivalism can also be detected in the built environment. In January 330 BCE the army of Alexander the Great had reached Persepolis, plundered the city, massacred its inhabitants, and sojourned in their homes for four months.[112] Then, at the start of spring, the bitterness of the Macedonians burst over the Achaemenid capital:[113] the palaces and treasuries of the royal terrace were picked clean of valuables, fires were ignited simultaneously within several of its hypostyle halls, and the cedared edifices of Persia's empire were felled by the conflagration.[114] When Alexander struck north on the hunt for Darius III, he left behind a smoldering devastation where the world's ceremonial center had gloried for almost two hundred years.[115]

At some point in the later third or (more likely) early second century BCE, a small part of the southwestern corner of the Persepolis terrace was cleaned of destruction debris: the toppled columns of a building begun by Xerxes and completed by Artaxerxes I were removed; smashed cornices, column fragments, staircase parapets, relief slabs, and smaller stones (including some from elsewhere on the terrace) were consolidated into a high rubble platform. A double-winged, sculpted staircase facade, dismantled from the palace of Artaxerxes III, was reemployed as a monumental facing (Figure 23). Another flight of steps was reconstructed from two different earlier staircases. Above, on top of this mound, old column bases were arranged into a 6 × 2 columnar podium entrance that led into a 4 × 4 columnar hypostyle hall, with more modest and restricted rooms beyond (Figure 24).[116] This palace, likely residence of the *frataraka* dynasts, was built up of not much more than reused Achaemenid masonry. Substantial elements from the edifices of Darius I, Xerxes, Artaxerxes I, and Artaxerxes III were gathered from across the terrace, united, and reerected in a precise emulation of Persepolis's distinctive imperial architecture. The building was raised at the most visible corner of the Persepolis platform, where the parapet was edged by the same or similar horned wall-crowns that are depicted on the *frataraka* coinage.[117] To its sides and back nothing remained besides the ruinscape of colossal wreckage, the material witness to Macedonian savagery.

Every visitor to this palace was obliged to climb the terrace's monumental staircase and pass among fallen mud bricks, shattered stones, ash heaps, weeds, rodents, and whatever odors the rubble still exhaled. Needless to say, it was open to the *fratarakā* to build anew and live elsewhere (at nearby

Figure 23. Palace H, Persepolis, with reused staircase facade of Artaxerxes III

Istakhr, say),[118] sidestepping this daily shaming by history. They chose, instead, to raise up what had been thrown down, albeit symbolically and on a small scale. If Alexander's destruction of Persepolis had been intended as a spectacle of closure for Persian rule, a mode of periodizing by ruination employed also by the Seleucids (see Chapter 5), then the physical reerection of at least part of the site implied the revival or reopening of an indigenous political history.[119] This was, inescapably, an architecture of resistance, prompting, perhaps, outrage or defiance or nostalgic rumination.

The disputed nature of our sources must caution against any hard political or chronological conclusions: we, perhaps like the inhabitants of Hellenistic Persis, will never be certain about the *fratarakā*'s degree of autonomy or subordination at any particular moment, and the striking of coinage was too infrequent, self-promoting, and politically idealizing a mode to capture the actual nature of interaction with the Seleucid empire. Certainly, these dynasts, modest by necessity, do not appear to have challenged the Seleucids for Asia or even to have claimed an authority beyond Persis. Yet the *fratarakā*'s aggressive cultural politics are brought into relief when their coin issues are compared with the

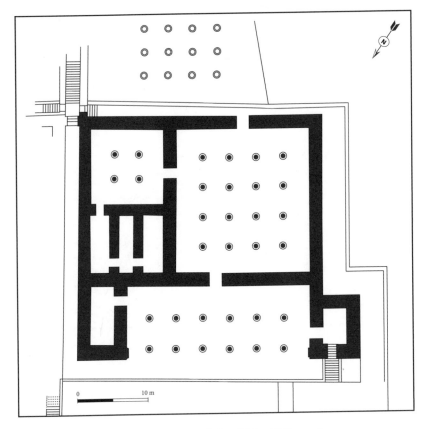

Figure 24. Reconstructed plan of Palace H, Persepolis

Seleucid numismatic types they overstruck (Figure 20e): Greek alphabetic script, names, and titles, Macedonian clean-shaven royal portraits, and Graeco-Macedonian deities and iconography were hammered away by the *fratarakā*'s Aramaic script, ancestral nomenclature, Achaemenid titles, Persian bodies, Iranian clothing, and local cult.[120] Together with their old-new palace, this amounted to an "anti-language" of rule,[121] a reframing of their contemporary Hellenistic environment according to the terms of the lost empire.

Armenia

Greater Armenia (Map 5) has not yet appeared in our account and would rise to major international significance only as the tear-line of the Roman and

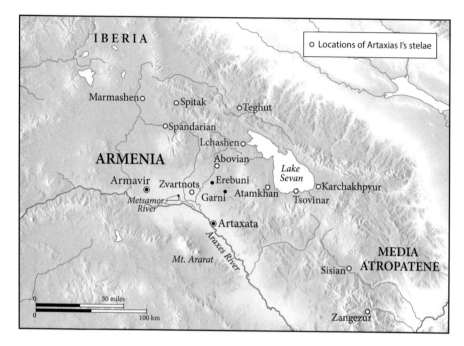

Map 5. Greater Armenia

Parthian partition of the Seleucid empire. Yet already in the first half of the second century BCE, its new Artaxiad monarchy presented itself as a local and rooted revival. The outlines of the reign of Artaxias I, the dynasty's founder, are known from two kinds of sources: first, classical writers, particularly Strabo, who made use of the historical account of Theophanes of Mytilene, companion and panegyrist of Pompey's eastern wars;[122] and second, the late antique Armenian history of Moses of Khoren (Movses Khorenatc'i), who cited for his Artaxias narrative the writings of Eusebius, a temple history of Ani, translated by the great Bardaisan of Edessa, and an oral epic of Artaxias's rise and conquests, the "Songs of Artašēs and His Sons."[123]

In the early first millennium BCE the upland plateau surrounding mount Ararat and the Araxes basin had been united into the kingdom of Biainili, better known by its Neo-Assyrian name of Urartu. This centralized state, along with its network of commanding ashlar-stone hill forts, suddenly and violently collapsed in the late seventh century BCE, perhaps falling to the Cimmerian invaders of Anatolia.[124] The region was absorbed into the successive Median and Persian empires.[125] Its Achaemenid hereditary governors, the line of Orontid or

Yervandid satraps, weathered the Macedonian invasion, apparently acknowl-
edging Alexander[126] and permitting his exploitation of the Hyspiritis gold
mines,[127] to regain a high degree of regional autonomy during and after the
Wars of Succession.[128] In all likelihood, for most of the early Hellenistic period,
Armenia hovered on or just beyond the northern horizon of practicable Seleucid control. For what it is worth, the medieval Georgian royal history, the
Kartlis Tskhovreba, reports that the client rulers of Armenia were dispatched by
a "king Antiokoz of Asurastan" to support the first kings of Iberia (Kartli), Parnavaz and his son Saurmag, both of whom appear as Seleucid vassals.[129]

At the end of the third century, the last of these Orontid dynasts was de-
posed by or with the support of Antiochus III,[130] who substituted Zariadres and
Artaxias as the *stratēgoi* (imperial governors) of Sophene and Armenia, respectively.[131] Artaxias served the Seleucid crown until Antiochus III's defeat at the
battle of Magnesia in 189 BCE. Breaking for Rome at this point, he was rewarded
by the Senate with recognition as an autonomous king and extended his dominion toward the Caucasus, the Caspian Sea, and Media Atropatene, earning a
reputation for aggressive rapacity.[132] In 165 BCE, Artaxias was reincorporated
into the Seleucid imperial structure during the eastward *anabasis* of Antiochus
IV,[133] but is found only half a decade later in alliance with Timarchus, the rebellious satrap of Media, against the Seleucid monarch Demetrius I.[134] Shortly
thereafter he passed on to his son Artavasdes a fully independent kingdom. In
other words, Greater Armenia in the first half of the second century BCE oscil-
lated between a precarious, grasping autonomy and violent resubordination, a
political dynamic that gave an anti-Seleucid charge to Artaxias's public acts of
monarchic self-fashioning. Three actions in particular—the foundation of an
eponymous capital, the marking of territorial divisions, and the striking of
coins—made potent appeals to a local, pre-Seleucid past.

First: Artaxias formalized his break from the Seleucid empire by establishing a new eponymous capital, Artaxata (or Artašat in Armenian, meaning
"the Joy of Artaxias").[135] The city was built along a chain of hills at the junc-
tion of the Araxes and Metsamor rivers, on the otherwise flat plain beneath
the peaks of mount Ararat and not far from the later monastery of Khor Vitap,
where Gregory the Illuminator, converter of Armenia, would be imprisoned.[136]
Archaeological excavation since the 1970s has revealed that Artaxata spanned
out from a high, central hill ("Hill 2" or the "Royal Hill") as a set of interlocking
curtain walls, fortified corridors, and a vast city wall (Figure 25). On the summit
of this radial rise and visible throughout the city and its hinterland were the

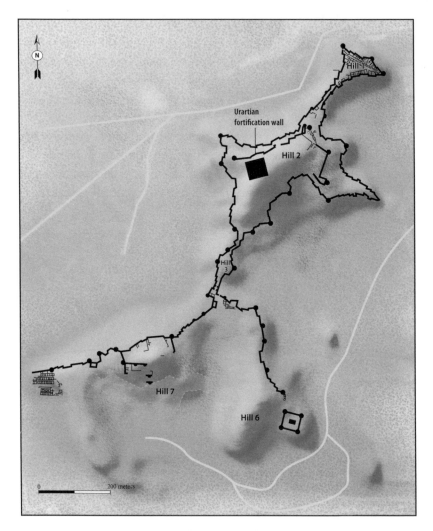

Figure 25. Plan of Artaxata citadel area, Armenia

imposing remains of an Urartian fortress, perhaps constructed by Argishti I in the eighth century BCE and likely one of the largest Urartian sites in the Araxes valley (Figure 26). Artaxias retained this ruin, already half a millennium old, restored its mud-brick walls on the Urartian foundations and according to Urartian methods of fortification, partly updated it with the addition of round towers, and reoccupied its interior as his own royal residence and administrative center.[137] Moses of Khoren reports that Artaxias trans-

214

Figure 26. Rebuilt Urartian wall, Artaxata

ferred to this new capital the cult of royal ancestors,[138] although it has also been proposed that this was in fact Artaxias's own creation.[139]

It is difficult to elicit intentions from mute walls, yet the selection of this site and the anachronistic mode of construction suggest that the relationship of Artaxias's new capital to the old Urartian fort was more than topographic.[140] However these ruins were comprehended, whatever tales had attached to their broken buttresses, Artaxias had chosen not to disguise or unmake them down to bedrock—as, in fact, the Urartians had done to the ruins of their own predecessors[141]—but to valorize their renewal. All is inference, of course, but Artaxata's seems a more ancient and unfinished time than that projected by the new cities of Seleucid imperial urbanism—Epiphania-on-the-Tigris (Amida, modern Diyarbakır) was on the Armenian marches[142]—as if the history of the land were a rampart for this new and threatened king to retreat behind. A couple of ancient sources allow us to flesh out, if only slightly, an engaged reverence for those ancient traces that could be found on Armenian hilltops: perhaps referring to Urartian ruins, Strabo speaks of *Iasonia* (monuments of the traveling hero Jason), some of which had been constructed by the rulers of the land;[143] and the *Babyloniaca* of Berossus, discussed in Chapter 4, adduces as testimony to the great flood and the landing of the ark of Xisouthros on Ararat that even in his day "a part of this ship still remains in the mountains of the Gordyeans of Armenia; some people scrape off and carry away the ship's asphalt to use as talismans"[144]—a rare glimpse of a vernacular and enchanted archaeology.

Second: about fifteen inscribed boundary markers of Artaxias I have been found across Greater Armenia, at Spitak in the north, Zangezur in the south, and around Lake Sevan near Artaxata. With various degrees of elaboration and slight lexical differences these inscriptions give the name, title, and occasionally regnal year of Artaxias, followed by a declaration of territorial demarcation. For instance, one of the longer stelae reads, "In the 10th year (*bšnt . . . 10*) of Artaxerxes, king (*mlk*), Orontid (*'rwndkn*), son of Zariadres, the good (*ṭby*), the bearer of the crown, united with *xšaθra* (*ḥšhrsrṭ*), vanquisher of all who encourage evil (*wnqpr*); Artaxerxes, Orontid, son of Zariadres, divided the land between the villages (*ḥlq 'rq' bny qry'*)."[145]

These stones made an argument, in language and in time. All of Artaxias's boundary stones were inscribed in Aramaic (see Figure 27). Lori Khatchadourian, whose lead I follow here, has emphasized that this represents a deliberate and public choice as to the new kingdom's linguistic alignment. The Aramaic is not quite correct—the determinative is missing, the form of the patronymic is improper, and there are hints of Middle Iranian pronunciation[146]—but it is resonant of the old, pre-Hellenistic, Achaemenid order.[147] Above all, these prominent markers were not in Greek, the administrative and prestige language of Artaxias's erstwhile and returning Seleucid masters. Certainly, Greek was known in the region and had been employed by the last Orontids: at Armavir, the earlier Orontid dynastic capital in the Araxes basin, seven Greek inscriptions were neatly carved onto rock face, one even listing out the months of the Seleucid Macedonian calendar.[148] Moreover, Artaxias's name and titles appealed to an Iranian religious and ethical tradition, not a Seleucid or Graeco-Macedonian one. His royal titles committed to the Zoroastrian or Mazdean dualism of good and evil; the *xšaθra*, to which he is allied, represents either abstract rightful power or its divine hypostasis, one of the six *Amaša Spantas*.[149] The king did not take or did not employ a Greek name. There may be some suggestion of a conscious linguistic policy in Strabo's statement that, during the reign of Artaxias, all the inhabitants of Greater Armenia came to speak the same language ("ὥστε πάντας ὁμογλώττους εἶναι").[150]

Figure 27. Boundary stone of Artaxias I

A couple of these boundary inscriptions are dated to Artaxias's tenth regnal year, using the Aramaic opening formula "in the year (bšnt)." In all likelihood, Artaxias had inaugurated this throne-year count at his elevation to autonomous monarchy in 189 BCE, after the battle of Magnesia. As we have seen in Chapter 3, casting off the Seleucid Era, with its horizon of memory and dynastic worldview, marked a public and chronographically formalized break from empire. It may be significant that Moses of Khoren immediately follows his very accurate report of these boundary stelae[151] with an account of Artaxias's reform of the calendar, "the cycles of weeks, months, and years,"[152] perhaps indicating a replacement of the Seleucid Macedonian time with a monarchic Armenian one. Artaxias's claim to Orontid ancestry, whether supposititious or not, asserted a link both to the satrapal dynasty he had replaced and, by the Orontid marriage to a daughter of Artaxerxes II, to the Achaemenid line.[153] Furthermore, Armenia was not a landscape scattered with inscriptions: other than the Greek texts at Orontid Armavir, perhaps a restricted sanctuary site, publicly displayed writing seems to have entirely disappeared after the fall of Urartu in the Iron Age. In such a context it is significant, as Khatchadourian has suggested, that Artaxias's boundary markers drew their shape—worked rectangular faces, rounded crowns, and base lugs—from freestanding Urartian cuneiform inscriptions that were still visible across the territories of Greater Armenia. While in no way textually comprehensible in the Hellenistic period, these monuments of the early first millennium may nonetheless have generated for Artaxias's new boundary stones the sense of ancient precedent and formal prototype. Put another way, while Artaxias's apportioning of territory was an assertion of his new sovereign prerogative, related to his sponsorship of settlers or judging of boundary disputes, the content and form of his inscriptions synthesized pre-Seleucid ways, names, and shapes, Orontid, Achaemenid, and Urartian. It is as if they are copies without an original, a kind of déjà vu in stone.[154]

Finally, briefly, crudely manufactured copper coins, apparently discovered at Artaxata, appear to be the first mint issues of Artaxias I. On the obverse, a bearded and diademed Artaxias wears a *kyrbasia*-type headdress, tied up at the top, with the Aramaic legend "'rthšsy mlk'" ("king Artašes"). The reverses vary, depicting the bare head of a male divinity, a bee, an eagle, and a dog.[155]

It seems probable that Artaxias I, like Adad-nādin-aḫḫē and the *fratarakā* of Persis, framed his breakaway monarchy in non- and pre-Seleucid ways: Aramaic script, appeal to Orontid and Achaemenid ancestry, archaizing forms, and ancient places. Who could have imagined that a century later his descendant, Tigranes the Great, would ascend the throne of Antioch? If only the "Songs of Artašēs and His Sons" had survived!

Judea

The Maccabean revolt in second-century BCE Judea is the most famous of our stories: the winter festival of Ḥanukkah (candles, dreidels, and donuts) is the sole contemporary commemoration of Hellenistic politics, and the Maccabees' casting off of the Seleucid yoke remains a potent paradigm for Zionist thinking.[156] Yet it is also the most complex and disputed of the Seleucid regional histories.

The details of the revolt are mainly known from two self-standing works of Hellenistic historiography. 2 Maccabees is an epitome of Jason of Cyrene's five-book history in Greek of events in and around Jerusalem from the reign of Seleucus IV down to 161 BCE, most likely composed shortly after the events that it recounts and somewhere in the eastern Mediterranean Jewish diaspora.[157] 1 Maccabees opens with the conquests and death of Alexander the Great and then narrates the Maccabean victories over the Seleucid empire and its local allies from the late 170s BCE down to the accession of Simon's son, John Hyrcanus, in 135 BCE. It was originally written in Hebrew in Judea toward the end of the second century BCE, but it survives only in a Greek translation.[158] These accounts are buttressed or undermined by passages in other historiographical works (Josephus, Diodorus Siculus, Tacitus, and Porphyry), the biblical book of Daniel, the Apocalypse of Animals from 1 Enoch, and a couple of Greek inscriptions. The incomparable thickness of this written evidence and the rapidly expanding archaeological data permit access, as nowhere else, to multiple perspectives, contradictory dynamics, and fine-grained distinctions. Scholarship is uncovering an entangled, often dispiriting history of imperial abuse and persecution, sacred war and sanctified massacre, elite factionalism and sectarianism, and oscillations of bitter struggle or grudging alliance between the Seleucids and the Maccabees, between the Maccabees and other Judeans, between the Maccabees and neighboring subject populations, between those other Levantine populations, between different lines of the Seleucid house, and between those Seleucid branches, the Ptolemies, the Parthians, and Rome.[159] No doubt, were we to have the evidence, things would be as complicated and confusing in the rest of the empire—and this must be remembered for the preceding regions and chapters.

A sentimental hostility to the Seleucids is already evident from shortly after Antiochus III's capture and incorporation of the southern Levant in the Fifth Syrian War at the opening of the second century BCE. Our first literary work from Seleucid Judea, the Hebrew wisdom book of Ben Sira, discussed in Chapter 4, already incorporates the author's heartfelt prayer for the vindication of Israel, asking God to deliver His people from the imperial overlords—"Come to our aid, God of the universe, and put all the nations in dread of You! Raise Your hand against the foreign folk, that they may see Your mighty deeds.... Smash the heads of the hostile rulers, who say, 'There is no one besides me!'"[160] Following Antiochus III's defeat by Rome, during the reigns of his sons Seleucus IV (r. 187–175 BCE) and Antiochus IV (r. 175–164 BCE), administrative reforms sought to normalize the former Ptolemaic province, better control its trade and taxes, and more directly exploit the resources of its sanctuaries.[161] It is fairly clear that Antiochus IV's deposing of the Zadokite high priest Onias III and his selling the high priesthood to Onias's brother Jason and then to the unrelated Menelaus,[162] his support for Jason's founding of a gymnasium and the conversion of Jerusalem (or a part of it) into a *polis* named Antioch,[163] and the consequent marginalization of certain established elites together triggered an anti-Seleucid uprising at the time of Antiochus IV Epiphanes's engagement in Egypt during the Sixth Syrian War (170–168 BCE).[164] In response, Antiochus IV launched a brutal campaign of punishment, plunder, colonization, and ultimately, and unprecedentedly, religious persecution. Some Jews acquiesced, others retreated in prayer to the wilderness, and still more, under the leadership of Mattathias of Modeïn and his sons Judas, Jonathan, and Simon, fought back with a holy zeal. Even after Antiochus V (r. 164–162 BCE) revoked his father's prohibition of Jewish cult and worship, the brothers Maccabee continued to consolidate their military dominance locally and regionally. The obvious weakness and divisions of the Seleucid house, the strength and apparent unity of the Maccabean culminated in king Alexander I Balas's authorizing Jonathan as high priest in 152 BCE, Simon's expelling the Seleucid garrison from the Jerusalem Acra, "lifting the yoke of the gentiles," and inaugurating his Year 1 in 142 BCE, and John Hyrcanus's (Simon's son) establishing a fully independent rule in Judea by 128 BCE for the first time since the exile to Babylon.

Given this history's complications and ramifications (which constitute an entire disciplinary subfield), here I will attempt only to elucidate certain temporal tendencies at play. For the markedly Hellenizing reforms of Jason, Menelaus, and their associates (founding a gymnasium and *ephebeia*, participating in the athletic games of Tyre, and self-politicizing as Antiochians) followed by Antiochus IV's prohibition of markedly Jewish practices (circumcision, proper diet, Temple cult and exclusivity) appear to have given a temporal charge to the Judean cultural landscape. By the logic of the Hebrew Bible, the Jewish way of life was legitimate and obligatory because it was ordained by God, not because it was old. Indeed, it depicted itself as a series of fairly recent interventions within history, a set of instituted changes quite unlike the organic, mythic origin, in *illo tempore*, of the polytheisms round about. Furthermore, with scriptural narratives of recurrent Israelite backsliding and apostasy, piety was not to be inherited from the fathers. Yet in Seleucid Judea, the cultural identifiers that had once been, simply, the way things were done ("their own way of life (τὴν ἑαυτῶν ἀγωγήν)," as king Antiochus V puts it)[165] became "ancestralized," that is, redescribed as traditional, ancient, and inherited. The Greek words for this deep, historical rootedness (πάτριος, πατρῷος, and προγονικός) hardly appear in this sense across all the books of the Septuagint;[166] but in 2 Maccabees alone, the short précis of Jason of Cyrene's history of the persecution and Judas's wars, there are fifteen occurrences.[167] The work lines up, against the voluntary or compelled cultural changes, "ancestral laws," "ancestral honors," "ancestral festivals," "ancestral language," "ancestral ways," "ancestral food," and "ancestral customs." By contrast, the high priest Jason "innovated (ἐκαίνιζεν)" customs contrary to the laws; Antiochus IV is characterized as an "inventor (εὑρετής)" of new evils, perhaps toying with contemporary theories of civilizing kingship.[168]

1 Maccabees dramatizes this temporal polarization. According to its narrative, Antiochus IV targeted by decree and violent erasure the "religion of the fathers (λατρείας πατέρων)" and "the covenant of our fathers (διαθήκη πατέρων)."[169] As discussed in Chapter 4, the king's officers, recognizing that Jewish practices derived from a textualized total history, tore up and burnt the scrolls of the Law ("τὰ βιβλία τοῦ νόμου") wherever they found them.[170] Antiochus IV established in their place a new set of laws and regulations: "the king sent scrolls by the hand of messengers (ἀπέστειλεν ὁ βασιλεὺς βιβλία ἐν χειρὶ ἀγγέλων) to Jerusalem and the cities of Judea," with instructions to stop Temple offerings, violate the Sabbath and festivals, defile the Temple,

build illicit altars and sanctuaries, sacrifice swine, and leave male babies uncircumcised. "Whoever disobeyed the king's word (τὸ ῥῆμα τοῦ βασιλέως) was to be put to death."[171] The royal orders, as depicted in this Greek translation of the lost Hebrew original, are a perversion of scripture. Antiochus delivers to Jerusalem not "letters" or "instructions" (ἐπιστολαί, λόγοι), the terms used for all other royal communications in 1 Maccabees,[172] but "scrolls" (βιβλία), a word employed only for biblical or historiographic texts;[173] "by the hand of his messengers (ἐν χειρὶ ἀγγέλων)" recalls prophetic delegation;[174] the "word" (ῥῆμα) of the king is the standard Septuagint term for God's own utterance.[175] It is as if Antiochus IV dispatches, through his ministering messengers, a counter-Torah, a textualized inversion of the key biblical commandments, sealed with the threat of punishment for disobedience, that abrogates a testament it makes old for a new and terrifying dispensation.[176]

The response of the Judean rebels, as framed by 1 Maccabees, further temporalizes the conflict. Narrative, diction, and historical allusion construct an image of Judas and his brothers not only fighting for the old ways, but even more, fighting in the old ways. That is, the Maccabean rebels appear as latterday Joshuas, judges, and Davids, reenacting modes of religious violence from the first conquest of the land. 1 Maccabees 5, for instance, pulls together a series of Maccabean campaigns across the Jordan river, into the Galilee, toward the south, and throughout the coastal plain.[177] The rival inhabitants of the southern Levant are anachronistically labeled "the nations round about (τὰ ἔθνη κυκλόθεν)," "the sons of Esau," and "the Philistines."[178] We are told that Judas, confronted by the enmity and wickedness of the sons of Baean, near neighbors to the south, drove them into their defensive forts, "'banned' them (ἀνεθεμάτισεν αὐτούς), and burned their towers along with everyone inside them."[179] As reported, this Idumean clan was "anathematized" by Judas's army—that is, put under the "ban" or ḥerem, the biblical consecration by total destruction with which Joshua wiped away Jericho and Saul most of the Amalekites.[180] Comparable violence was launched against the cities of Bozrah, Alema, Ephron, and Carnaïn (where the gentiles, seeking refuge in the temple of Atargatis, were burned to death within their pagan sanctuary).[181] At Azotus (Ashdod, near Gaza), Judas "destroyed their altars and burned the graven images of their gods and plundered the towns."[182] Jonathan, his brother and successor, would return to Azotus to burn its inhabitants, who had taken refuge in the temple of Dagon.[183] These temple sackings or destructions fit the normative instructions of Exodus and Deuteronomy and recall the historical accounts of Joshua, Saul, Hezekiah, and Josiah.[184] There is an emotional or moral revival, too: in one episode, following his massacre of the men of Ephron and the plundering and razing of their city, Judas marches his army and local Jewish men, women, and children back to Jerusalem over the bodies of their slain enemies.[185] This is a delight in destruction that recalls the book of Judges. Indeed, the victorious rule of Jonathan is framed explicitly in these archaizing terms: "Jonathan dwelled in Michmash. And Jonathan began to judge the people."[186]

Judeans had not launched such extremes of sanctified violence against "the peoples round about" for centuries, since at least the fall of the house of David and probably long before, if ever. If these Maccabean campaigns actually happened, as I think likely at least in outline, they renewed forms of warfare understood to go back to the origins of the people in the land.[187] Military actions were molded to a biblical script, turning participants into imitators and reenactors of the first conquest. Note that 1 Enoch's Apocalypse of Animals, contemporary with the early Maccabean wars (for date and details, see Chapter 5), figures the militant Judas as a butting ram;[188] the only other historical actors allegorized by this animal are Jacob, Saul, David, and (briefly) Solomon,[189] which constitutes for Judas the same kind of "first conquest" typology. Even if unhistorical or exaggerated, the accounts in 1 Maccabees function as image makers and memory shapers, framing the Hasmonean emergence as a kind of martial Israelite revival. Group violence, like other ritual,[190] seems to lend itself to this "time-machine effect," that dizzying feeling of being whisked back, momentarily conjuring into sensory presence an archetypal conquest and reexperiencing its primal intimacy. Similarly, modern jihadists emulate the idol smashing of early Islam, which in turn reached back to foundational tales of Ibrahim;[191] and we have seen, above, that the *frataraka* Vahbarz likely framed his defeat of Seleucid settlers in the mode of Achaemenid anti-Greek victory imagery.

Two spectacles from the opening years of the Maccabean revolt—the gathering for war at Mizpah (modern Tell en-Nasbeh, about twelve kilometers north of Jerusalem) and Judas's rededication of the Temple—are especially rich in this reiterative temporality, concatenating antecedents and echoing, rhyming, and even correcting old history. With a vast Seleucid army bearing down on Judea, under instructions to colonize and distribute the land and "to banish even the memory of [the Jews] from the place (ἆραι τὸ μνημόσυνον αὐτῶν ἀπὸ τοῦ τόπου),"[192] Judas assembled his followers at Mizpah. The reason, according to 1 Maccabees, was that "formerly (τὸ πρότερον) Israel had a place of prayer in Mizpah":[193] when faced with the militarization of the Seleucid horizon, the rebels took their stand in the Judean past. Mizpah had come to prominence toward the very beginning and the very end of the Israelite scriptural-national history. According to the book of Judges, Mizpah had been a mustering point and site of prebattle worship for the Israelite tribes in their wars against the Ammonites[194] and the Benjaminites.[195] Later, the judge and prophet Samuel twice convoked the Israelites at Mizpah, the first time to repent and fast for Israelite apostasy and thereby secure a battlefield victory over the Philistines,[196] the second to declare Saul Israel's first king.[197] The Enochic Apocalypse of Animals, composed around the time of Judas's assembly at Mizpah, has high regard for Samuel's actions at the site.[198] At the other end of the monarchic history, when Nebuchadnezzar of Babylon had conquered and razed Jerusalem, Mizpah was turned into the land's replacement administrative hub and the residence of Gedaliah, governor of the Judean remnant. In a culminating horror, perhaps four years later, this Gedaliah was murdered at Mizpah, along with the Judeans and Babylonians stationed with him, an assassination still mourned by Jews with an annual fast.[199]

Thus, it was at this place of ancient prayer—a site resonant of an Israel endangered, the congregation of its warriors, the traumas of Judean factionalism, and the occupation of Jerusalem by others (Jebusites or Babylonians)—that Judas assembled his men. 1 Maccabees reports that the Jewish rebels ritually mourned, consulted the Torah scrolls, brought forward the priestly garments, first fruits, tithes, and Nazirites (self-consecrated Jews); then Judas, observing the rules of war in the book of Deuteronomy, sent home all who were building homes, recently married, planting vineyards, or fearful of battle;[200] finally, in emulation of Moses, he divided the fighters into battalions of thousands, hundreds, fifties, and tens.[201] The accounts of 2 Maccabees and Josephus, though less detailed, concur.[202] The actions at Mizpah, at the very place where the Israelite state had been born and where it perished, are a remarkably concentrated performance of ancestral devotion and religiously licit warfare.

The ceremony worked. The Maccabees defeated the massed Seleucid army at Ammaus and entered Jerusalem to restore the desecrated Temple on 25th Kislev, 164 BCE. Our ancient sources insist on the organizing metaphor of Seleucid ruination and Maccabean reconstruction. Indeed, this amounts to a slogan of the whole rebel movement: "Let us restore the ruination of our people (ἀναστήσωμεν τὴν καθαίρεσιν τοῦ λαοῦ)!"[203] So the Temple, despite its use by the Seleucid colonists and their local allies, is depicted as a wasteland—desolate, burned, overgrown with weeds and bushes, its priestly chambers broken.[204] The Maccabean forces cleared the sanctuary of its Seleucid abominations, purified it, "constructed a new altar like the former one (ᾠκοδόμησαν θυσιαστήριον καινὸν κατὰ τὸ πρότερον),"[205] and outfitted the Temple with new cultic instruments. Similarly, if less significantly, some years later Jonathan would "reconstruct the ruins (ᾠκοδόμησεν τὰ καθῃρημένα)" of Beth-Bassi in the Judean desert, perhaps a post of the old king Uzziah.[206] Furthermore, Judas's resanctification was explicitly framed as a reenactment of the Second Temple's founding—for, if Moses's Tabernacle and Solomon's First Temple had shown how to begin the cult, the returnees from Babylon modeled, more relevantly, how to begin it again. The so-called Second Epistle, a festal letter attached to the beginning of 2 Maccabees, urged the Jews of Egypt to join in celebrating the new festival of Ḥanukkah, "the Days of Tabernacles and the Days of Fire, as when Nehemiah, the builder of the Temple and the altar, brought sacrifices."[207] The letter goes on to depict Judas as a more general imitator of Nehemiah, rebuilding the altar, reigniting the sacred flame, and reconstituting the library that Nehemiah had first gathered.[208] Finally, the new Maccabean festival of dedication was modeled on the festival of Tabernacles (Sukkoth);[209] according to Jewish tradition, this had been the occasion on which Solomon celebrated the dedication of his First Temple[210] and Zerubbabel and Joshua the altar of the Second Temple,[211] Nehemiah completed the wall of Jerusalem,[212] and Ezra and Nehemiah renewed the covenant.[213]

It is enormously significant that, of all possible models and genres, the author of 1 Maccabees chose to emplot these Jewish victories in the language and style of biblical historiography—in particular, of the conquest-of-the-land books of Joshua, Judges, and Samuel.[214] Just as the actions of the Maccabean rebels were reenactments, imitations, and revivals of archaic Israelite models, so, a generation later, they were given appropriate form in the literary classicism of Jerusalem's court historian. The Hebrew original of 1 Maccabees has been lost to us, but it constituted part of a more general revival of this language in Seleucid Judea.

Much as Adad-nādin-aḫḫē in southern Babylonia, the Persian *fratarakā*, and Artaxias I of Armenia advertised their indigenous nomenclature at the moment of self-emancipation from Seleucid rule, so Mattathias and his sons

Judas, Jonathan, and Simon used good Hebrew names.[215] By contrast, Jason, Menelaus, and Alcimus, the Seleucidizing enemies of the Maccabean brothers, preferred Greek names; Josephus reports that the high priest Joshua, brother of Onias III, "changed his own name (αὐτὸν μετωνόμασεν)" to Jason.[216] Such a self-consciousness as to linguistic affiliation is evident in a number of contemporary sources. We have already seen, in Chapter 5, that the language shifts in the book of Daniel—Hebrew to Aramaic and back to Hebrew—thematized Judea's incorporation into the global language of Near Eastern empire and then the reclaiming of the ancestral tongue at the moment of persecution and martyrdom.[217] Contemporary theological works developed the idea that Hebrew was not only the old language of the Judean monarchy, biblical literature, Moses, and the patriarchs, but also the original, archetypal, and sanctified language of God and His creation.[218] Furthermore, at several places in 2 Maccabees Hebrew was violently hurled at representatives of the Seleucid empire. Judas led the war cry "in the ancestral language (τῇ πατρίῳ φωνῇ)" against the forces of Gorgias.[219] Later, when the Maccabean troops came across the corpse of the Seleucid general Nicanor, they blessed God "in the ancestral language."[220] This victory Hebrew was followed immediately by cutting out Nicanor's tongue ("τὴν γλῶσσαν"),[221] suggesting the physicalization of linguistic difference; Daniel Schwartz notes that a later liturgical song for the festival of Ḥanukkah rejoices that "the sword has cut down every Greek tongue."[222] 2 Maccabees' famous martyrology of the pious Jewish mother and her seven sons shows to the full such weaponization of Hebrew. Ordered to eat pig on pain of death by king Antiochus IV Epiphanes, fictitiously conjured into presence for this encounter, all the brothers refused and were subjected to ghastly torture: the second eldest son responded, simply, "No!" in the ancestral language ("τῇ πατρίῳ φωνῇ 'Οὐχί'");[223] the third in line stuck out his own tongue to be cut off;[224] and their noble mother urged all in Hebrew to prefer an undefiled death to earthly rewards.[225] In this latter, fascinating confrontation, the persecutor king, unable to understand the mother's indigenous speech, assumed that he was being mocked.[226] Hebrew operates here as a kind of hidden or disguised transcript, audibly and provocatively beyond, before, and after the authority of the empire, even when this was at its most brutal and totalizing extension.[227]

Hebrew's ideological force was also given material form. Our very earliest surviving archaeological evidence for Hasmonean rule are bronze and lead coins minted by Simon's son, John Hyrcanus, in whose reign 1 Maccabees was composed. These coins, of little value in themselves and for circulation only within Judea, abandoned the Seleucid and Ptolemaic numismatic model of Greek legends, divine images, and personal portraits for nonfigural or vegetal imagery and a legend inscribed in Hebrew with John's name and "the council of the Jews."[228] The Hebrew used on these coins was not the "square" Aramaic alphabet—the familiar Hebrew of today—that had been adopted by Jews for almost all Hebrew writing since the Babylonian exile. Rather, it was the paleo-Hebrew script, closely derived from the early Iron Age Phoenician alphabet and used in the kingdoms of Israel and Judah during the First Temple period. Other than very rare appearances on coins and stamp seals at the end of the Achaemenid kingdom or shortly after its fall, John's Hasmonean coinage represented this ancient script's first and comprehensive resuscitation by state authorities.[229] We can be certain that this "neo-paleo-Hebrew" was understood as sanctified by age, for the contemporary Dead Sea Scrolls from Qumran use it only for the chronologically "early" biblical books (the Torah and Job) or for the name of God in Hebrew texts otherwise copied in the square Aramaic script.[230] In contrast to the Aramaic alphabet, which underwent strong stylistic development from the seventh to the second centuries BCE, this paleo-Hebrew script hardly changed,[231] as if its words, buried for centuries and frozen in time, had suddenly been brought back to light. Despite the fact that most second-century BCE inhabitants of Judea spoke Aramaic and / or Greek, the Hasmonean deployment of the Hebrew language in general and the paleo-Hebrew script in particular made charged and visible claims for a revived autonomy, legitimacy, and deep, pre-Seleucid antiquity.

Let me unite these theses. Recent, powerful interpretations of the second-century BCE Jewish resistance to the Seleucid empire have located the conflict in a fiscal or political "bottom line"—taxes, soldiers, institutions—that treats any cultural, ethnic, and religious tension as epiphenomenal, anachronistic, or invented.[232] Even the facticity of Antiochus's persecution has been denied, on the grounds of a supposed historical implausibility[233] and adherence to preexisting narrative patterns.[234] Such analyses have been a helpful corrective to credulously literal readings of our sources and an overdetermined, ultimately Christian opposition of Hebrew and Hellene. Yet their privileging of a "common-sense" pragmatic and institutional historical causation goes against the evidence of our sources and at the expense of affect, identity, and belief.[235] For I find it hard to doubt the deeply dislocating consequences of Seleucid imperialism and colonization, the loathings of others (and self) it could prompt, and the temporal quality it gave to cultural difference and change. Antiochus IV's persecution, in particular, is too firmly attested in non-Hasmonean sources to be a legitimizing invention of this new dynasty. To take just one instance, Diodorus Siculus draws on a Seleucid court historian to depict the Friends of Antiochus VII Sidetes (then besieging John Hyrcanus in Jerusalem) urging the king to follow the example of his great-uncle and "to wholly destroy the people, or, failing that, to abolish their customs and compel them to change their way of life (ἄρδην ἀνελεῖν τὸ ἔθνος, εἰ δὲ μή, καταλῦσαι τὰ νόμιμα καὶ συναναγκάσαι τὰς ἀγωγὰς μεταθέσθαι)."[236]

Undoubtedly, the political history of mid-second-century BCE Judea was, in toto, as much a tale of a single family's capture of cultic and political leadership within a Seleucid administrative framework as of a massed, anti-imperial uprising; and all political movements contain contradictions and hypocrisies, manifold diversity, and inventions masked as traditions.[237] Yet in the (partial) light of this book's hermeneutic, the reforms of the high priests Jason, Menelaus, and their followers can be seen as an attempt to more fully enter the imperial time, to pull the Jerusalem community within the Seleucid horizon, and Antiochus's persecution as a doing away with the indigenous past. There is, accordingly, a coherence to the Maccabean response, even as its ambitions progressed and its character altered: the style of warfare, the modes of historical memory, the rhetoric of reconstruction, and the linguistic revivalism all point to a kind of self-ancestralization. However we characterize this movement—a synthetic vision that saw an affinity between early Israel and the Hasmonean age; a "tiger's leap" back to the remote past of the Hebrew people in the promised land; a paradoxical proximity of present and foreshortened past; a "secondary" culture of imitation and quotation at the meeting point of linear and cyclical, transcendent and imminent historicality—we can detect in second-century Judea, as in contemporary Babylonia, Persis, and Armenia, a deliberate and politically potent desynchronizing of the indigenous and imperial temporalities.[238]

The Greek Exception

From the inception of Hellenistic history as a modern scholarly field, the Seleucid empire's Greek communities have been walled off from discussions of local or indigenous resistance. Such categorical exclusion derives from the framing of this period in ethnic terms. The programmatic identification of "synthesis" (Droysen's *Verschmelzung*) as the Hellenistic world's governing principle required the opposing of "Greek" and "Oriental" as two discrete objects of historical process. The Seleucid empire, taken as a political vehicle for Greekness, was held to share not only an ethnic community with the ancient Greek city-states of its western fringe but also an ethical alliance.[239] A scholarly comfort with empire and, without doubt, a certain chauvinism lined up kings and Greeks against the peoples of the east. Accordingly, the only methodologically comparative examination of resistance to the Hellenistic monarchies, Samuel Eddy's groundbreaking and still influential *The King Is Dead*, considered revolts in Iran, Judea, Egypt, and the non-Greek parts of Anatolia. Eddy argued that these subject populations, armed with revealed religions of social justice, theologies of native kingship, and memories of lost grandeur, targeted not so much the imperial state's overbearing exploitation, but rather its seductive and threatening Greekness.[240]

We must instead, at least provisionally, take indigeneity as a universal subject position within the Seleucid empire. The Greeks of the old Anatolian coastal *poleis* and new inland colonies, just like the Babylonians, Persians, Armenians, and Jews, were in various degrees subordinated to the Seleucid monarchs and at many times hostile to them. If the epigraphic corpora that constitute the bulk of our source data on these Greek communities show little animosity to the Hellenistic monarchs, it must be remembered that these were public transcripts, built up from a limited repertoire of stock euphemisms, intended to formulate political incorporation and domination as enthusiastic complicity and beneficial reciprocity.[241] Only rarely, and mostly in narrative sources, does an alternative viewpoint break through, one more closely aligned to the noninstitutional, affective, ancestralizing evidence from the Near East.

Most striking is an episode in the local history of Nymphis of Heraclea Pontica, a Megaran colony on the northern coast of Bithynia. Nymphis's history, preserved thirdhand, ran from the city's foundation down to the accession of Ptolemy III Euergetes in 246 BCE. The historian tells how, shortly after Seleucus I's victory over Lysimachus at the battle of Corupedium in 281 BCE, his city enfranchised its occupying garrison, razed the citadel's walls, and, in a state of "glorious fearlessness (λαμπρὰς ἀδείας)," dispatched an embassy to this most recent conqueror.[242] Yet Seleucus disparaged and threatened the Heracliote envoys, "until one of them, Chamaeleon, not shrinking at all from the threats, said, 'Heracles is *karrōn*, Seleucus! ('Ηρακλῆς κάρρων, Σέλευκε)'—

karrōn means 'stronger' in Doric. But Seleucus did not understand (οὐ συνῆκεν) what had been said and, still angry, turned away."[243] The scene, a confrontation between a Hellenistic monarch and the representative of a threatened community, is a kind of encounter common to provincial or indigenous storytelling, giving narrative opportunity for an unusually direct, unmediated meeting with the imperial state in the person of the king or emperor.[244] As we have already seen, dialogues of this kind carry the responsibility of political dignity for the subject community. The episode turns on Seleucus's failure to understand ambassador Chamaeleon's speech. The use of the Doric dialect, a disguised speech without physical consequence, is defiantly vernacular, asserting in the face of transregional empire Heraclea's local identity, early history, and extra-imperial kinship network; note also that Seleucus's royal title is not used.

The phrase "Heracles is stronger ('Ηρακλῆς κάρρων)" has meaningful content beyond the obvious reference to the city's name. The formula—a divinity or local hero followed by the comparative "stronger"—was used across the Greek world to ward off demons, ghosts, and bad luck. So, Theophrastus's Superstitious Man exclaims, "Athena is stronger! (Ἀθηνᾶ κρείττων)" to avert the ill omen of an owl's hooting;[245] and the equivalent formula "Heracles is stronger! ('Ηρακλῆς κρείττων)" is inscribed onto apotropaic gemstones.[246] Moreover, the Doric phrase "Heracles is stronger ('Ηρακλῆς κάρρων)" is independently cited by Apollonius Dyscolus from a mime of the fifth-century BCE Syracusan poet Sophron.[247] We know from Athenaeus that Chamaeleon, Heraclea's unshakable ambassador, was a Peripatetic philosopher known for his writings on Greek comedy. Thus, it is likely that Chamaeleon was deploying against Seleucus a cultural reference, perhaps even adopting a prescripted comic frame. When Nymphis tells us that Seleucus "did not understand what had been said," this may have the secondary force of the king not getting the reference.[248] Thus, we have a piece of local Greek historiography that crafts an encounter of cultural delegitimization, characterizing Seleucus I as an alien monarch, and perhaps also a demonic and philistine one. The scene comes remarkably close to the martyrology of the Hebrew-speaking mother and her seven sons in 2 Maccabees, discussed above.

Elsewhere, we see equally disruptive speech directed at imperial authority. About a century later, for example, the cities of Lampsacus and Smyrna refused to submit to Antiochus III and sent ambassadors to meet with the king, his representatives, and Roman officials. We are told that these envoys spoke with such *parrhesia*, "liberated speech," ("μετὰ παρρησίας διαλεγομένων") that Antiochus III interrupted them, unable to bear the slight in the presence of Romans.[249] Livy reports that, in an earlier encounter, the envoys asked how they could be expected to suffer the postponement of their freedom with a calm demeanor (*"aequo animo"*).[250] The vehemence of the ambassadors and the indignation of the king attest a breach in the mannerly dance of ritualized deference that served as a source of cohesion and definition-making for Hellenistic kingdoms, where the king could get angry but subordinates had to bite their tongues and regulate visible emotions. "Radical rudeness" is a well-attested strategy of destabilization in modern imperial contexts, disrupting a power-serving consensus and drawing conflicts into the open.[251] Episodes of this kind must have been far more frequent than our evidence indicates. Such *parrhesia* was a diagnostically ancestral style of speaking and itself an assertion of freedom and autonomy.[252]

Several Greek and Hellenized Phoenician cities in Cilicia and the Levant attempted to take control of their pasts and to assert pre-Seleucid political identities, as I discussed in my previous book: employing noncolonial names, inventing or emphasizing pre-Seleucid histories, and displaying archaizing iconography on their semiautonomous coinage.[253] Similarly, it has recently been argued that, in the second half of the third century BCE, the Carian city of Mylasa and its nearby sanctuary of Labraunda experienced a "Hecatomnid revival"—the copying, reemergence, and preservation of the coins, names, and architecture, respectively, of Caria's great fourth-century dynasty.[254]

Yet even if Greeks could join with their fellow Seleucid subjects in charging their pasts against the empire, they seem not to have generated either sealed total histories or apocalyptic eschatology. Perhaps the very frontier dynamics of the Asia Minor coastlands that so heightened *polis* agency vis-à-vis the Hellenistic monarchies also inhibited an easy sense of imperial sequence and historical periodization. Perhaps Greek communities, already possessing or constituting much of what fell within the Seleucid horizon, did not experience as acutely the temporal displacement we detect elsewhere. Perhaps their disproportionate incorporation into the court and colonial elite too closely identified imperial with personal or communal interests. Indeed, it is only with the total dominance of Rome that the Greek world would come to experience an analogous distantiation of their pasts and the end-of-regime fantasies and nostalgic pathologies that follow. Polybius, the earliest of the extant Greek historians to apprehend this new world, is least alien to the kind of total history we see in the Seleucid empire, with his uniting of regions, his concern for periodization and temporal interweaving, his sense of events leading up to a definite end in the "fifth kingdom" of Rome, his prognosis of its fall, and his importing, through the idea of *Tychē*, of a new element of determinism.[255]

Conclusion

Our regional stories were articulated to one another. Not only are the political situations interwoven—Artaxias I allied with the rebel Timarchus, the Hasmoneans took lessons from Elymean and Persian resistance,[256] Adad-nādin-aḫḫē likely drew on Urukean expertise, the revolt of Tarsus dragged Antiochus Epiphanes from Judea, and so forth. But also, despite all the bottomless distinctions of history, social structure, beliefs, and language, there was a basic and remarkable convergence in these local responses to a weakening Seleucid state. From the Levant to Iran, from Armenia to southern Babylonia, we witness reenactment, rebuilding, and renewal, archaizing violence, ancient spaces, vernacular languages, and ancestral titles.

Whatever the degree of incorporation into or independence from Seleucid political structures in each case, this culture of resurrection depended upon and reacted to the distantiation and alienation of indigenous pasts brought about by the Seleucid dynasty. Put as schematically as possible, if the empire worked in various ways to "break off" pre-Seleucid history, then in a second movement, as a response to a call, this was undone or invalidated by subject elites pressing for ever-greater autonomy. The very Seleucid horizon that worked hard to dissociate past from present was transformed into the locus of their enlivening encounter. By putting the past back into the present, by opening it once more to the future, these indigenous populations reentered history. Perhaps this could be characterized as a "fusion of horizons" (*Horizontverschmelzung*) or, in the terms of Ankersmit again, as a move from knowledge to desire.

It is a commonplace of contemporary criticism that nostalgia requires modernity.[257] But much as the Seleucid temporal regime could operate in ways analogous to and anticipatory of the modern, so the empire's subjected peoples seem to have experienced something approaching the romantic valorization or chthonic poetics of the exhumed and tangible past.[258] Svetlana Boym has helpfully contrasted two modes of nostalgia: a reflective nostalgia of estranged irony and historical rumination that "lingers on ruins" (similar to what I have termed "total history 1") and, more germane here, a restorative nostalgia of public action and aggressive recreation that "manifests itself in total reconstructions of monuments of the past."[259] The self-archaizing of second-century BCE southern Babylonia, western Iran, Armenia, and Judea was not a ceding of the historical stage to the Seleucid overlords; rather, it was an explosive challenge to their hegemony. The irruption of pre-Hellenistic worlds introduced a radical alterity that punctured the illusion of imperial time. If the apocalyptic eschatology of the previous chapter constituted a redemption from history, then these *Altneuland* rebellions amounted to a redemption through history.

From at least the first half of the second century BCE, the Seleucid empire and its Near Eastern subjects inhabited times of state that, though coexistent, were neither contemporary nor commensurable.[260] The temporal landscape of the mid- to late Seleucid empire, like its political landscape, was ever more multifarious, protean, and riven with conflict. For a brief moment, as the massive tide of Macedonian imperialism began to recede, the inhabitants of Seleucid Asia could look across the sand and see what of their own pasts might be worth salvaging before the next wave began to pound their shore. By contrast, the Seleucid state in Syria, ringed by these new-old kingdoms, denied any history before itself and so was condemned to an unregeneration.

Conclusion

R. Naḥman opened his discourse with the text, "Therefore, fear not, O Jacob, My servant" (Jer. 30:10). This speaks of Jacob himself, of whom it is written, "And he dreamed, and behold, a ladder was set upon the earth, and the top of it reached to heaven, and, behold, the angels of God ascending and descending on it" (Gen. 28:12).

R. Samuel b. R. Naḥman said, ". . . The Holy One, blessed be He, showed our father (Jacob) the prince of Babylon ascending seventy rungs, the prince of Media fifty-two rungs, and the prince of Greece one hundred and eighty rungs. But as for the prince of Edom (Rome), he kept ascending until Jacob did not know how many rungs. Thereupon, our father Jacob was afraid. He thought: 'Is it possible that this one will never be brought down?' But the Holy One, blessed be He, said to him: 'Fear not, O Jacob, My servant. Even if he ascends and sits beside me, I will bring him down from there.'"

LEVITICUS RABBAH 29:2

This midrash, or exegetical commentary, attributed to the late-third- or fourth-century CE rabbi Samuel b. Naḥman, unites in one place several of this book's themes.[1] The dream of the patriarch Jacob at Bethel, of a ladder reaching to heaven, here becomes a vision of comparative chronography, historical periodization, temporal anxiety, and apocalyptic promise. Four angels, guardian princes of empires, as in the book of Daniel, ascend and descend these rungs of historical time. Each has his allotted, enumerated period of rule: Babylon for seventy years, the Medes (and Persians) for fifty-two, the Macedonians for one hundred and eighty—but the angel of Edom continues to rise, and Jacob is afraid.

The Seleucid empire is barely remembered, and the compresses of years have healed all wounds. Yet as the midrash demonstrates, its temporalities, Seleucid and subjected, outlived their host.

First, the kingdom's declining agony (its interdynastic wars, the absorption of its territories by Parthia and Rome, and its final dissolution by Pompey in 64 BCE) can be seen to have perfected the Seleucid Era's tendency to transcendence. A temporal system that had always worked in tension with monarchy and in contradiction to imperial realities—oblivious to enthronement, a growing number for a shrinking territory, an idealized unity as kinglets multiplied—now, finally, could cut its earthly anchor. Henceforth, Era years would roll on like emptied jars, a dating of convenience all across Asia, for late antique Syrian mosaics, Yemenite Jewish marriage contracts, Arabic astronomy, Chinese Nestorian inscriptions, and so forth.

In turn, the kings of Syria were succeeded in their last lands by the Romans, Edom in the rabbinic coding. Rome, quite unlike the Seleucid state, was a rooted Republic and Empire, closely bound to the traditions of its ancient city and so undrawn to the more radical abstractions and simplifications of Seleucid statecraft. Under the Caesars, dating by events, offices, and reigns returned to prominence in the coastal Near East, even as the densest employment of civic era counts remained in old Seleucid territories.

And, finally, history failed to meet its deadline. The midrash neatly shows the logics of apocalyptic deferral, the tempering of an eschatological urgency by the promise of a future, but not imminent, redemption: *sed noli modo.* This was a strategy of stabilization and integration—a mithridatism by years—that would be essential for Judaism and the Church alike.[2]

So what is left? To some extent, no doubt, all empires regulate the times of their subjects, all kingships fashion infinitude, all insurgents work for political closure, all worshipers make heaven sovereign, and all historians seek meaning. Yet the new systems of time and history introduced under the Seleucid watch brought a heightened prominence, self-consciousness, and interconnectedness to these movements. The experience of inhabiting the Seleucid empire was one of being in a discrete period of history, a rupture with the past that most states preferred to obscure. The historical had become empirical.

It is tempting to consider this Seleucid time as just another of the countless ways in which humans comprehend historical duration, and of course, this is undeniable. Yet the imperial temporality had also made visible, had exposed to a more immediate perception—at least in part, and in the face of all the cultural systems generated to deny it—a terrifying truth: the irreversibility, loss, and indifference of time. In doing so, the Seleucid empire made not only a new earth but also, by necessity, a new heaven. It is in this sense, in the tension of living between incompatible histories, between an empty time of state and the call of an ending, of balancing number with meaning, that the Seleucid east opened the very age to which we still belong.

Abbreviations

Notes

Bibliography

List of Maps, Illustrations, and Tables

Index

ABBREVIATIONS

References to classical authors are according to the *Oxford Classical Dictionary* (3rd edition); references to cuneiform sources follow the abbreviations listed in the *Chicago Assyrian Dictionary*, volume A.1 (1964), xxiv–xxxiv, and the *Cuneiform Digital Library Initiative* (http://cdli.ox.ac.uk/wiki/abbreviations_for_assyriology); references to epigraphic sources follow the abbreviations listed in the *Supplementum Epigraphicum Graecum* (*SEG*). In addition:

AD   Abraham Sachs and Hermann Hunger. 1988–1996. *Astronomical Diaries and Related Texts from Babylonia.* Vienna.

AJP   *American Journal of Philology*

BASOR   *Bulletin of the American Society of Oriental Research*

BCH   *Bulletin de correspondance hellénique*

BCHP   Irving Finkel and Bert van der Spek. 2012–. *Babylonian Chronicles of the Hellenistic Period* [published online at www.livius.org/cg-cm/chronicles/chron00.html]

BNJ   Ian Worthington, ed. 2012–. *Brill's New Jacoby.* [published online at http://referenceworks.brillonline.com/browse/brill-s-new-jacoby]

BSA   *Annual of the British School at Athens*

BSOAS   *Bulletin of the Society of Oriental and African Studies*

CA   *Classical Antiquity*

CIS   *Corpus Inscriptionum Semiticorum*

CP   *Classical Philology*

CQ   *Classical Quarterly*

FGrHist   Felix Jacoby. 1923–. *Die Fragmente der griechischen Historiker.* Berlin.

IEJ   *Israel Exploration Journal*

IEOG   Canali de Rossi. 2004. *Iscrizioni dello estremo oriente Greco.* Bonn.

IGIAC   Georges Rougemont. 2012. *Inscriptions grecques d'Iran et d'Asie centrale.* London.

IGLS   Louis Jalabert, René Mouterde, and Jean-Paul Rey-Coquais, eds. 1929–. *Inscriptions grecques et latines de la Syrie.* Paris.

JANES   *Journal of the Ancient Near Eastern Society*

JAOS   *Journal of the American Oriental Society*

JBL   *Journal of Biblical Literature*

JCS	*Journal of Cuneiform Studies*
JEA	*Journal of Egyptian Archaeology*
JHS	*Journal of Hellenic Studies*
JNES	*Journal of Near Eastern Studies*
JJS	*Journal of Jewish Studies*
JSJ	*Journal for the Study of Judaism in the Persian, Hellenistic, and Roman Period*
KAI	*Kanaanäische und aramäische Inschriften*
NABU	*Nouvelles assyriologiques brèves et utilitaires*
NC	*Numismatic Chronicle*
OGIS	Wilhelm Dittenberger. 1903. *Orientis Graeci Inscriptiones Selectae.* Leipzig.
RC	C. Bradford Welles. 1934. *Royal Correspondence in the Hellenistic Period.* London.
REG	*Revue des études grecques*
SEG	Angelos Chaniotis, Thomas Corsten, Nikolaos Papazarkadas, and R. A. Tybout, eds. 1923–. *Supplementum Epigraphicum Graecum.* Leiden.
TAPA	*Transactions of the American Philological Association*
VT	*Vetus Testamentum*
ZA	*Zeitschrift für Assyriologie und verwandte Gebiete*
ZAW	*Zeitschrift für die alttestamentliche Wissenschaft*
ZDPV	*Zeitschrift des deutschen Pälastina-Vereins*
ZPE	*Zeitschrift für Papyrologie und Epigraphik*

NOTES

INTRODUCTION

1. *BNJ* 1 F300 = Hdt. 2.142-143.
2. On this famous episode, see, e.g., Dillery 2015, 119-122; Moyer 2002; Bertelli 2001, 89-94; Calame 1998; Froidefond 1971, 137-139 and 145-151; Heidel 1935, 92-97.
3. On the career and tradition of Seleucus I, see, e.g., Kosmin 2014a, 31-37, 59-61, and 79-87; Capdetrey 2007, 25-50; Mehl 1986.
4. Kosmin 2014a.
5. Thonemann 2013, 4, 12 (drawing on Boone 2003, 11-33).
6. See, e.g., Strootman 2017; Plischke 2014; Capdetrey 2007; Sherwin-White and Kuhrt 1993.
7. On the Maccabean revolt, see, e.g., Gruen forthcoming; Eckhardt 2016; Honigman 2014. On the Attalids, see, e.g., Chrubasik 2013. On the *fratarakā*, see, e.g., Strootman 2017, 194-197; Engels 2013.
8. The formula originates in a famous 1862 speech of the German-Jewish socialist Ferdinand Lassalle and was influentially appropriated to the minimal state theories of Nozick 1974, 25-27.
9. Eddy 1961.
10. References to the more significant synthetic studies (which are listed in the bibliography) can be found throughout the book.
11. For a discussion of this possible "Greek exception," see Chapter 6.
12. Momigliano 1987, 105.
13. For an introduction to current debates on the category of "indigeneity," see, e.g., Trigger and Dalley 2010.
14. For these categories, see Moyn and Sartori 2013, 4-20. Cuneiform historiography from Babylonia incorporates events in Bactria, Thrace, Armenia, and the Egyptian frontier, and 2 Maccabees, the epitome of Jason of Cyrene's account of the Jewish revolt, knows of parallel anti-Seleucid violence in Cilicia and Persia (see, e.g., Kosmin 2016).
15. For ancient times, see, e.g., Schiffman 2011; Feeney 2007; and D. Allen 1996. For French Revolutionary time, see, e.g., Shaw 2011; E. Zerubavel 1977. For modern imperial times, see, e.g., Prasad 2013; Barrows 2011; Banerjee 2006. For Zionist time, see, e.g., Chowers 2012 and Ohana 2012. For American time, see, e.g., T. Allen 2008. For medieval time, see, e.g., Le Goff 1960. For factory time, see, e.g., Thompson 1967. For now time, see, e.g., Hartog 2015; Guyer 2007; Harootunian 2007; Jameson 2003.
16. E. Bloch 1977. See also Harootunian 2007; Koselleck 2004, 93-100, 246-247, and 265-268.

17. For the particular example, see Genovese 1976, 285-324; Thompson 1967.

18. See, e.g., Ricoeur 2004, 209-216.

19. Koselleck 2002, 149. Koselleck's contrast between years, on the one hand, and hours, days, weeks, and months, on the other, is fully drawn in a discussion of the calendrical reforms of the French Revolution (ibid., 148-153). On these, see also Shaw 2011; E. Zerubavel 1977.

20. Aristotle, *Pol.* 1252a-b, as developed by Agamben (1998, 1-12 and 181-184) and Arendt (1958, 22-28).

21. Note that even those literary genres or worldviews, such as the Babylonian Astronomical Diaries or apocalyptic eschatology, that may seem to attempt to reconcile these two orders (durational time and periodic time) in fact subordinate one to the other—historicality to calendricality in the first case, and calendricality to historicality in the second.

22. Löwy 2005; Mosès 1994; Eliade 1969; Rosenzweig 1921, 335-523. See also Hammer 2011.

23. Note, though, the efforts of Bickerman (1944-1945, 381-383), Momigliano (1970), and, most recently, Briant (2017) to pull the origins of Hellenistic scholarship back into the eighteenth century CE.

24. See, e.g., Hooker forthcoming; Courtray 2011; Botha 2007; Barnes 1994.

25. Jerome, *in Dan.* prol. 94-95. From the third to fifth centuries CE alone, we have the commentaries of Hippolytus, Origen of Alexandria, Porphyry of Tyre, [Ephrem the Syrian], Aphrahat, Jerome, Theodoret of Cyrrhus, and Polychronius of Apamea.

26. August. *Conf.* 11.17-22.

27. See, e.g., Fabian 1983, 34.

28. Collingwood 1946, 252-253 and 261-263.

1. THE SELEUCID ERA AND ITS EPOCH

1. τί γὰρ Σωκράτει πρὸς σοφίαν καὶ Θεμιστοκλεῖ πρὸς δεινότητα συντελεῖται παρὰ τῶν χρόνων (Eunapius Sardianus F1 [Müller], quoted in Momigliano 1966, 16).

2. ITT 4, 8001 = MVN 7 #397.

3. Mari Text #62 = LAPO 16 90.

4. Thuc. 2.2.1.

5. On Adad-guppi and her autobiographical inscription, see Longman 1991, 97-103; Beaulieu 1989, 68-79; Gadd 1958, 46-56.

6. Gadd 1958 H1, B 1.29-34, trans. Longman III 1991: 226, adapted. For other Mesopotamian multiple-reign calculations of chronographic depth and their problems, see Na'aman 1984.

7. On the inscription's chronological difficulties and choices, see Beaulieu 1989, 139 and 143; Gadd 1958, 70-72.

8. In the Old Babylonian period, year-name formulae could be extended into a numbered succession: Ur-Ninurta of Isin numbered the four years from the year in which he constructed the wall named Imgur-Enlil; Sumu-la-El of Babylon named five years in relation to the year in which he destroyed

the city of Kish; and most spectacularly, Rīm-Sîn of Larsa counted the last half of his sixty-year reign from the year in which he had conquered Isin. On these proto-eras, see Charpin and Ziegler 2013, 61-62; van de Mieroop 1993, 48; Hallo 1988, 177-178. On the supposed Nabonassar Era, see Chapter 4. More germane are certain Phoenician city eras, but several are likely to be regnal or minting dates. Where truly present, these eras seem to be associated with the removal of local city monarchies (see Stylianou 1991; Millar 1983, 61-63). On the political force of the epigraphic dating formula [*year number*] *št l'm* [*city name*], "in the . . . year of the people of . . . ," see Sznycer 1975, 52-56. On the Era of Sidon, see *KAI* 60, discussed by Lipiński 2004, 171-172; Wheatley 1995, 438, and 2003; Baslez and Briquel-Chatonnet 1991, 235-237. On the Era of Lapethus, see *KAI* 43, discussed by Lorber 2007, 115-116; Lipiński 2004, 172; Huß 1977, 134-136; Sznycer 1975, 53-54; Honeyman 1941. On the Era of Citium, see *KAI* 40, discussed by Lipiński 2004, 172; Hauben 1973, 261-262. A Carthaginian inscription, *CIS* 5632 dated to "the twentieth year of the magistracy of the *šōfṭīm* in Carthage," may belong in the mid-fifth century BCE (Sznycer 1981, 295-296; see also Krahmalkov 1976). Intriguingly, the eras of the Cypriote cities of Citium and Lapethus begin at around the same time as the introduction of the Seleucid Era count; perhaps they are part of the same moment of conscious chronographic design that followed the liquidation of the Argead dynasty of Macedonia. I am grateful to Jeff Chu for bibliographical assistance.

9. Boiy 2001; Hunger, Sachs, and Steele 2001, 84-87, #34. Note that an eclipse report is written, at a right angle to the rest of the tablet, at the bottom of its reverse side; see Aaboe, Britton, Henderson, Neugebauer, and Sachs 1991, 35n21. On the Saros cycle, see Britton 2007, 88-92.

10. Of course, the Saros Tablet was not composed to reveal these differences; it is illuminating precisely because it demonstrates assumed temporal frameworks.

11. Space-saving abbreviations of individual kings' names are known, but only successive Seleucid monarchs were replaced by an unchanging syllable. See, e.g., the Saros Canon (BM 34597), which abbreviates Artaxerxes III Ochus as *ú* (for *Umasu*), Artaxerxes IV Arses as *ár*, Darius III as *da*, Alexander the Great as *a*, Philip III as *pi*, and Antigonus Monophthalmus as *an*.

12. While ancient astronomers and mathematicians had long been able to imagine vastly distant future years of celestial periods—see, e.g., the Keskintos inscription from Hellenistic Rhodes, *IG* XII,1 913, with A. Jones 2006—these were problems of scholarly calculation and not historicality. For a concise contrasting of these domains, see M. Bloch 1992, 23-24. With the Seleucid Era, for the first time, we are dealing with both counting from an unchanging, governmentally fixed epoch date and an unprecedented social pervasiveness (on which, see Chapter 2).

13. Note, as a rule-proving exception, that a unique proto-almanac (BM 33066 = Strassmaier Cambyses 400) contains records of various astronomical phenomena, with reference to planetary movements in Cambyses's nonexistent ninth regnal year. This erroneous date is merely one or two years in advance of the tablet's composition (see Britton 2008, 8-10 and 30-31). Otherwise,

the observation of Aaboe, Britton, Henderson, Neugebauer, and Sachs holds: "We do not know of a single astronomical cuneiform text in which a regnal year exceeds the natural reign of the king before the introduction of a continuing year count in the Seleucid Era" (1991, 2).

14. Knausgaard 2014, 152. I am struck, but do not know how far to press the point, that from the same Seleucid imperial environment the second-century BCE heliocentric astronomer, Seleucus of Seleucia-on-the-Erythraean Sea, was among the first to propose the infinity of the cosmos (Aëtius, *Platica* 2.1.7, as reconstructed by Mansfield and Runia 2009, 300–305, with a possible Arabic citation discussed in Pinès 1963).

15. The following schema is reconstructed from Boiy 2011, 2010, 2009, 2007, 73–89, 2002a, 2002b, 2001, 2000, and 1998; Anson 2005; Grzybek 1992, 190–192; Aaboe, Britton, Henderson, Neugebauer, and Sachs 1991, 30–31; McEwan 1985; Joannès 1979–1980; Oelsner 1974.

16. Only a couple of tablets are attested for years 1 and 2 of Alexander IV (BM 78948, CT 49 13, and CT 49 27). See Boiy 2000, 120, for a possible year 4 of Alexander IV (*AION Suppl.* 77 87).

17. A couple of exceptional documents continued dating by the year of Antigonus's generalship—i.e., a continuation of stage 2—after Seleucus's return to Babylonia in stage 3: TBER 88 (AO 26765), a ration list from the Esagil archive, for year 8 of General Antigonus, and BM 105211, a rental contract from Larsa, for his year 9.

18. See Capdetrey 2007, 27.

19. *BCHP* 3 column 4, lines 3–4.

20. See van der Spek 2014b, 328–329; Oelsner 1974, 136n33; Stolper 1990. The lacuna's restoration is confirmed by the stability and frequency of this dating formula.

21. The last tablet dated, posthumously, by Alexander IV is CT 49 25 (Adarru of Alexander IV's year 11 = February–March 305 BCE); the first published tablet dated by Seleucus I is CT 4 29d (3rd Nisannu 8 SE = 16th April 304 BCE). See Boiy 2011, 8; McEwan 1985.

22. *Babylon King List* (BM 35603) Obv. 6, discussed by Sachs and Wiseman 1954, 204–206.

23. With our current state of knowledge, we do not know whether Antigonus restarted his year count at his coronation. Certainly, his successors, once in control of the Macedonian homeland, dated by regnal years in the traditional manner.

24. These, found in Appian, Diodorus Siculus, Justin, Libanius, and Malalas, are thought to derive from an encomiastic and novelistic biographical tradition of the empire's eponymous founder, propagated by the court. See Ogden 2017; Kosmin 2014a, 94–100; Primo 2009, 29–35; Fraser 1996, 37–46.

25. Diod. Sic. 19.90–91. On this narrative, see Ogden 2017. 68–98; Primo 2009, 181–190; Capdetrey 2007, 35–38. For the oracles that hailed the success of Seleucus's return to Babylon, see Diod. Sic. 19.90.3–4, discussed by Bearzot 1984.

26. BM 35920 2.2–4. See van der Spek 2014b, 340–342. Based on my personal observation in the Arched Room of the British Museum on March 29, 2016,

column 1, which is very fragmentary, seems to have dealt with astronomical observations. This confirmed to my satisfaction van der Spek's readings. Indeed, the beginning of the sign kám, the determinative for ordinal
numbers, is visible after ud.1.

27. There is some confirmation for this early springtime return: a cuneiform
tablet from Borsippa, dating May 13, 311, is dated by the year of Antigonus,
while a week later, on May 20, 311, another tablet is dated by the regnal year
of Alexander IV. See Boiy 2004, 125–126.

28. Chaniotis 2011, 161–162; Levenson and Martin 2009, 333–336.

29. Thonemann 2005.

30. We have no testimony of Babylonian involvement in this process, but
the evident control of local cultic and calendrical knowledge makes the
participation of knowledgeable Babylonians highly probable. See, for instance, Antiochus I's offering to Sîn "according to the instructions of a certain Babylonian" (*ina qí-bi šá 1-en* [lú]dumu.e[ki]) (*BCHP* 5, line 8, discussed by
Kosmin 2013c, 210–211).

31. For reconstruction, see Zgoll 2006, 30–41; Bidmead 2004; Linssen 2004, 78–
86; Sommer 2000; Dirven 1997, 102–106; Berger 1970; Falkenstein 1959.

32. Pongratz-Leisten 1994, 17–18 and 71–75; Wyatt 1987; Talmon 1966, 36–45;
Haldar 1950.

33. Bidmead 2004; Pongratz-Leisten 1994, 11–19 and 71–75.

34. Seri 2012; Frahm 2010, 14–20; Vanstiphout 1992.

35. On the close relationship between the *Chaoskampf* and kingship in west Semitic mythology, see Wyatt 1996, 117–218.

36. K 1356 described the images and reproduced the texts adorning the doors
of Sennacherib's *bīt akīti* (see Lambert 1997). The temple and its cella, respectively, were named in Sumerian "The House Where the Sea Is Put to
Death" and "The House Which Routs the Sea" (George 1996, 376n32).

37. See, e.g., Zgoll 2006, 43–45 and 47–60; Black 1981, 50–51.

38. Pongratz-Leisten 1994, 177–178. George notes that the Ištar Gate, through
which the *akītu* procession reentered Babylon, bore the ceremonial name
nēreb šarrūti, "the entrance of kingship" (1996, 366).

39. *Babylon King List* (BM 35603) Obv. 3, discussed by Sachs and Wiseman 1954,
203–204. On this passage, Del Monte says: "L'assenza di un re era stata da
sempre avvertita a Babilonia come periodi particolarmente nefasti e di dis-
ordini per il paese" (2001, 145). Note that Sennacherib's rule in Babylon was
referred to as a "kingless" period because of his neglect of the gods (see
Brinkman 1979, 247n115).

40. See Wiesehöfer 2013b, 47–48; Silverman 2012, 188–191; Stausberg 2002,
169–170.

41. Weissert 1997.

42. Cyrus Cylinder l.17 (K 2.1 Schaudig), alluding to *Enūma eliš* 6.126, as noted
by Schaudig 2001, 555n906. I am grateful to Johannes Haubold for this
reference.

43. On the possibility that the anchor serves to mark Seleucus I's service as
admiral of Ptolemy I's fleet, see, e.g., Jacobson 2015, 18; Babelon 1890, vii–
viii. On Seleucus's maritime experience, see Diod. Sic. 19.12.5–13.2 (on the

Tigris), 19.58.5 and 62.4–6 (in the eastern Mediterranean), and 19.60.3–4 and 68.3–4 (in the Aegean).

44. Houghton and Lorber 2002, #88–100 and 105–108, discussed by Ogden 2017, 48–50. The anchor was erased from certain of these Seleucid coins Houghton and Lorber consider the possibility "that the erasure of Seleucus' personal badge was a hostile act, effected by the Antigonids in the course of their campaigns in Babylonia" (2002, 1.43).

45. Bidmead 2004, 163.

46. Kosmin 2014b, 184–185.

47. Kämmerer and Metzler 2012, 228.

48. D. Brown 2000, 227–239. See also Steele 2013 on the significance of this passage for the so-called astronomical fragments of Berossus's *Babyloniaca*.

49. Sahlins 1985, 125 (see also 104–135 and Sahlins 1981).

50. Stern 2012, 234–235 and 239–243; Assar 2003, 174–175.

51. See Bennett 2011, 187–222.

52. *Babylon King List* (BM 35603) Rev. 14, "[mu.148.kám] gan *it-te-eš-me šá* ¹An l[ugal nam^meš]," discussed by Sachs and Wiseman 1954, 208–209.

53. "καὶ ἀπέθανεν ἐκεῖ Ἀντίοχος ὁ βασιλεὺς ἔτους ἐνάτου καὶ τεσσαρακοστοῦ καὶ ἑκατοστοῦ" (1 Macc. 6:16). On the 1 Maccabees' internal dates, see J. Goldstein 1976, 23–25; Bickerman 1944, 73–74.

54. The battle of Gaza (Diod. Sic. 19.81–85; Plut. *Demetr.* 5.2–3; App. *Syr.* 54; Just. *Epit.* 15.1.6–8), according to the low and eclectic chronologies of the Diadoch Wars (outlined in Manni 1949, 53–61 and defended by Boiy 2007, 111–129 and 145), occurred in autumn 312 BCE. See also Meeus 2012, 88–93; Wheatley 2003. This victory of Ptolemy and Seleucus, exiled from his satrapy, over Demetrius Poliorcetes opened the way for Seleucus's heroic return to Babylon. We have no precise details about the battle's dating, but if it fell near 1st Dios, the SE(M) and SE(B) epochs could be understood as marking an extended moment of victorious return. Porphyry of Tyre (*BNJ* 260 F32.4) makes explicit the link between the battle of Gaza, Seleucus's return to Babylon, and the beginning of his rule.

55. See Chapter 3.

56. See, e.g., Ma 2003, 179–183.

57. On the institution of the *bīt šarri* (é lugal) or *basilikon*, see Clancier and Monerie 2014, 205–206; McEwan 1981a, 138–139.

58. *BCHP* 12 Obv., lines 3–8.

59. Pace Mitsuma, who unpersuasively suggests that the word a^meš-*šú, mārīšu,* "can be interpreted as referring to [Seleucus III's] desired and expected 'offspring'" (2013, 742–743).

60. *CAD* s.v. *māru* 1c, following van der Spek 1985, 557–561; McEwan 1981a, 161–162. However, van der Spek (2016) suggests a cult of Seleucus II and his children.

61. The -*šú* of dumu^meš-*šú* is the third-person singular pronominal suffix. For reasons of political history—the Third Syrian, or Laodicean, War—it is unlikely that Antiochus II's children were present in the region.

62. Bidmead 2004, 101.

63. *AD* -204 C Rev. 14–18.

64. *Ḫarû* rituals are well known from the Neo-Assyrian and Neo-Babylonian periods (Zgoll 2006, 34–37; Linssen 2004, 119; Kessler 2002), but a *ḫarû ša šatti* is not, to the best of my knowledge, found elsewhere.

65. In the publication of the Astronomical Diaries, it is simply translated as "the Akītu temple."

66. Raḥimesu #23, dating to 93 BCE, lists a series of supplementary payments for sheep offerings. In lines 6–12 sacrifices in the Temple of Day One are funded for a month, from 15th Adarru to 15th intercalated Adarru. In lines 13–17 funds are given for sacrifices at the *bīt akīti*, the Main Gate, and the Gate? of Nabû on the 11th Nisannu—that is, at the conclusion of the *akītu* festival. The recipient of both payments is the same butcher, a certain Urak, son of Bēl-eṭir. See McEwan 1981b, 132–136; Jursa 2002, 118n18; Mitsuma 2008; and Bergamini 2011, 30. Pace van der Spek 1998, 225 and 234–235; Boiy 2004, 9 and 85–86. Note that George raises the possibility that Babylon's é ud.1.kám is the *bīt arḫi*, "House of the Day of the New Moon" (2000, 280).

67. No such temple appears in George 1993, the standard catalogue of Mesopotamian sanctuaries. Note, however, that half a millennium earlier, according to cuneiform tablet IM 67543, edited and discussed in Postgate 1974, one of the rooms of the Neo-Assyrian Nabû temple at Nimrud (Kalhu) was named the é ud.7[/8].kám, "Temple of Day Seven [or Eight]" (the reading is uncertain). I am grateful to Eleanor Robson for this reference. More significantly, there is possible reference to an é ud.1.kám *šanat* in Nabonidus's Ebabbar cylinder from Sippar (1.31). George (1985–1986, 22n38), followed by Schaudig (2001, 386n466), suggests that this refers to Nabonidus's relocation of the cult of Šamaš to the supposed é ud.1.kám *šá-na-at*, "the Temple of Day One of the Year," during his reconstruction of Sippar's Ebabbar temple. Beaulieu reads it instead as a date: "I took the hand of Šamaš my lord and, on the first day of the year in which I caused him to dwell in the Ebabbar, I made excavations all around the cella" (1989, 7). Note that the alternative version of the text, now housed in Istanbul's Archaeological Museum (AH 81-4-28, 4, text 2 in Schaudig 2001), lacks the Sumerogram é (*bītu*, "temple"), giving the simple date "on the first day of the year." Even if Nabonidus's Ebabbar inscription can be considered evidence of a sixth-century BCE "Temple of Day One of the Year" in Sippar, we lack any evidence of such a cult site in first-millennium Babylon or, indeed, of anything approaching the prominence it receives in the Seleucid period.

68. *AD* -187 A Rev. 4–18.

69. *AD* -171 B Rev. 1–8 + Upper Edge 1–2.

70. *AD* -170 A Obv. 13–14; see also *AD* -187 A Rev. 6 and *Ludlul bēl nēmeqi* 2.27–28.

71. *AD* -158 C Flake 6; perhaps also *AD* -163 B Obv. 17.

72. *AD* -137 D Rev. 27–28, -136 C Rev. 12–13, -133 B Obv. 23–26, -132 A Rev. 2–4, -129 A$_2$ Obv. 18–19, -126 A Obv. 7–9 + Rev. 1, -126 B Rev. 6–8, -124 B Obv. 4–5, -111 B Rev. 8–10, -107 C Rev. 17–18, -105 A Obv. 13–14, B Obv. 13, -77 A Obv. 27; fragmentary, but probably: -132 D$_2$ Rev. 13–15, -129 A$_1$ Obv. 13–15, A$_2$ Rev. 14–16, -82 B Rev. 2, -78 Flake 13, -77 A Rev. 30–31. On the possible location of the ká.dumu.nun.na, see George 1992, 93, 308, and 397.

73. *AD* -126 A Rev. 4–5.

74. *AD* -132 B Obv. 1.

75. Kosmin 2013b, 71–72; Schuol 2000, 294–295.

76. The incident follows immediately on the *politai*'s receiving news of Arsaces's defeat of his Elamite enemy Pittit.

77. *AD* -124 B Rev. 14–16, "itu bi ud.15.kám ^{giš}gu.za lugal šá gim giš.ḫur lú [x] šá iṣ-ṣi x x [x x] Pi-it²-ti-ti [. . .] šá mu-šú i'-'a-ma-na-a-a tu-ru-nu-us šá pa-na-ma² ¹As-pa-a-si-né-e [luga]l ta é.gal lugal šá ina e^{ki} ti-ú šul-lu-man-nu ana ^{d}en id-din-nu ^{lú}pa-ḫat e^{ki} u ^{lú}pu-li-te-e šá ina e^{ki} . . . ^{meš}-nim-ma ⌐x x x x x x x⌐ ^{giš}ig^{meš} é ud.1.kám bad^{meš}-nim-ma ⌐x x⌐ . . . ⌐x⌐-ḫu-su e-pu-šu nu ku₄^{meš} ^{giš}gu.za lugal mu-a-ti šul-lu-man-nu šá ^{d}en ta é ud.1.kám è-ú² at-ta-aḫ-šú-nu ti-ú² ("That month, the 15th, the throne of the king, which, like the drawing of a man . . . of wood . . . Pittit . . . whose Greek name is 'thronos' . . . , which, previously, [kin]g Hyspaosines had taken from the palace of the king in Babylon and dedicated as an honorific present to Bel, the *paḫat* and *politai* who are in Babylon . . . and . . . the doors of the Temple of Day One they opened and . . . made. They did not enter. That throne of the king, an honorific present to Bel, they brought out from the Temple of Day One. They took it next to them"). On this episode, see Mitsuma 2008; van der Spek 2001, 451–453; Schuol 2000, 37–39. Van der Spek (2009, 111, and 2005, 404) and Monerie (2014, 104) note that in this cuneiform passage the Greek word for throne (θρόνος) was employed in Akkadian transliteration (*tu-ru-nu-us*).

78. Note that Hyspaosines may have been informed about cultic details by Itti-Marduk-balāṭu, an astronomer in his employ from Babylon's Esagil temple (see Schuol 2000, 31–34).

79. Malalas, *Chronographia* 8 p. 202 ll.17–19 (Dindorf).

80. Note that this precise narrative, eliding the Battle of Ipsus with Seleucus's return to Babylon, is present in Paus. 1.16.1: "After the death of Alexander, Seleucus, in fear of Antigonus, who had arrived in Babylon, fled to Ptolemy, son of Lagus, and then returned to Babylon (τελευτήσαντος δὲ Ἀλεξάνδρου Σέλευκος Ἀντίγονον ἐς Βαβυλῶνα ἀφικόμενον δείσας καὶ παρὰ Πτολεμαῖον φυγὼν τὸν Λάγου κατῆλθεν αὖθις ἐς Βαβυλῶνα); on his return, he overcame Antigonus's army and killed Antigonus himself (κατελθὼν δὲ ἐκράτησε μὲν τῆς Ἀντιγόνου στρατιᾶς καὶ αὐτὸν ἀπέκτεινεν Ἀντίγονον)." The same implication is present in the warning of the Chaldeans to Antigonus in Diod. Sic. 19.55.7.

81. On Malalas's use of inscriptions, see Downey 1935.

82. Malalas's lengthy account of Seleucus's foundation of the Syrian Tetrapolis concludes with an attribution to Pausanias of Antioch: "The most wise Pausanias has written many others things poetically (πολλὰ δὲ καὶ ἄλλα ὁ αὐτὸς σοφώτατος Παυσανίας ποιητικῶς συνεγράψατο)" (*BNJ* 854 F10).

83. Malalas, *Chronographia* 8 p. 202 ll.8–9 (Dindorf).

84. Amm. Marc. 21.15.2. See Downey 1961, 38–39.

85. It has been argued that the year 1581/1580 BCE may be the starting point of an alternative Olympiad numbering, according to a bronze discus of 241 CE from Olympia (*IvO* #240/1) that equates the 255th Olympiad with the 456th (see Christesen 2007, 508–513; Lämmer 1967).

86. The Greek acrophonic numerical system is used, in contrast to the Seleucid Era's alphabetic system (on which, see Chapter 2).
87. Parian Marble (*BNJ* 239 A29).
88. Parian Marble (*BNJ* 239 B8).
89. See Rotstein 2016, 10-11 and 89; Hazzard 2000, 162-163.
90. Rosenberger 2008, 230-231.
91. The Soter Era was proposed by Hazzard (2000, 25-30 and 161-167). Note that alternative interpretations for Hazzard's data have been proposed by Lorber (2007). Chaniotis comments that "the author of the Parian Chronicle explicitly explains his retroactive system: εἴως ἄρχοντος ἐμ Πάρωι [μὲν Ἀστ]υάνακτος, Ἀθήνησι δὲ Διογνήτου; if he wanted to propagate the new [Soter] era, he missed his only chance to explain this to his readers" (*SEG* 53 871).
92. Parian Marble (*BNJ* 239 B16).
93. Diod. Sic. 20.5.5, where it is held to portend misfortune. For the astronomical calculation, see the commentary on *BNJ* 239 B16.
94. For the force of the vocabulary, see Scolnic 2014b.
95. We can also ask, did the development of alternative era counts by library scholars—for works of historical chronography rather than contemporary statecraft—represent a competitive engagement with the Seleucid Era epoch? Take, for instance, a fragment of *On the Kings of Judea* by the Hasmonean ambassador and historian Eupolemus (*BNJ* 723 F4 = Clem. *Strom.* 1.21.141.4): "He says that all the years from Adam (ἀπὸ ᾽Αδὰμ) up to the fifth year of the kingship of Demetrius, the twelfth year of Ptolemy, who reigned over Egypt, add up to 5,419 years; and from the time that Moses led the Jews out of Egypt until the above-mentioned time (ἐπὶ τὴν προειρημένην προθεσμίαν), add up to [2],580 years." The *annus Adami*, the Exodus era, and the use of regnal years for the Seleucid monarch Demetrius I Soter undermine the Seleucid Era epoch and count. On Eupolemus, see, e.g., Wacholder 1968. Similarly, the third-century, Alexandria-based polymath Eratosthenes was the first to fully enumerate the Olympiads, establishing a systematic, annually successive era-type system (see Baron 2013, 24-27; Clarke 2008, 64-70; Christesen 2007, 173-179; Dunn 2007, 33-34; Feeney 2007, 84; Bickerman 1968, 75-76). What could be more unlike the *akītu*-aligned Babylonian origin of Seleucid time than this Ptolemaic scholar's opening of years in the most exclusively Greek event of all? The Adamic and Olympiad datings, even if not invented to be hostile to the Seleucid Era epoch, are made to anticipate and run contrapuntally to it. Of course, in a clear contrast to the impact of the Seleucid Era count (see Chapter 2), they remained restricted to libraries and armchairs, neither institutionalized by the state nor pervading social spaces. I cannot improve on the methodological observation of Spengler: "Die Olympiadenrechnung ist keine Ära wie etwa die christliche Zeitrechnung, und außerdem ein später, rein literarischer Notbehelf, nichts dem Volke Geläufiges. Das Volk besaß überhaupt kein Bedürfnis nach einer Zählung, mit welcher man Erlebnisse der Eltern und Großeltern festlegen konnte, mochten einige

Gelehrte immerhin sich für das Kalenderproblem interessieren. Es kommt hier nicht darauf an, ob ein Kalender gut ist oder schlecht, sondern ob er im Gebrauch ist, ob das Leben der Gesamtheit danach läuft" (1923, 13n1).

96. White 1987, 20.

97. See Goldberg 2016, 125, 141, and 148–149, and 2000, 275. Note that al-Biruni's *Chronology of Ancient Nations* mistook the Seleucid Era for an era of Alexander the Great; his later works, including the *Canon Mas'udicus*, corrected the error (see Stowasser 2014, 57–58). Savasorda, in his 1122 *Seder ha-Ibbur*, 3.8, p. 99 (Filipowski) correlated the introduction of the Seleucid Era with the "end of prophecy" (on which, see Chapter 4). Note that Sasanian chronographers identified the Seleucid Era epoch with the birth of Zoroaster (see Lewy 1944).

2. A GOVERNMENT OF DATING

1. E.g., Lysippus's *Kairos* (see Posidippus 142 A-B = *Anth. Pal.* 16.275).

2. Within the civic environment, Gell. *NA* 3.3.5 preserves a lament, attributed to Plautus's *Boeotian Women* but deriving from a Greek context, of the spread of sundials: "You know, when I was a boy, my stomach was the only sundial, by far the best and truest compared to all of these. It used to warn me to eat, wherever—except when there was nothing. But now what there is isn't eaten unless the sun says so. In fact the town's so stuffed with sundials that most people crawl along, shriveled up with hunger."

3. Plut. *Demetr.* 41.7–8; Ath. 12.535f–536a. Demetrius and his father, Antigonus Monophthalmus, were also woven into the sacred *peplos* of Athena (Diod. Sic. 20.46.2; Plut. *Demetr.* 12.3).

4. Ma #18 = *SEG* 41 1003 II ll.50–59.

5. Callixenus of Rhodes (*BNJ* 627 F2 = Ath. 5.196a–203e); Polyb. 30.25–26.

6. See, e.g., Evans 2016; A. Jones 2016, 34–42.

7. Appadurai 1996, 133.

8. Tod 1950.

9. This was first observed in print by Jeanne and Louis Robert (1967), but it has not been extensively discussed. The most significant exception is the Tyrian coinage of Antiochus III, which continued the province's Ptolemaic numbering practices. From Seleucus IV we see the standard Seleucid numerical ordering (see, e.g., Houghton, Lorber, and Hoover 2008, 275).

10. *OGIS* 244, line 42.

11. *OGIS* 257, line 18.

12. *SEG* 36 973 = Errington 1986, line 2; see also Leschhorn 1993, 31.

13. 1 Macc. 1:54; 2:70; 3:37; 4:52; 6:20; 7:1; 9:3 and 54; 10:1, 21, 57, and 67; 11:19; 13:41 and 51; 14:1 and 27; 15:10; and 16:14.

14. Joseph. *AJ* 12.321.

15. This is confirmed by, e.g., 1 Maccabees, where the only Seleucid year date not ordered in the Seleucid Era mode (1:10) was expressly identified, as "the year of the kingdom of the Greeks" ("ἐν ἔτει ἑκατοστῷ καὶ τριακοστῷ καὶ ἑβδόμῳ βασιλείας Ἑλλήνων"); elsewhere, the ordering of numerals sufficed.

16. This may offer further evidence for the role of Seleucus I in the era's design, as the ordering of alphabetic tens and units would have been a necessity from 11 SE (302/301 BCE). It already appears in 276/275 BCE at Karatepe, only half a decade after the Seleucid conquest of western Asia Minor (*TAM* 5.2 881, lines 2-5: "βασιλευόντων Ἀντιόχου | καὶ Σελεύκου τοῦ Ἀντιόχου | ἑβδόμου καὶ τριακοστοῦ ἔτους, μη|νὸς Ὑπερβερεταίου"). The form had penetrated to the minor Phrygian villages of Neon Teichos and Kiddioukome by 267 BCE (*I.Laodikeia am Lykos* 1= *SEG* 47 1739 ll.1-3: "βασιλευόντων Ἀντιόχου καὶ [Σ]|ελεύκου πέμτου καὶ τεσαρακο|στοῦ ἔτους, μηνὸς Περιτίου"). For a rule-proving exception, see Chapter 3 on the year 205 SE in Maresha.

17. E.g., Doyen 2014, 278-289; Finkielsztejn 2014, 2003, 1999, and 1998a; Kushnir-Stein 2005b and 2002.

18. E.g., *SEG* 48 1852; Finkielsztejn 2006 and 1998b, 87-103; Kawkabani 2003; Hartal 2002, 98-101.

19. E.g., Houghton, Lorber, and Hoover 2008, 275-290; Kushnir-Stein 2001; Mørkholm 1991, 31-33; Seyrig 1950.

20. E.g., *SEG* 58 1756; Wörrle 2000.

21. E.g., *SEG* 53 1914; *TAM* 5.2 881; *IGIAC* 70; Milik 2003.

22. As suggested in *IEOG* 442-445, with reference to Pfrommer 1993.

23. E.g., *OGIS* 244 and 257; *IGIAC* 66.

24. E.g., *SEG* 36 1280, 37 1232; *MAMA* 6 154; *I.Stratonikeia* 1030.

25. E.g., *IEOG* 428; *TAM* 2.1 42; Overlaet, Macdonald, and Stein 2016; John Peters and Thiersch 1905, 37-75.

26. E.g., *SEG* 55 1113; *OGIS* 245; Adiego, Debord, and Varinlioğlu 2005.

27. E.g., *MAMA* 4 75.

28. E.g., the Saros Tablet, discussed in Chapter 1.

29. E.g., Sachs 1952.

30. E.g., E. Eshel 2010, 72, no. 66.

31. E.g., *IGIAC* 13-16 and 76.

32. E.g., note the demotic documents, dated by the Seleucid Era, from Antiochus IV's second invasion of Egypt in 168 BCE, P. Tebt. Suppl. 640 and P. Stan. Green. dem. 016, discussed by Fischer-Bovet 2014, 227-228.

33. On the domestication of modern national standard times, see, e.g., Edensor 2006.

34. McCarthy 2005, 56; I am grateful to John Ma for the reference.

35. For Sardis, *OGIS* 225, lines 23-24. For Antioch-by-Daphne, Joseph. *BJ* 7.55, discussed by de Giorgi 2016, 48 and 58. For Babylon, Hicks 2016, 20. For Uruk, Westh-Hansen 2017, 160-162; Hicks 2016; Lindström 2003; Rostovtzeff 1932. For Nippur, Gibson 1994. For Seleucia-on-the-Eulaeus (or Susa), *SEG* 7 15= *IEOG* 192= *IGIAC* 17, line 16. For Kedesh, Dura-Europus, and Seleucia-on-the-Tigris, see below.

36. Hicks 2016, 179.

37. Archives A and B of the "Great House" of Seleucia-on-the-Tigris (McDowell 1935); temple and domestic archives at Uruk (Lindström 2003; Rostovtzeff 1932); the site of the so-called "Parthian Fortress" at Nippur (Gibson 1994).

38. On the form, see Hicks 2016, 41-42 and 193.

39. See, e.g., Messina and Mollo 2004, S6-7271.

40. Some sense of the strong identities at play comes from the private archives of earlier Mesopotamia, which were usually placed in the innermost room of the house, under the floor of which family members were buried (see Pedersén 1998, 242).

41. See, e.g., Radner 2008. Instructions for distribution survive in a letter to Esarhaddon (*SAA* 18 163): "(Concerning) the message which the king, the lord, sent and the golden signet rings which they gave to me, saying, 'Send them to Babylon!' One signet ring I have already given to Mušēzib-Marduk, but the signet ring for Šuma-iddin, the prelate of Dēr, is still with me."

42. *OGIS* 56, line 23 (Canopus Decree). See also Pfeiffer 2004, 105–106; *P.Ryl* 9 14.13, discussed by Vittmann 1998, 2:501.

43. Note, for comparison, that private seal rings from Hellenistic Uruk were replaced on average every eight years (Wallenfels 1996, 119).

44. From the combination of seal impressions, Hicks estimates that about two dozen administrators were regularly active in Seleucia-on-the-Tigris's archive building (2016, 216).

45. Jeremiah 22:24, on Jehoiachin; Haggai 2:23, repurposing the figure for Zerubbabel (see O'Kennedy 2014, 532–535).

46. Platt 2006.

47. App. *Syr.* 56; Just. *Epit.* 15.4.3–6. Cf. Philip II's dream of sealing Olympias's womb in Plut. *Alex.* 2.4.

48. *BRM* 2 10 = S-3718. See Wallenfels 1996, 115–116; Mollo 1997, 99; McEwan 1982.

49. Note that 2 Macc. 4:10 aligns the reconfigured, Hellenized subjectivity of the renegade Jews with the sealing practices of the imperial archive: high priest Jason's construction of a gymnasium, enrollment of an *ephebeia*, and registration of Jerusalemites as Antiochians (see Chapter 6) are figured as the community's being moved beneath the "Greek stamp" ("πρὸς τὸν Ἑλληνικὸν χαρακτῆρα τοὺς ὁμοφύλους μετέστησε"). See also Arr. *Epict. diss.* 4.5.15–17 for a similar use of imperial engravings for discussions of personal character. Furthermore, it is possible that another conceptual frame was activated by imperial administrators wearing Era dates—the use of bodies to tell the time (Hannah 2009, 75–79; Bilfinger 1886, 10–15). We have good evidence that body parts could be "read" like timepieces. Thus, an epigram in the Palatine Anthology rather unkindly turns a face into a sundial—"If you put your nose to the sun and open your mouth wide you'll show the hour to all passers-by" (*Anth. Pal.* 11.418). Old and New Comedy have guests invited to dinner at the time when the footfall of their shadow reached ten, twelve, or twenty feet, with gluttons too quick to overestimate their own shadow's length (Ar. *Eccl.* 651–652; Ath. 1.8b–c = Eubulus F117; Ath. 6.243a = Men. F304). In the good old days, dinnertime was marked by a stomach's rumbling (Gell. *NA* 3.3.5). A late antique house from Antioch-by-Daphne shows a parasite, gazing expectantly at a sundial, with the inscription "τρεχέδειπνος" ("running to dinner"; Pamir and Sezgin 2016). I am grateful to Christopher Jones for this reference; see also A. Jones 2016, 19–21. Counting on one's fingers was, of course, ubiquitous (Menninger 1969, 34–35).

50. The following description is derived from Coqueugniot 2014, 2012, and 2011; Messina 2006, 57-58; Leriche 1996; F. Brown and Welles 1944; and Hopkins 1934.

51. The best precedents for this form are Mesopotamian, from the archive rooms of the sanctuary of Nabû at Dūr-Šarrūkin (Khorsabad) and the sanctuary next to the ziqqurat at Sippar (see Coqueugniot 2013, 48; Pedersén 1998, 155).

52. The following description is derived from Hicks 2016, 82-87; Invernizzi 1996; and Messina 2006.

53. For a reconstruction of the situation at the time of the archive's destruction, see Hicks 2016, 170-174.

54. Nicolaou 1997 and 1979.

55. Note that in the nonroyal archives found in the Skardhana quarter of Delos, the only dated impression (SP 10) was from a Phoenician-Greek bilingual seal giving the inscribed Seleucid Era year number 185 (L επρ', 128/127 BCE; see Boussac 1992, 16-17, and 1982, 444-446).

56. See Pedersén 1998.

57. This was a fairly standard location for public archives in the Greek world (e.g., Athens, Cyrene, Messene, and Pella), but most fell under the protection of a tutelary deity (see Coqueugniot 2013, 32-34 and 43-44).

58. Derrida 1995, 10.

59. Aneurin Ellis-Evans helpfully terms the archive building "an anti-stoa (in every sense)" (personal correspondence).

60. On French public archives, Pomian comments, "Le lien qu'elles maintiennent avec le passé est subordonné à leur orientation vers l'avenir" (1992, 225).

61. See, e.g., Appadurai 2003; Stoler 2002.

62. Derrida 1995, 27-28.

63. The neologism is from Koselleck 2002, 87.

64. Details are from Hicks 2016, 79-82; Hopkins 1972, 28-32; McDowell 1935, 4-14. See also Coqueugniot 2013, 140-142.

65. McDowell 1935, 11.

66. Note, however, the important caution of Hicks that some of these *bullae* could have been accidentally broken in the fire (2016, 180-181).

67. For the archaeological details, see Herbert and Berlin 2003.

68. *P. Cairo Zen.* I 59004; Westermann, Keyes, and Liebesny 1940, #61.

69. 1 Macc. 11:63-74.

70. The *tabularium* of Heraclea was destroyed during the Social War (Cicero, *Arch.* 4.8). Jerusalem's archives were burned at the beginning of the revolt (Joseph. *BJ* 2.427; see Goodman 1982, 418-419). Cyrene's *nomophylakeion* may have been destroyed during the Jewish revolts of 115-117 CE (Maddoli 1963-1964, 42). The destruction of the Dyme archive was associated with opposition to the Roman regime imposed on Achaea (Syll.³ 684 ll.6-7, 22).

71. Kallet-Marx (1995, 149-150) observes that the perpetrators of the Dyme fire were members of the political elite and that the motivation imputed to them by Q. Fabius Maximus is political, not financial.

72. For the use of a fetus or child corpse in malicious magic, see Wilburn 2012, 95–101. Note that Plin. *HN* 28.20.1 mentions cutting up stillborn babies for hostile purposes.
73. For such instances of bodily dismemberment, see, e.g., Judges 1:5–7, 1 Samuel 5:1–5 (of the Dagon statue), 2 Samuel 4:5–12, and 1 Kings 13:2.
74. Löwy 2005, 92–93.
75. Daniel 8:26, 12:4, and 9; 1 Enoch 89:71 and 90:20.
76. On *Kulturtechniken,* see Siegert 2015.
77. For the metrological reforms, see Doyen 2014, 278–297; Finkielsztejn 2007. Not a single lead weight can be dated to the Ptolemaic period; in all likelihood, they were removed or recast when the regime changed (Finkielsztejn 2014, 65n6).
78. Finkielsztejn 2010b and 1999; *SEG* 49 2068.
79. Finkielsztejn 2006; *SEG* 56 1907.
80. Perhaps Demetrius I's refounding of Staton's Tower (Caesarea) or Dora (see G. Cohen 2006, 201–203).
81. Seyrig 1950, 52–53. On occasion, the royal authority was far more explicit. E.g., a weight of 95/94 BCE bears on its obverse the name of Seleucus VI ("βασιλέως Σελεύκου"), an anchor, and the king's epithets ("Ἐπιφανοῦς Νικάτορος"). On its reverse are the Seleucid Era year date 218 ("ἔτους ηισ΄"), the name of the *agoranomos* ("ἀγορανομοῦντος Δημητρίου"), and the one-mina weight unit ("μνᾶ") (Weiss and Ehling 2006; *SEG* 56 1856).
82. Houghton, Lorber, and Hoover 2008, #1804.
83. See, e.g., Johnstone 2011, 49–54.
84. See, e.g., Finkielsztejn 2001, 17; *Sifrei Deuteronomy* 294 (discussed by Sperber 1977); [Aristotle] *Ath. Pol.* 51.2; *IG* II² 1013.
85. Ismard 2017, 40–42; Lang and Crosby 1964.
86. Monachov 2005; Garlan 1993.
87. Walthall 2011.
88. Giddens 1990, 80–88.
89. Kockelman and Bernstein 2012.
90. Daniel 2:8.
91. Newsom, locating this within the Babylonian tradition of apotropaic rituals, proposes that this 'iddānā' represents a defined period of time rather than time as duration (2014, 69–70).
92. On the "redeeming time" of Eph. 5:16, see Janzen 2012. 348–391. Sertorius, because of storms in the mountains, pays protection money to barbarians to secure his passage, quipping that he was buying time ("καιρὸν ὠνεῖσθαι"; Plut. *Sertorius* 6.6). Plutarch treats this as a witty apothegm rather than a common idiom. It is possible that time's commodification was thematized in Eubulus's lost comedy *Clepsydra*, in which a courtesan used a water clock to regulate her dalliances ("οὕτω δ᾽ ἐκλήθη αὕτη ἡ ἑταίρα, ἐπειδὴ πρὸς κλεψύδραν συνουσίαζεν ἕως κενωθείη," Ath. 13.567c–d).
93. Portier-Young 2011, 180.
94. For a study of the contemporary "buying time" idiom and its spread across the global South, see Mueller 2016.
95. For "assessed," see Newsom 2014, 178; Wolters 1991a, 169–170.

96. Dan. 5:25–28.

97. Old Greek, Theodotion, Josephus, the Vulgate, and Jerome give just three— *mᵉneʾ tᵉqel û)arsin*—or their Greek and Latin equivalents. But the number four is of central thematic relevance to the book of Daniel, signifying a completed totality from the statue of Nebuchadnezzar's dream in chapter 2 to the imperial beasts of chapter 7.

98. This was first observed by Clermont-Ganneau (1887) and is now generally accepted.

99. See Joseph. *AJ* 10.243 (explicating for a Greek-speaking audience): "μάνη· τούτῳ δὲ ἔλεγεν Ἑλλάδι γλώσσῃ σημαίνοιτο ἂν ἀριθμός, ὥσπερ τῆς ζωῆς σου τὸν χρόνον καὶ τῆς ἀρχῆς ἠρίθμηκεν ὁ θεὸς περισσεύειν ἔτι σοι βραχὺν χρόνον." Note that in the Hellenistic Levant, the standard weight was the mina (μνᾶ), appearing in its denominations in the inscriptions found on the weights (Finkielsztejn 2014, 68).

100. Note that Seleucus I appears as a master of weights in the letter accompanying his votive dedications to the temple of Apollo at Didyma (*OGIS* 214, lines 25–29): "τῶν δὲ ἀφεσταλμένων χρυσωμάτων καὶ ἀργυρωμάτων εἰς τὸ ἱερὸν ὑπογέγραφα ὑμῖν τὴν γραφήν, ἵνα εἰδῆτε καὶ τὰ γένη καὶ τὸν σταθμὸν ἑκάστου." There follows the list of votives, each with its weight in drachmas, minas, or talents.

101. Wolters 1991a, 173–174. For other wordplay in this chapter, see Wolters 1991b.

102. *b. Ta'an.* 21b, quoted in Newsom (2014, 176) and J. Collins (1993, 251).

103. Neither the Hebrew original nor the Greek translation survive, but complete translations from the Greek survive in Arabic, Armenian, Ethiopic, Georgian, Syriac, and Latin (which I used), and substantial fragments survive in Coptic. For a brief introduction to the text and its transmission, see Stone 2013, 2–8.

104. 4 Ezra 4:36–37. It is generally accepted that this passage is drawn from an earlier source document: this is the only appearance in the book of an angel other than Uriel, the term archangel is not used elsewhere in the apocalypse, and the passage is framed as a citation. On the rhetorical calculation of end-times here, see Chapter 5 and Yarbro Collins (1996, 64).

105. On the influence of state bureaucracy for modeling Syro-Palestinian divinity, see, e.g., Handy 1994.

106. Frost comments: "That a kingdom should be weighed in a balance and found wanting is the very stuff of prophecy; that it should be divided is a pronouncement of divine judgement in the very manner of an Isaiah; but that it should be *numbered* is the thought of an apocalyptist alone" (1952, 186).

107. On the "publication clauses" of royal letters, all sent to officials or groups of people, rather than *poleis*, see Bencivenni 2014, 148–149, and 2010, 153–161.

108. See, e.g., Virgilio 2011, 34–37; Bencivenni 2014, 159–160.

109. See, e.g., Ma #9, 10, 23, 24, 29; *I.Laodikeia am Lykos* 1 = *SEG* 47 1739; *I.Stratonikeia* 1030; *IGIAC* 13, 14, 17, and 20; *IGLS* 1261; *TAM* 5.2 881; *MAMA* 6 154.

110. *SEG* 33 1184 = Ma #23 = *Amyzon* 15b.

111. See, e.g., Boffo 2013, 223–225, especially 224n63; Savalli-Lestrade 2010, 133–135; Buraselis 2010, 426–430; Capdetrey 2007, 215–216; Ma 2002, 27 and 147–148; Leschhorn 1993, 34. Note that Plut. *Demetr.* takes as diagnostic of a

shameful subordination the Athenians' replacement of the civic archon with the new priest of Antigonus Monophthalmus and Demetrius Poliorcetes as the annual eponym for dating public and private documents (10.4 and 46.2).

112. See, e.g., a decree from the neighboring city of Telmessus, subordinate to the Ptolemies: "[βα]σιλεύοντος Πτολεμαίου τοῦ Πτολε|[μα]ίου καὶ Ἀρσινόης θεῶν ἀδελφῶν, ἔτους | [ἑβδ]όμου μηνὸς Δύστ[ρ]ου, ἐφ᾽ ἱερέως Θεοδό|[το]υ τοῦ Ἡρακλείδου, δευτέραι, ἐκκλησίας κυ|[ρί]ας γενομένης ἔδοξε Τελμησσέων | [τ]ῆι πόλει" (OGIS 55, lines 2-7).

113. SEG 7 17 = IGIAC 13, line 1.

114. OGIS 175, line 12.

115. On the taxonomy and terminology, see Capdetrey 2006, 105-110, and 2007, 335-341. Ma #1, 2, 4, and 5; Yon 2015, 92-101; Isaac 1991; SEG 41 1556; IGIAC 76 and 422; RC 44; IEOG 422; 2 Macc. 11:16-21, 27-33, and 34-38. We also find in this category RC 45, a decree from Seleucia-in-Pieria.

116. See Bertrand 2006, 90-92.

117. Capdetrey 2007, 339-340; Bencivenni 2014, 158-164, and 2011; Ceccarelli 2013, 298-300. Note that royal letters could be publicly recited. See Ceccarelli (2013, 308) on the readings at Telmessus (SEG 28 1224) and Miletus (Milet I, 3 139b); and AD -140 A Rev. 5-6, -132 D$_2$ Rev. 15, -124 B Rev. 17, -119 A Rev. 19, -87 C Rev. 31 for their recitation before the politai in the theater of Babylon.

118. SEG 57 1851. For readings of the dates, see Cotton and Wörrle 2007, 193-194; Piejko 1991; Bertrand 1982; Fischer 1979; Landau 1966.

119. SEG 57 1851, lines 2-3: "σύνταξ[ον ἀνα]γράψαν[τας] ἐν στήλαις λιθ[ίναις - - - - - -] | [τὰ]ς ἐπισ[τ]ολὰ[ς ἀνα]θεῖναι ἐν [ταῖ]ς ὑπαρχούσαις [σοι κώμαις."

120. See Fischer 1979, 137.

121. Recall the characterization of Seleucus I as a master of time, explored in Chapter 1. For other cases, see Ma #2 (Sardian decree and letter of Laodice); Joseph. AJ 12.258-264 (on which, see Bertrand 2006, 97-103; Bickerman 1937); perhaps also Isaac 1991. For the incorporation of royal instructions into civic archives, see Boffo 2013.

122. See, e.g., the classic statements of Giddens (1990, 19-21) and Thompson (1967), though note Sauter (2007) on time discipline without industrialization.

3. DYNASTIC TIME

1. Kantorowicz 1957. See also Kahn 2009; Landauer 1994.

2. Alexander's monarchy already contains hints of such separation. See the distinction between office and officeholder in Diod. Sic. 17.114.2: "ἐπεφθέγξατο Κρατερὸν μὲν γὰρ εἶναι φιλοβασιλέα, Ἡφαιστίωνα δὲ φιλαλέξανδρον."

3. See Kantorowicz 1957, 409-419. Perhaps better is the hybrid form "the king is dead, the kingship lives for ever" (Fortes 1967, 9).

4. I borrow this phrase from Bazzana (2015, 272). This was not a necessary practice; note that the Babylon King List (BM 35603) gave both Seleucid Era dates and abbreviated royal names. Alternatively, the scribe of the Saros

Tablet could have used ki.min, the Sumerian logogram signifying the repetition of the previous line, like our "ditto."

5. Daniel 7:7-8.
6. Borsippa Cylinder (BM 36277) 1.13–14.
7. Haubold 2013, 136; da Riva 2008, 64 and 68.
8. Schaudig 2001, 510 (Stele 3.2 I.31).
9. *Uruk King List* (IM 65066) Obv. 7.
10. Berossus *BNJ* 680 F7c = Euseb. (Arm.), *Chron.* 13, 18–15, 4 (Karst); F8a = Joseph, *AJ.* 10.219; F9a = Joseph, *Ap.* 1.146.
11. Kosmin 2014b, 189–190; Haubold 2013, 136–137.
12. Polyb. 11.34.11–12. See also Kosmin 2014a, 35–36.
13. A clay *bulla* from Seleucia-on-the-Tigris suggests the presence in the city of the royal court in 211 BCE. Stamp Alk 25 reads, exceptionally, "of the king's house" (ἀλικῆς | Σελευκείας | βρ´ | βασιλισης οἴκου) (see Hicks 2016, 113).
14. For Maresha as a Seleucid outpost, see 2 Macc. 12:35. For Hasmonean aggression, see 1 Macc. 5:66; Joseph. *AJ* 13.257.
15. Korzakova 2010, #1, 2, 3, 33, 34, and 37; Finkielsztejn 1998a, 34–38. The reversed numbering is characterized by Finkielsztejn as "noticeable, but not fully understandable" (2010a, 185).
16. On the symbolic force of the Macedonian shield, see Liampi 1990, 160–168.
17. On such *gematria*, playing on the phonetic reading of alphabetic numerals, see, e.g., Berossus (*BNJ* 680 F1b = Syncellus *Chron.* 29 M, where Tiamat's alternative name, *Omorka* (Ὁμόρκα), is "equal in numerical value (ἰσόψηφον)" to the moon (σελήνη): according to Greek alphabetical numbers (see Table 4), each adds up to 301 (see Dillery 2015, 236). See, also, the "isopsephic" poems of Leonides of Alexandria (Luz 2010, 247–250 and 254–270).
18. Proposed by Finkielsztejn 2010a, 186. Did the material of the market weight—lead, as used in scratched curses and imprecations—encourage such aggression?
19. The bibliography is vast. See, e.g., Frazer 1890; Hocart 1927; Fortes 1967; Feeley-Harnik 1985; Sahlins 1985. The logic culminates in the requirement of royal regicide, although the enacted reality of this has been questioned. See, e.g., Evans-Pritchard 1948, 20–21 and 30–35.
20. Eliade 2005, 80.
21. Concerning Eliade's enthusiastic participation in the Romanian Iron Guard, see Ginzburg 2010, 316–320.
22. *OGIS* 90, lines 18, 26–27, and 12–14. On the Ptolemaic amnesty decrees and their temporal ideology of resetting socioeconomic conditions, see Bazzana 2015, 171–176.
23. Polyb. 18.55.3.
24. Polyb. 25.3.1–3; judgment at Polyb. 25.3.5. Note also that Herodotus observed that acceding Spartan and Persian kings alike canceled debts and forgave tribute (Hdt. 6.59).
25. See Tuplin 2009, 120.

26. This is the force of the letters of Antiochus V, ending his father's persecution, in 2 Macc. 11:22-26 and 27-33.
27. Pace Honigman 2014, 86-87.
28. See, e.g., Tadmor 1999, 56; Liverani 1990, 56-57.
29. *OGIS* 54, lines 10-13: "καὶ ἐλεφάντων Τρωγλοδυτικῶν καὶ Αἰθιοπικῶν, οὓς ὅ τε πατὴρ αὐτοῦ καὶ αὐτὸς πρῶτο(ι) ἐκ τῶν χωρῶν τούτων ἐθήρευσαν καὶ καταγαγόντες εἰς Αἴγυπτον κατεσκεύασαν πρὸς τὴν πολεμικὴν χρείαν."
30. *Anth. Pal.* 16.6 and 9.518.
31. E.g., *I.Erythrai* 31 = *RC* 15, lines 9-10 and 23-24: "ἔτι δὲ καὶ τὴμ προαγωγὴν ἐν ἧι γέγονεν ἡ πόλις ἐπὶ τῶν πρότερον βασιλευσάντων . . . οἱ ἡμέτεροι πρόγο[νοι] ἔσπευδον ἀεί ποτε περὶ αὐτῆς"; *RC* 22 = *OGIS* 227, lines 2-3: "τῶμ προγόνων ἡμῶν καὶ τοῦ πατρὸς πολλὰς καὶ μεγάλας εὐεργεσίας κατατεθειμένων εἰς τὴν ὑμετέραμ πόλιν"; *IEOG* 422 = *SEG* 35 1476, lines 13-14/19-20: "κατὰ τ[ὴν τοῦ βασ]ιλέως αἵρεσιν καὶ τῶν προ[γόνων] αὐτοῦ"; *SEG* 37 859 = Ma #31 B II ll.8-9: "ἀνακεκομισμένων ἡμῶν τῶι βασιλεῖ τὴν πόλιν ἐξ ἀρχῆς ὑπάρχουσαν τοῖς προγόνοις αὐτοῦ"; Polyb. 18.51.5: "κατὰ δὲ τοὺς τῶν αὐτοῦ προγόνων περισπασμοὺς"; Joseph. *AJ* 12.150: "μαρτυρουμένους δ' αὐτοὺς ὑπὸ τῶν προγόνων"; Liv. 35.16.6: "*bello superatas a maioribus*"; Diod. Sic. 34/35.1.3: "ὑπέμνησαν δὲ αὐτὸν καὶ περὶ τοῦ προγεγονόμενου μίσους τοῖς προγόνοις πρὸς τοῦτο τὸ ἔθνος"; App. *Syr.* 6: "καὶ Θρᾴκην ἐᾶν ἀεὶ τῶν προγόνων αὐτοῦ γενομένην."
32. E.g., *SEG* 41 1004 = Ma #19 B l.11: "καθά]περ καὶ οἱ πατέρες καὶ [αὐτός . . .]."
33. E.g., *RC* 22 = *OGIS* 227, lines 2-3: "τῶμ προγόνων ἡμῶν καὶ τοῦ πατρὸς πολλὰς καὶ μεγάλας εὐεργεσίας κατατεθειμένων εἰς τὴν ὑμετέραμ πόλιν"; *RC* 45 B, lines 4-5: "μετὰ πάσης εὐνοίας τῶι τε πατρὶ ἡμῶν καὶ τῶι ἀδελφῶι καὶ ἡμῖν"; Polyb. 28.20.8: "ἐπὶ τὴν τελευταίαν κατὰ πόλεμον Ἀντιόχου τοῦ πατρὸς ἔγκτησιν"; 2 Macc. 9:23: "θεωρῶν δὲ ὅτι καὶ ὁ πατήρ."
34. E.g., *SEG* 37 1010 = Ma #4, lines 40-41: "καθὰ καὶ ἐπὶ τοῦ πάππου ἡμῶν"; *IGIAC* 80bis, line 6: "ὑπό τε τοῦ παπποῦ ἡμῶν."
35. E.g., *RC* 44, lines 1-3: ["ἀ]δελφῶι γεγενημένον ἐν τιμ[ῆι κα]ὶ πίστ[ει καὶ] τῆς εἰς ἡμᾶς καὶ τὰ πράγματ[α] αἱρέσεως"; *RC* 45 B, lines 4-5: "μετὰ πάσης εὐνοίας τῶι τε πατρὶ ἡμῶν καὶ τῶι ἀδελφῶι καὶ ἡμῖν."
36. *IGIAC* 80bis.
37. *SEG* 37 1010 = Ma #4, lines 40-41.
38. On kinship terminology generating an "unperishing present," see Geertz 1973, 378-379.
39. E.g., *OGIS* 222, line 20: "[ἀκολουθήσει τῆι τ]ῶν προγόνων αἱρέσει"; *OGIS* 234 = *F.Delphes* III.4.163 = Ma #16, line 22: "κὰτ τὰν τῶν προγόνων ὑφάγησιν"; *SEG* 41 1003 = Ma #18 II, line 95: "[τῶν ἐ]πωνύμων πόλεων τῶν τοῦ βασιλέως προγόνων"; Ma #26 B I, lines 10-11: "προγονικὴν αἵρεσιν διατηροῦντος εἰς πάντας [το]ὺς Ἕλληνας"; *SEG* 41 1556, lines 9-11: "τῶ]ι πάππωι ἐν τοῖς κατὰ τὴν ναυτικὴν χρε[ίαν . . . τὸν πατ]έρα σου προάγον εἰς τοὺς κατ' Αἴγ[υπτον τόπους]." The so-called Lehmann Tablet (MMA 86.11.299 Obv. 3-4) has the *šatammu* of the Esagil temple of Babylon announce to the sanctuary assembly that Antiochus II continued the policy of his father Antiochus I and grandfather Seleucus I: "King Antiochus made a transaction *about fields,*

namely, that everything which his father Antiochus and his grandfather Seleucus, the kings, had done . . . (*mim-ma šá ¹An-ti-'-uk-su* ˡᵘad-šú u ¹Se-lu-ku ˡᵘad.ad-šú lugal*ᵐᵉˢ i-pu-š[ú-x]*)." Conversely, Daniel 11:24 turned such arguments by dynastic precedent against Antiochus IV, accusing him of engaging in behavior which "his fathers and his fathers' fathers have never done *(lō'-ʿāśû ᵃbōṭāyw waᵃbōṭ ᵃbōṭāyw)*" (see Chapter 5).

40. *OGIS* 223 = *RC* 15 = *I.Erythrai* 31, lines 21–27.
41. See Welles's commentary at *RC* 15 (page 82).
42. Cf. Polyb. 10.28.3, necessarily deriving from a Seleucid court historian, where information on the privileges granted by the (untitled) Achaemenid kings toward the upkeepers of the Hyrcanian *qanats* is presented as an argument of the local inhabitants ("περὶ δὲ τούτων ἀληθὴς παραδίδοται λόγος διὰ τῶν ἐγχωρίων").
43. 2 Macc. 9:23.
44. Joseph. *AJ* 12.150. The authenticity of this letter has been much discussed (see, e.g., G. Cohen 1995, 212–213; Marcus 1943, 473–481 and 494–496).
45. Diod. Sic. 34/35.1.3–4.
46. See, e.g., Montiglio (2005, 139) on διέξειμι in Herodotus.
47. Polyb. 5.83.6. Note that this is paralleled by the opposing Ptolemy.
48. Kosmin 2014a, 31–76.
49. Primo 2009, 24–28.
50. *BNJ* 163 T1 = Suda s.v. Σιμωνίδης· Μάγνης <ἀπὸ> Σιπύλου, discussed by Primo 2009, 87–88.
51. *BNJ* 45 T3 = Ath. 4.155a–b; T4a = Polyb. 18.47.1–4; T4b = Polyb. 18.49.2–18.50.3; T5 = Liv. 34.57.1–6. See also Olshausen 1974, 191–192.
52. *BNJ* 45 F3 = Strabo 13.1.27, possibly in the context of Hegesianax's *Troica*.
53. *BNJ* 162 T1 = Diog. Laert. 5.83.
54. Primo 2009, 104–105.
55. *BNJ* 164 T1 = Ath. 15.53.697d: "Μνησιπτολέμου . . . ποτε τοῦ ἱστοριογράφου τοῦ παρὰ Ἀντιόχῳ τῷ προσαγορευθέντι Μεγάλῳ πλεῖστον ἰσχύσαντος υἱὸν γενέσθαι Σέλευκον."
56. *BNJ* 164 F1 = Ath. 10.40.432b–c, with the commentary of Michel Cottier, contra Primo 2009: 90. A Delian proxeny decree (*BNJ* 164 T3 = *IG* XI.4 697) for "the historian Mnesiptolemus" may indicate his political importance at the Seleucid court.
57. *BNJ* 853 F3 = Ath. 3.98.124d–f, discussed by Primo 2009, 106.
58. *BNJ* 165 F1 = Euseb. *Praep. evang.* 9.35.1 (452b–c). See Bar-Kochva 2010, 458–468.
59. See Kosmin 2014a, 211–218.
60. On the status of Berossus, see Chapter 4. The Alexander historian Aristos of Salamis (*BNJ* 143) is likely a late Hellenistic writer, not the lover of Antiochus II (Ath. 10.51.438d), contra Meißner 1992, 465.
61. See *I.Erythrai* 205, lines 74–76, discussed by Kosmin 2014b, 178–180.
62. See, e.g., van Nuffelen 2004; Müller 2000, 532–536.
63. For Antioch-in-Persis, *I.Magnesia* 61 = *OGIS* 233 = *IGIAC* 53 = *IEOG* 252. For Seleucia-on-the-Tigris, *SEG* 51 1914 = *IEOG* 76; discussed by van Nuffelen 2001. For Seleucia-in-Pieria, *OGIS* 245 = *SEG* 35 1521. For Dura-Europus, *PDura*

25. For Scythopolis, *SEG* 8 33. For Samaria, *SEG* 8 96. For Teos, *OGIS* 246. On the local variations, see Buraselis 2010, 429–430.

64. Contra Rostovtzeff 1935, 62–65.

65. *OGIS* 245 = *SEG* 35 1521, lines 11–19.

66. All this is well known (see, e.g., A. Stewart 1993, 229–262). Theoc. *Id.* 17.26–27 emphatically aligns the lineages of Alexander and Ptolemy I: "ἄμφω γὰρ πρόγονός σφιν ὁ καρτερὸς Ἡρακλείδας/ἀμφότεροι δ᾽ἀριθμεῦνται ἐς ἔσχατον Ἡρακλῆα." I was fortunate first to read Theocritus's encomium under the guidance of Albert Henrichs (see Kosmin 2017a). For the purported kinship through Philip II, see Satyros *BNJ* 631 F1 = Theophilus, *ad Autolycum* 2.7; Curt. 9.33.22; Paus. 1.6.2; Ps.-Callisthenes 3.32; Aelian F285; with Collins 1997.

67. In Satyros's reconstruction (*BNJ* 631 F1 = Theophilus, *ad Autolycum* 2.7). For alternative genealogies, see the *BNJ* commentary.

68. Moyer 2011, 87–88; Jenni 1986, 4–6.

69. See, e.g., Jasnow 1997, 100–103.

70. Gorre 2009.

71. Plut. *Pyrrh.* 26.6. See Andronicos 1984, 55–62; Lane Fox 2011, 505–506.

72. See Wilson 2015; Hintzen-Bohlen 1990, 138–139: Walbank 1967, 258n3.

73. Theopompus *BNJ* 115 T31 = Photius, *Bibliotheca* 121a35: "διὸ καὶ Φίλιππος, ὁ πρὸς ῾Ρωμαίους πολεμήσας, ἐξελὼν ταύτας καὶ τὰς Φιλίππου συνταξάμενος πράξεις, αἳ σκοπός εἰσι Θεοπόμπῳ εἰς ις᾽ βίβλους μόνας." See Walbank 1967, 258–259.

74. *SEG* 56 625, lines 8–11: "κατὰ τὴν ἱστορίαν, ἣν ὁ βασιλεὺς εἰσηγεῖται περὶ τοῦ πράγματος," discussed by Chaniotis 2010, 304 (#71); Hatzopoulos 2007, 694–695 (#358).

75. See, e.g., Müller 2009, 77–80; Kosmetatou 2000, 45–46, and 1995. For Heraclid ancestry, see Nicander's *Hymn to Attalus* F104 (Schneider): "Τευθρανίδης, ὦ κλῆρον ἀεὶ πατρώιον ἴσχων,/κέκλυθι μηδ᾽ ἄμνηστον ἀπ᾽ οὔατος ὕμνον ἐρύξῃς,᾽Ἄτταλ,᾽ ἐπεί σεο ῥίζαν ἐπέκλυον Ἡρακλῆος/ἐξέτι Λυσιδίκης τε περίφρονος, ἣν Πελοπηίς/Ἱπποδάμη ἐφύτευσεν ὅτ᾽ Ἄπιδος ἤρατο τιμήν."

76. L. Robert 1973, 478–485; Hintzen-Bohlen 1990, 140; Schalles 1985, 127–135.

77. Paus. 1.25.2. See also A. Stewart 2004, 181–236.

78. For a description of the site, see Sanders 1996. For identifications of Commagenian Antiochus I's ancestors, see Dörner 1996; Brijder 2014a and 2014b.

79. See, e.g., Holt 1984. It is unlikely that these coins were displaying an imputed kinship, as they depicted the different descent lines of kings who had fought and killed off one another.

80. On the Hegira era and its making of a preceding "age of ignorance" see, e.g., Nixon 2004, 431; Trombley 2000.

81. App. *Syr.* 3; Strabo 11.7.2 and 11.9.2; Diod. Sic. 31.19a; Polyb. 18.51.5.

82. Cf. Polybius's and Livy's preservation of the programmatic vocabulary of Seleucid reconquest as discussed in Ma 2002, 29–30.

83. Diod. Sic. 31.19a.

84. Polyb. 18.51.5. Similarly, see Diod. Sic. 31.19.5 on the independence of Ariarathes of Cappadocia ("Ἀντιγόνου δὲ καὶ Σελεύκου περισπωμένων").

85. 1 Macc. 15:33. See Kosmin 2014a, 254–255.

86. Strabo 14.5.2. The arguments may well derive from Posidonius (see Engels 2011; Strasburger 1965, 42–43 and 50). For a reconstruction of Posidonius's account of Seleucid decline, see Engels 2011, 182–185; Malitz 1983, 257–302.
87. OGIS 219 = I.Ilion 32 = SEG 49 1752, lines 2–12.
88. Memnon BNJ 434 F1 9.1.
89. Ma aptly described it as "the Great Idea to which Antiochos devoted his life" (2002, 32).
90. 1 Macc. 15:3.
91. Gabrielsen 2008, 23. See also Portier-Young 2011, 136–138.
92. For the thesis and its interpretation, see Löwy 2005, 23–29.
93. 1 Macc. 13:31–32; Diod. Sic. 33.28; App. *Syr.* 68; Liv. *Per.* 55; Just. *Epit.* 36.1.7; Oros. 5.4.18.
94. For a narrative of Diodotus Tryphon's career and self-presentation, see Chrubasik 2016, 154–161; Ehling 2008, 164–192.
95. Houghton, Lorber, and Hoover (2008, 338) contrast this with the victorious or divinizing cult names Nicator, Soter, Callinicus, Megas, Euergetes, and Theos Epiphanes and the dynastic emphasis of the epithets Philopator, Eupator, Theopator, and Philadelphus.
96. E.g., 1 Enoch 90:9: "And I saw till a great horn of one of those sheep branched forth, and their eyes were opened." For a discussion of the horns of Daniel's visions and of the Apocalypse of Animals' savior, see Chapter 5.
97. Ariel, Sharon, Gunneweg, and Perlman 1985.
98. Kushnir-Stein 1997, 90: "L βξρ'," 162 SE (151/150 BCE).
99. Gera 1985; SEG 35 1535. Obverse: "Τρύφωνο(ς) | νίκη." Reverse: "ꓶ L [bird] ε' | Δω(ριτῶν) π(όλεως)· Ῥοῦ | γεῦσαι" (confirmed by personal observation). The alternative reading of Fischer 1992, based on Diodotus's supposed devotion to Zeus, is doubtful (see Sève 1993). For the siege, see 1 Macc. 15:10–14 and 25; Joseph. *AJ* 13.7.2.
100. For slingshot bullet inscriptions, see Ma 2010, 167–173; Pritchett 1991, 43–53; Robinson 1941, 418–439.
101. The Greek translator of the Hebrew original probably took *hkhn* as ἀρχιερεύς and *hgdl* as ὁ μέγας (see Babota 2014, 243n75).
102. 1 Macc. 13:41–42.
103. 1 Macc. 14:43.
104. *Megillat Ta'anit* (ms Parma) 18. For introduction to the text, see Goldberg 2016, 193; Noam 2006; Lichtenstein 1931–1932; Zeitlin 1918.
105. SEG 57 1838 E, lines 11–12, discussed by Gera 2009, 137–138; C. Jones 2009, 103–104.
106. See Sievers 1990, 110–112; J. Goldstein 1976, 478–479. This does not mean, at least initially, the removal of the Seleucid Era year number, which was retained for the Simon decree (see 1 Macc. 14:27). On dating by the provincial high priesthood, see Buraselis 2010, 426–430.
107. On prophetic allusion in 1 Maccabees, see J. Goldstein 1987, 71–81.
108. See, e.g., Aesch. *Pers.* 176–199; Nabopolassar 4, lines 18–21 (Langdon). The idiom is discussed by Machinist (1983, 734–735).
109. See, e.g., Isaiah 9:4, 10:24–27, and 14:24–27; Nahum 1:13; Jeremiah 27:2 and 30:4–8; and Ezekiel 34:27–28—all perhaps deriving ultimately from Leviticus

26:13 ("I the Lord am your God who brought you out from the land of the Egyptians to be their slaves no more, who broke the bars of your yoke and made you walk erect").

110. Jeremiah 27:2.

111. Jeremiah 27:4-6: "This is what the Lord Almighty, the God of Israel says: . . . With My great power and outstretched arm I made the earth and its people and the animals that are on it, and I give it to anyone I please. Now I will give all your countries into the hands of My servant Nebuchadnezzar, king of Babylon; I will make even the wild animals subject to him."

112. Jeremiah 30:8, from the so-called Book of Consolation. For a lexical analysis of this passage, see Becking 2004, 135-164. The prophet continues: "Instead, they shall serve the Lord their God and David the king, whom I will raise up for them" (Jeremiah 30:9)—a provocative implication in the Hasmonean context—to exchange gentile vassalage for Israel's legitimate service to God and his monarchic representative. Note that the responsibility of Israel's deliverance, assigned to the Hasmonean family in 1 Macc. 5:62 ("that seed of men to whom had been granted the deliverance of Israel through their agency"), directly echoes God's appointment of David according to 2 Samuel 3:18 ("Deliverance of My people of Israel is through the agency of David My servant, *bᵉyaḏ dāwiḏ ʿaḇdi hôšiaʿ ʾeṯ-ʿammi yiśrāʾel*"; see J. Goldstein 1987, 80.

113. See, e.g., 1 Macc. 14:12: "καὶ ἐκάθισεν ἕκαστος ὑπὸ τὴν ἄμπελον αὐτοῦ καὶ τὴν συκῆν αὐτοῦ, καὶ οὐκ ἦν ὁ ἐκφοβῶν αὐτούς," which alludes to Zechariah 3:10 ("In that day—declares the Lord of Hosts—you will be inviting each other to the shade of vines and fig trees") and Micah 4:4 ("In the days to come . . . every man shall sit under his grapevine or fig tree with no one to disturb him"). Carr aptly describes the portrayal of Simon's rule in 1 Maccabees as "virtually messianic in effect" (2011, 156). See also Philonenko 1992, 95-96; Boccaccini 1991, 157.

114. It was employed in 1 Maccabees to fix events of panimperial significance, such as monarchic accessions and deaths, major military expeditions, the arrivals of dynastic competitors in Syria, and interdynastic marriage (1 Macc. 1:10, 20, 3:37, 6:16, 20, 10:57, 67, 14:1, and 15:10). The only events dated precisely (by day, month, and year) are the desecration and rededication of the Jerusalem Temple, since it is important to the author's chronology that these occurred at the same season and day (1 Macc. 1:54 and 4:52-54). For other significant Jerusalem events, some mentioned in the *Megillat Taʿanit*, see 1 Macc. 6:20, 9:3, 9:54, 10:1, and 13:51.

115. 2 Macc. 10:3. See Honigman 2014, 136.

116. The significance of this year number is shown by Josephus's two accounts of the same episode. Joseph. *BJ* 1.53 reports that Simon liberated the Jews after 170 years of Macedonian rule ("τῆς Μακεδόνων ἐπικρατείας μετὰ ἑκατὸν καὶ ἑβδομήκοντα ἔτη Ἰουδαίους ἀπαλλάττει"). This was not, of course accurate—the number uses the Seleucid Era date for the full succession of Macedonian rulers who had commanded the region since the Battle of Issus in 333 BCE, including Alexander, Perdiccas, Antigonus, and the first five Ptolemies. The error was corrected in Joseph. *AJ* 13.213 in a way that fully brings to the surface the origin-privileging quality of the Seleucids' dynastic tem-

porality, even for a province that had been dominated by the Seleucids only since Antiochus III at the beginning of the second century BCE: "ἡ δὲ ἐλευθερία καὶ τὸ ἀνείσφορον τοῖς Ἰουδαίοις μετὰ ἑβδομήκοντα καὶ ἑκατὸν ἔτη τῶν Συρίας βασιλέων ἐξ οὗ χρόνου Σέλευκος ὁ Νικάτωρ ἐπικληθεὶς κατέσχεν Συρίαν ὑπῆρξεν."

117. Jeremiah 25:11, 12, and 29:10 (subsequently cited by Zechariah 1:12 and 7:5, 2 Chronicles 36:21 and 22, and Daniel 9:2 and 24–27). See Applegate 1997.

118. Daniel 9:2, 24–27. See Fröhlich 1996, 47; Dimant 1993; Wacholder 1975.

119. Apocalypse of Animals (1 Enoch 85–90). See Gore-Jones 2015, 276; Olson 2013, 99–102.

120. See, e.g., Olson 2013, 99–103; Berner 2006; Werman 2006.

121. Alexander Jannaeus Type 5. See Dąbrowa 2010, 137; Shachar 2004, 7; Naveh 1968; Kindler 1968.

122. Joseph. AJ 13.356.

123. SEG 50 1479 = I.Gadara 1, discussed by Wörrle 2000.

124. Eumenes I retained the use of the Seleucid Era into the 260s BCE (OGIS 266). See, e.g, Chrubasik 2013, 94. On Attalus I's self-coronation, see Polyb. 18.41.7; Strabo 13.4.2; Paus. 1.25.2, discussed by Chrubasik 2016, 26–34, and 2013, 95–96; Virgilio 1993, 30–31; R. Allen 1983, 195–199; Bickerman 1944, 76–77. For Attalid dating on inscriptions, brick stamps, and coins, see, e.g., OGIS 268 (see L. Robert 1962, 36n6); OGIS 314; I.Perg #652–729; Kleiner 1972. For Artaxias I's inscriptions, see Chapter 6.

125. Lorber 2015, 66–67; Duyrat 2005, 225–231.

126. Meadows 2009. Note that all attested Pergean year dates (from α' to λγ') follow the standard, non-Seleucid, reverse alphabetical order (see McIntyre 2007).

127. Hoover 2009, 293–297; Kushnir-Stein 2005a; Bickerman 1968, 72–73; Seyrig 1950.

128. See Kushnir-Stein 2005a, 159.

129. See, e.g., Ziegler 1993 on the eras of Cilicia (Mopsuestia, Epiphania, Alexandria-by-Issus, Mallos, and Soloi), which originated either in Lucullus's victory over Tigranes and the restoration of citizens from forced settlement at Tigranocerta, or in Pompey's pirate settlement. Perhaps the footprint of the Seleucid temporal regime can be seen in the fact that the densest concentration of these Roman imperial era dates came from the territories formerly under the scepter of the Syrian kings.

130. It is perhaps for this reason that the Roman republic introduced an era system for conquered Macedonia, following Metellus's victory over Andriscus-Philip VI, in 148 BCE, implying a liberation from the national monarch they had defeated. On the Macedonian Era, see Tod 1918/1919 and 1919/1920–1920/1921.

131. Coloru 2009, 159–160; Assar 2003, 176; Lerner 1999, 13–29; Bickerman 1983, 782; Bivar 1983, 28–29. The Arsacid Era may have been introduced retrospectively; it does not seem to have been employed for Parthian administration before the reign of Phraates I or Mithridates I.

132. Overtoom 2016, 993–994; Olbrycht 2013, 64–65; Coloru 2009, 160; Lerner 1999, 17; Brodersen 1986; Wolski 1956–1957, 52. Other, less likely, suggestions include

the secession of Andragoras, Seleucid satrap of Parthia (Bivar 1983, 29, and 1969, 31); the insurrection of the Parni (Assar 2003, 176); and the coronation of Tiridates, Arsaces I's supposed brother (Gardiner-Garden 1987, 13-14 and 18; Tarn 1951, 65; Bickerman 1944, 79-83, interpreting Nisa ostracon #1760).

133.　Isidore of Charax *BNJ* 781 F2.11. For the connection between sacred fire and legitimate Persian kingship, see Diod. Sic. 17.114.4-5; Boyce 1979, 87.

134.　See Overtoom 2016, 993-995; Wiesehöfer 1996, 132. Arsaces I and his Parni conquered Parthia only in the later 230s BCE, following the rebellion of the Seleucid satrap Andragoras. The nomad king's successful resistance to Seleucus II's *anabasis* of reconquest was, according to Just. *Epit.* 41.4.10, "ever since commemorated as the beginning of their independence *(quem diem Parthi exinde sollemnem velut initium libertatis observant)*." It is possible that Justin, Pompeius Trogus, or his source here erroneously conflates the Arsacid Era epoch with the later *defectio Parthorum*. Alternatively (reconciling the traditions with Wolski 1959, 237), perhaps this victory was the occasion for establishing the year count retrospectively.

135.　Just. *Epit.* 41.5.5-6: *"Arsaces . . . decedit, cuius memoriae hunc honorem Parthi tribuerunt, ut omnes exinde reges suos Arsacis nomine nuncupent."*

136.　For the political chronology of these post-Seleucid kingdoms of Central Asia and northwestern India, see E. Errington and Curtis 2007, 43-71.

137.　Pace Coloru 2009, 169. On Diodotus's graduated independence, see Capdetrey 2007, 124-126.

138.　*IGIAC* 93; Clarysse and Thompson 2007; Rapin 2010, 241-242; Coloru 2009, 198. The fragmentary parchment refers to the employment of Scythian mercenaries, a previously unknown Graeco-Bactrian settlement or colony of Amphipolis, king Antimachus I, and the year thirty—neither king Antimachus I nor II approached a reign of thirty years. For Antiochus III's recognition of king Euthydemus and his son, see Polyb. 11.34.9-10. Note that the appointment of Euthydemus's son, Demetrius, as coregent may have encouraged, as in the case of Antiochus I, the dynastic temporality of a continuing year count.

139.　*IGIAC* 92; Rea, Senior, and Hollis 1994.

140.　It was employed in its 279th year at Surkh Kotal (SK2) and Dasht-e Nawur (DN1 = *IEOG* 319ᵃ) by the Kushan king, the grandfather of Kanishka I (see below), Vima Taktu. The occasion of its epoch is unclear. If inaugurated by Antimachus I, it may have marked his establishment of a joint monarchy with Eumenes and Antimachus II. Other proposals include that it was inaugurated by kings Demetrius I, Agathocles (Bopearachchi 2007, 47-50, and 2008, 49 and 52), and Eucratides I (E. Errington and Curtis 2007, 55; Cribb 2005, 214; Bernard 1985, 99-103).

141.　Errington and Curtis 2007, 51 and 53.

142.　This was already the case on a receipt from the Aï Khanoum treasury, dated to year 24 (κδ') of, most likely, Eucratides I (*IGIAC* 117 = *IEOG* 329). See Bernard 1985, 99.

143.　Falk and Bennett 2009; Fussman 1985. The reliquary's second inscription— "These relics of the Lord are (deposited) in veneration of all the Buddhas, (as a pious donation) of Aprakhraka, son of Heliophilus; (year) 172, inter-calated Gorpiaeus, day 8"—allowed the Azes Era epoch to be fixed at 47/46 BCE, not 58/57 BCE as previously thought: the intercalary *(embolimos)* Macedonian month Gorpiaeus, following the Parthian not Seleucid calendar, occurs with a periodicity of nineteen years. Cuneiform data have shown that the Parthian king Orodes II retarded the Macedonian calendar year by one month in 48 BCE, meaning that the Seleucid intercalary months of Xandicus and Hyperberetaeus were replaced with the Parthian intercalary months of Dystrus and Gorpiaeus. For this "Parthian slip," see Bennett 2011, 126–129 and 198–222; Assar 2003.

144. *IEOG* 318ᵃ; Sims-Williams 2004; Sims-Williams and Cribb 1995–1996. On the elimination of Greek, attested on Kanishka's coins, see Sims-Williams and Cribb 1995–1996, 110–111.

145. 1 Macc. 14:27; *AD* -132 B Obv. 1.

146. On markedness in code-switching, see Myers-Scotton 1993.

4. TOTAL HISTORY 1: RUPTURE AND HISTORIOGRAPHY

1. Plut. *Alex.* 37.3.

2. Aeschin. 3.132.

3. See, e.g., Gonzalez 2013, 21–24; Mathys 2000, 52–54; Albertz 1994, 567–568; Delcor 1951. Compare, e.g., the fate of the Phoenician island-city of Tyre ("Tyre has built herself a stronghold; she has heaped up silver like dust, and gold like the mud in the streets. But the Lord will take away her possessions and destroy her power on the sea, and she will be consumed by fire" [Zechariah 9:3–4]) or of the old Philistine Pentapolis ("Ashkelon will see it and fear; Gaza will writhe in agony, and Ekron too, for her hope will wither. Gaza will lose her king and Ashkelon will be deserted. A mongrel people will occupy Ashdod, and I will put an end to the pride of the Philistines" [Zechariah 9:5–6]) with the Alexander Historians' accounts of the destruction and repopulation of Tyre and Gaza (Diod. Sic. 17.40–47; Arr. *Anab.* 2.18–27; Plut. *Alex.* 24.3–25.2; Curt. 4.2.1–4.21, 6.7–31). See, however, the doubts of J. Collins (1995, 32).

4. Demetrius of Phalerum F82a (Fortenbaugh and Schütrumpf) = Polyb. 29.21.

5. Visscher (2016) noted the treatment of the battle of Gaugamela as a moment of epochal significance in the Babylonian Astronomical Diaries.

6. See, e.g., the two leading English-language textbooks, Marc van de Mieroop, *A History of the Ancient Near East, ca. 3000-323 BC* (Oxford, 2007), and Amélie Kuhrt, *The Ancient Near East c. 3000-330 BC* (London, 1995). Note that both authors explicitly state the significant continuities after Alexander's conquests. On the use of textbooks and syllabi for disciplinary modeling, see Smail 2005.

7. The bibliography is enormous and regionally focused. See, e.g., Sherwin-White and Kuhrt 1993.

8. For a general discussion of Hellenistic survey archaeology, see Alcock, Gates, and Rempel 2003 and Alcock 1994.

9. See, e.g., Murray 1969—an appropriately scathing review of Carl Schnei-der's *Kulturgeschichte des Hellenismus* (Munich, 1967).

10. See, e.g., McKenzie 1994; Sherwin-White and Kuhrt 1993; Briant 1990.

11. Mairs 2014, 39–46; Shaked 2004; Eph'al and Naveh 1996.

12. On this text, see Aperghis 2004, 117–135.

13. See, e.g., Mitchell 2017; Vlassopoulos 2013.

14. Note, however, the caution of Tuplin (2009).

15. Diod. Sic. 19.12–14, discussed by Bosworth 2002, 109–114.

16. Diod. Sic. 19.90–92, 100.3–7; Plut. *Demetr.* 7.2–3; Polyaenus *Strat.* 4.9.1. See also van der Spek 2014b; Wheatley 2002.

17. *BCHP* 3 Rev. col. 4, lines 24 and 37.

18. Prices rose similarly for sesame, cress, dates, mustard, and wool. See Pirn-gruber 2017, 34–35; Aperghis 2004, 78–86; van der Spek 2000a.

19. Discussed in Kosmin 2014a, 183–221.

20. See the strangely brilliant discussion of negative numbers in Spengler 1923, 71–91.

21. Rosenzweig comments on the Christian epoch, "Sie duldet, daß alles, was davor liegt, als verneinte, gewissermaßen als unwirkliche Zeit erscheint" (1921, 427).

22. Burman (1981), for example, has argued that when the Gregorian calendar was introduced to the Solomon Islands, and the nature of time changed, a sharp distinction was drawn between "custom" and "new ways," the water-shed being the arrival in 1904 of Methodist missionaries. Similarly, ac-cording to Koselleck, the French revolutionary calendar opened up and announced a new time: "What is really new about it is the idea of being able to begin history anew by accounting for it in terms of a calendar" (2002, 152; see also 148–53).

23. Compare, e.g., the works by van de Mieroop and Kuhrt mentioned earlier with Marc van de Mieroop, *A History of Ancient Egypt c. 3400 BC–AD 395* (Ox-ford, 2011) and Ian Shaw, ed., *The Oxford History of Ancient Egypt c. 7000 BC–AD 395* (Oxford, 2000).

24. Ginzburg 2010, 318. Ginzburg discusses a constellation of works—including Walter Benjamin's *Über den Begriff der Geschichte*, Max Horkheimer and The-odor Adorno's *Dialektik der Aufklärung*, and Marc Bloch's *Apologie pour l'histoire ou Métier d'historien*—that emerged from the sense of civilizational collapse brought about by National Socialism.

25. See, e.g., Dillery 2015; Ristvet 2015, 190–194; Haubold 2013, 143–177; and the contributions to Haubold, Lanfranchi, Rollinger, and Steele 2013.

26. *BNJ* 680 T1 = Euseb. (Arm.), *Chron.* 6, 14 (Karst), T2 = Tatian, *Oratio ad Graecos* 36.

27. *BNJ* 680 T2 = Tatian, *Oratio ad Graecos* 36.

28. Based on a variant reading: "τῷ μετὰ Σέλευκον τρίτῳ" (Euseb.).

29. This is attested on cuneiform administrative documents from 258 to 253 BCE (see van der Spek 2000b, 439). The Antiochus II date has recently been sup-ported by Bach (2013, 157–162).

30. Abydenus *BNJ* 685 F7 = Euseb. (Arm.), *Chron.* 25, 26–26, 8 (Karst): "Thus do the Chaldeans regard the kings of their land, from Alorus up to Alexander;

they do not themselves pay any attention to Ninus or Semiramis." Furthermore, according to Abydenus *BNJ* 685 F1a = Euseb. (Arm.), *Chron.* 19, 18–25
(Karst) and F1b = Euseb. *Praep. Evang.* 9.41.6, reproducing Berossus's account,
the wall of Babylon, first constructed by Bel, "lasted up to the rule of the
Macedonians (τὸ μέχρι τῆς Μακεδονίων ἀρχῆς διαμεῖναν)." Burstein (1978,
17n23) places this statement in the *Babyloniaca*'s first book. See also Diod.
Sic. 2.31.9, discussed by Boncquet 1987, 191–192, on Chaldean astronomical
record keeping "from earliest times until the crossing of Alexander to Asia"
(see below).

31. See Kosmin 2014a, 37–53. Note that the *Aegyptiaca* of Hecataeus of Abdera
(*BNJ* 264) incorporated Ptolemaic episodes into Egyptian history.

32. On the implication of an uninformed, external audience in Greek local historiography, reframing the esoteric as exoteric, see Tober 2017.

33. Dillery 2015, 352.

34. *BNJ* 680 F1b = Syncellus, *Chron.* 49, 19 (3–4) (Dindorf), trans. Dillery 2015, 74,
adapted; F1a = Euseb. (Arm.), *Chron.* 6, 8–9, 2 (Karst).

35. *BNJ* 680 F1b = Syncellus, *Chron.* 49, 19 (5–6) (Dindorf).

36. On Berossus and canonization, see Lange 2004, 53; Lambert 1957, 9.

37. Burstein 1978, 7.

38. See Dillery 2015, 146–147; Burstein 1978, 20nn50–51. Note that the Akkadian
^{l}ud-*napišti*(zi) is an etymologizing interpretation of the Sumerian name
Ziusudra (see George 2003, 152–153).

39. *BNJ* 680 F4a = Euseb. (Arm.), *Chron.* 10, 17–12, 16 (Karst); F4b = Syncellus,
Chron. 53, 19 (14–15) (Dindorf).

40. *Erra Epic* 4.50: "As for Sippar, the primeval city, through which the Lord of
the countries did not let the deluge pass because she was the darling of his
eyes (*šá* ^{uru}*si-par uru ṣa-a-ti šá* ^{d}*en kur.kur ina a-qar pa-ni-šu abu-bu la
uš-bi-'u-šu*)."

41. Berossus may be punning on the city's meaning in Aramaic, where the
verbal root *spr* is used for writing and scribal recording of all kinds; see Knobloch 1985.

42. Dillery 2015, 256; Lang 2013, 54; Haubold 2013, 156–160.

43. See Dillery 2015, 70; Lang 2013, 54; Kosmin 2013c, 210–211.

44. Haubold 2013, 160. Note that Dillery (2015, 70 and 140) adduces a Babylonian parallel (from Lambert and Millard 1969, 137), but this refers only to
the recitation of the beginning and end of *Atrahasis*.

45. The "closed archive" formulation is from Haubold 2013, 156.

46. In *Jubilees* 1–2, Moses, on the peak of mount Sinai to receive the tablets,
copies down the direct speech of the Angel of the Presence, as recited from
celestial documents.

47. Dillery notes, "In terms of his own scale, then, Berossus spent about as much
time on the first year or so of human history as he spent on the next
432,000 years" (2015, 72).

48. See the discussion in Dillery 2015, 77–78. Note that a Late Babylonian text,
the so-called Theogony of Dunnu (BM 74329), uniquely provides day and
month dates for each stage of its divine succession myth (e.g., Obv. 20: "In
the month Chislev, on 16th day, he took the overlordship for himself (*i-na*

^{iti}*kislimi* [gan.gan.è] ud.16.kám en-*ta* ù lugal-*ta il-qú*-[*ú*]"). Lambert and Walcot (1965) propose that the reference is cultic, not historical—that is, that the dates refer to events in the city of Dunnu's festive calendar.

49. On *illud tempus*, see Eliade 1969.
50. Dillery 2015, 76–77; contra Burstein 1978, 13n6.
51. On this location for Oannes's revelation, see Dillery 2015, 136.
52. *BNJ* 680 F4a = Euseb. (Arm.), *Chron.* 10, 17–12, 16 (Karst); F4b = Syncellus, *Chron.* 53, 19 (14) (Dindorf).
53. The Sumerian Flood story, *Atrahasis*, and *Gilgamesh* (see the interesting discussion in Dillery 2015, 77–78). Note, though, a recently republished tablet from Ugarit that seems to date the flood hero's release of birds to the first day of a month (see Darshan 2016). I am grateful to Stephanie Dalley for this reference. Similarly, the flood is the only event in the book of Genesis that receives precise dates in terms of days and months (see Day 2011, 212–213; Dalley 1989, 6).
54. *BNJ* 680 F21 = Sen. *Q Nat.* 3.29.1: "*Berosos, qui Belum interpretatus est, ait ista cursu siderum fieri. adeo quidem affirmat, ut conflagrationi atque diluvio tempus assignet.*"
55. *BNJ* 680 F2 = Ath. 14.44.639c.
56. On the chronological problems of reconciling the surviving fragments with cuneiform data, see Burstein 1978, 33–35.
57. Both forms are attested (see *BNJ* 680 F7c = Euseb. (Arm.), *Chron.* 13, 18–15, 4 (Karst); F9a = Joseph. *Ap.* 1.145–153; and F10 = Euseb. (Arm.), *Chron.* 15, 11–20 (Karst).
58. On Herodotus's temporal space taking its shape from the flow of *logoi*, see, e.g., Schiffman 2011, 38–47.
59. *BNJ* 680 T3 = Joseph. *Ap.* 1.131.
60. *BNJ* 680 F1a = Euseb. (Arm.), *Chron.* 6, 9–9, 2 (Karst).
61. Diod. Sic. 2.31.9. Alternative totals, from highest to lowest, are 490,000 years, per *BNJ* 680 F16b = Plin. *HN* 7.56.193; 480,000 years, per Julius Africanus, *Chronographiae* F15 (Wallraff); 470,000 years, per Cic. *Div.* 1.19.36 and 2.46.97; and more than 400,000 years per Chaeremon *BNJ* 618 F7 = Michael Psellos, Πρὸς τοὺς ἐρωτήσαντας πόσα τῶν φιλοσοφουμένων λόγων 443–444. Berossus is accepted as the common source for these figures. Diodorus's textual frames—the origins of astronomy and Alexander's conquest—are precisely the boundaries of Berossus's history, from Oannes's cultural teachings to the fall of the Achaemenids (see Eck 2003, 161; Boncquet 1987, 191–192). On Cicero, see Wardle 2006, 202; van der Horst 2002, 164–165. Note that Julius Africanus seems to have known Berossus in some form (see F34 (Wallraff): "τοῦ δὴ Ναβουχοδονόσορ μνημονεύει Βηρωσσὸς ὁ Βαβυλώνιος").
62. For comparison, note that, according to Hecataeus of Abdera, the Egyptian priests computed 23,000 years from the reign of Helios to Alexander's crossing (*BNJ* 264 F25 = Diod. Sic. 1.26.1: "οἱ δ᾽ ἱερεῖς τῶν Αἰγυπτίων τὸν χρόνον ἀπὸ τῆς Ἡλίου βασιλείας συλλογιζόμενοι μέχρι τῆς Ἀλεξάνδρου διαβάσεως εἰς τὴν Ἀσίαν φασὶν ὑπάρχειν ἐτῶν μάλιστά πως δισμυρίων καὶ τρισχιλίων"), and the third-century Samian historian Duris counted one thousand years from the capture of Troy to Alexander's invasion of Asia

(*BNJ* 76 F41a = Clem. Al. *Strom.* 1.21.139.4: "ἀπὸ Τροίας ἁλώσεως ἐπὶ τὴν Ἀλεξάνδρου εἰς Ἀσίαν διάβασιν ἔτη χίλια"). Burkert (1995, 143) detects here chiliastic speculation.

63. Certainly, there was a marked increase in the astronomical record from the eighth century BCE, after which appear the earliest Astronomical Diaries, Babylonian Chronicles, and eclipse reports. Ptolemy's *Royal Canon* and *Almagest*, probably using the second-century BCE astronomer Hipparchus, turned Nabonassar's accession year into the epoch of a continuous count of years, the so-called Nabonassar Era. This, like the similar Era of Philip III Arrhidaeus, was merely an invention of astronomers for mathematical calculation and had no wider currency in the world; it does not appear in cuneiform sources. See, e.g., Depuydt 1995, 101; Rochberg-Halton 1991, 108–111; Hallo 1984–1985, 149–151; Kugler 1907–1924, 2:368.

64. *BNJ* 680 F16a = Syncellus, *Chron.* 388, 5 (Dindorf).

65. See Burstein 1978, 22n66.

66. *BNJ* 680 F3a = Euseb. (Arm.) *Chron.* 4, 8–6, 8 (Karst), discussed by Burstein 1978, 22n67; Schnabel 1923, 22–25.

67. On the placing of these caesurae, see Tuplin 2013, 185.

68. Note, though, a comparable instance in distant China in 213 BCE: Li Ssu presented a memorandum to the First Emperor advocating the burning of all the historical records of the states vanquished by Qin, adding that "those who dare criticize the present by means of the past shall be executed, they and their families." On this extraordinary episode, see, e.g., Petersen 1995.

69. 1 Macc. 1:56–57. On this passage, and the Torah scroll as synecdoche, see S. Schwartz 2001, 59–60.

70. We should not entirely resist the inescapable National Socialist comparison. As Alon Confino has argued in his troubling *A World without Jews*, the Torah burnings of Kristallnacht (November 9, 1938) were attempts at historical annihilation and establishing a periodizing distinction: "In burning the Bible the Nazis laid the foundation for a revolutionary idea of time that could legitimize their expansionist and genocidal plans of the future. . . . It showed the Jews not only as an ancient people of the Bible but also as a people who already belonged to the past and had no place in the present" (2014, 141 and 149).

71. 2 Macc. 2:14. This passage has been at the center of discussions on biblical canonization (see, e.g., van der Kooij 1998, 24–26).

72. Note that Burstein (1978, 13n3) considers Chaeremon's observation (*BNJ* 618 F7 = Michael Psellos, Πρὸς τοὺς ἐρωτήσαντας πόσα τῶν φιλοσοφουμένων λόγων, 443–444), that astronomical wisdom "was inscribed on baked tablets so that neither would fire take them away nor overflowing water ruin them (ἐν ὀπταῖς πλίνθοις ταῦτα ἐγγράψασθαι, ἵνα μήτε πῦρ αὐτῶν ἅπτοιτο μήτε ὕδωρ ἐπικλύσαν λυμαίνοι)" to derive from Berossus.

73. Boym 2001, 307.

74. Ed. pr. van Dijk 1962, 44–52, with suggested emendations in van Dijk 1963 217, Klotchkoff 1982, and Lenzi 2008.

75. On Hellenistic Uruk and its Bīt Rēš, see, e.g., Baker 2014 and 2013, 56–61; Röllig 1991.

76. Pearce and Doty (2000) have identified a second, identically named Anu-bēlšunu, grandson of the first, to whom they assign the Uruk List of Kings and Sages, on the ground of Anu-bēlšunu the elder's otherwise lengthy scribal career. This in no way alters the interpretation of the tablet's historical positioning.

77. *NCBT* 1231. See also Pearce and Doty 2000, 337; Beaulieu and Rochberg 1996.

78. See, e.g., Dillery 2015, 66–72; Ristvet 2015, 192; Tuplin 2013, 183. See also Reiner 1961, 9–10.

79. *BNJ* 680 F5a = Euseb. (Arm.), *Chron.* 12, 17–13, 18 (Karst).

80. Note that the exceptionally long entry for Enmerkar and his *apkallu* (not *ummânu*), Nungalpirigal, seems to provide an aetiological myth for certain cult objects at Uruk (see Lenzi 2008, 161). There are slight traces of a ruled line beneath this entry.

81. On the mythological foundations of this dynasty, see, e.g., Woods 2012.

82. Note that Uruk had been well treated by the Neo-Assyrian kings, who sponsored its Ištar cult and renovated the Eanna temple (see, e.g., Krul 2014, 16; Clancier and Monerie 2014, 208n105).

83. See, e.g., Lambert 1962.

84. Beaulieu 2000, 4.

85. The Neo-Assyrian library catalogue, published by Lambert (1962), contains the entry "'The Series of Gilgamesh,' from the mouth *(ša pî)* of Sîn-lēqi-unninni (éš.gàr ^{d}gilgameš : šá pi-i ^{Id}30-li-qí-un-nin-ni)" (VI.10). From approximately the same time Sîn-lēqi-unninni was claimed as a lineal ancestor by Urukean scribes (see Beaulieu 2000).

86. See Beaulieu 2004, 316–317; Hallo 1963, 175.

87. For another indigenous reworking of contemporary court historiography, see Pseudo-Daniel, discussed in Chapter 5.

88. This is preceded by the broken sign [-*i*]*š*, in all likelihood indicating an adverb. Restoration is difficult: Klotchkoff (1982, 153) suggested *paniš*, meaning "before" ("Before Nicarchus"), and Lenzi (2008, 141n12) suggested *ediš*, meaning "alone" ("(But) Nicarchus is alone").

89. *YOS* 1.52, lines 1–3. The rendering of Nicarchus into cuneiform was enormously varied: *ni-iq-ar-ku-su, ni-iq-ar-qu-ra, ni-iq-ar-qu-su, ni-iq-ar-qu-ú-su, ni-iq-ar-ra-su, ni-iq-ár-ra-su, ni-i-qí-ar-qu-su, ni-iq-qar-su,* and *ni-qí-ar-qu-su* (see Monerie 2014, 40, 72–73, and 154–155; Lenzi 2008, 163n91).

90. Genesis 41:45; Daniel 1:7.

91. For the phenomenon of double names in Seleucid Babylonia, see Monerie 2014; Sherwin-White 1983, 214–218. For the biblical parallels, see Newsom 2014, 46–47.

92. Hallo (1963, 175–176) suggests that the names of the antediluvian *apkallū* encode the incipits of cuneiform literary series.

93. Klotchkoff 1982, 154.

94. Lenzi 2008, 164–165.

95. Krul 2014, 67–69; Beaulieu 2000, 11; van Dijk 1962, 50.

96. J. Smith 1982, 47. See also van de Mieroop 2016, 219–224; Neusner 1990.

97. Lenzi 2008, 162; contra what Eco characterizes as "the *etcetera* of the list" (2009, 81).

98. See the similar conclusion of Krul: "By placing Nikarchos alone at the end, the List of Kings and Sages expresses the notion that the relationship between rulers and advisors was different under Seleucid rule than it had been throughout history" (2014, 69). Pace Stevens (2016, 74–77), who sees in the Uruk List of Kings and Sages an integration of the Seleucid regime.

99. The details of this canonization process are, of course, hotly debated. See, e.g., Fantalkin and Tal 2012; Carr 2011 and 1996, 25–35; Lange 2008; McDonald 2006; Reed 2005c; Schniedewind 2004; Zeitlin 1974.

100. E.g., Sir. Prologue 1–2, 8–10, and 24–25; 2 Macc. 15:9; 4DibHam^a (4Q504) 1–2. iii.12–14; 4QapocrJer C^e (4Q390) 2.i.5; 1QS i.2–3 and viii.15–16.

101. On the possible composition or editing of Ezra-Nehemiah in the Hellenistic period, see, e.g., Eckhardt 2017.

102. On the rare deep counts of years, from the arrival of the Israelites into Egypt (Exodus 12:40–41) or their exodus (1 Kings 6:1), see, e.g., Miano 2010, 56–58.

103. On this understanding of prophecy, see, e.g., Carr 2011, 159–160; Hall 1991, 22; Neusner 1988, 17–18. For instance, 1 Maccabees 7:16–17 cites Psalm 79:2–3 as prophecy; Joseph. *Ap.* 1.8 notes that "prophets, who followed Moses (οἱ μετὰ Μωυσῆν προφῆται)" wrote down "what was done in their lifetimes from the death of Moses until that of Artaxerxes (ἀπὸ δὲ τῆς Μωυσέως τελευτῆς μέχρι τῆς Ἀρταξέρξου . . . τὰ κατ' αὐτοὺς πραχθέντα συνέγραψαν)." On the end of prophecy, see Gonzalez 2013, 32–41; Nihan 2013; Cook 2011; Jassen 2011; S. Cohen 2006, 167–168 and 186–192; Sommer 1996; Greenspahn 1989; Blenkinsopp 1983, 226–229. Note that when later prophets did arise this represented not so much a continuation of direct revelation but a popular revival of a tradition seen as dormant; see Sommer 1996, 34–36.

104. Zechariah 13:4–5. See also Zechariah 11:8–9, where the prophet resigns as shepherd of the flock. On these passages, see, e.g., Gonzalez 2013, 32–34.

105. Psalm 74:9. See, e.g., Donner 1973, 44–45; Kraus 1989, 96–97.

106. 1 Macc. 9:27; see also 4:46 and 14:41, discussed by Philonenko 1992; J. Goldstein 1976, 12–13.

107. LXX Daniel 3:38: "οὐκ ἔστιν ἐν τῷ καιρῷ τούτῳ . . . προφήτης."

108. See Newsom 2014, 114; J. Collins 1993, 202–203.

109. *Seder 'Olam* 86b, in a comment on Daniel 11:3–4, and citing Proverbs 22:17. For this passage, see Goldberg 2016, 120, 123–129, and 273n11; Gafni 1996, 32–34.

110. See Goldberg 2016, 131–132, and 209, and 2000, 275, citing the *Sefer ha-Ibbur* 3.8, p. 99 (Filipowski) of the eleventh- to twelfth-century CE astronomer-rabbi Abraham bar Hiyya, also known as Savasorda.

111. *T. Soṭa* 13:2; *y. Soṭa* 24b; *b. Soṭa* 48b; *b. Sanh.* 11a; *b. Yoma* 9b. Note the parallel "end of miracles" in *Midrash Tehillim* 22:10: "R. Benjamin bar Japheth taught in the name of R. Eleazar: 'As the dawn ends the night, so all the miracles ended with Esther.'"

112. Malachi 4:5. See, e.g., Blenkinsopp 1983, 241–243. Similarly, the Community Rule of the Qumran sect will hold until the appearance of the eschatological prophet (1QS ix.11), and the stones of the desecrated Jerusalem Temple altar will be reserved until a prophet should come to instruct them (1 Macc.

4:46). See, e.g., Jassen 2011, 2:588–589; Knibb 1999, 2:381–382; J. Goldstein 1976, 285.

113. Ben Sira 49:10: "May the bones of the Twelve Prophets send forth new life from where they lie, for they comforted the people of Jacob and delivered them with confident hope." After this statement, Ben Sira mentions no further prophetic figure in his historical review (on which, see below).

114. Amos 1:1; Haggai 1:1. P. Davies 2000, 67–78. B. Jones (1995) has suggested, on the basis of the different positions of Joel, Obadiah, and Jonah in the Masoretic Text and Septuagint versions and ancient manuscripts, an original Book of Nine, as a triple sequence of prophets coordinated with each empire: Hosea-Amos-Micah, Nahum-Habakkuk-Zephaniah, and Haggai-Zechariah-Malachi. Note that the Qumran text 4Q76 (4QXII^a), in which Jonah seems to follow Malachi, does not disrupt the chronological order.

115. See Blenkinsopp 2013, 131–132; Di Lella 2006, 155–156; Yarbro Collins 1996, 78–80 and 127–134; types of heroic ancestors (Ben Sira 44:3–7), ages (4 Ezra 14:11–12), rungs (*Ladder of Jacob* 1), heavenly Jerusalem (Revelations 21:12–14).

116. On apocalyptic pseudonymy and its various interpretations, see, e.g. Hartog 2014; Portier-Young 2011, 38–45 and 217–220; Reed 2005a, 54–55; Najman 2003, 4–17; J. Collins 1993, 55–57; Bickerman 1988, 201–203; Meade 1986, 77–101; Hengel 1974, 112 and 205–206; Willi-Plein 1977, 63–64; Metzger 1972; Plöger 1968, 23.

117. By contrast, J. Collins (1977) has noted that in early Christianity, where authoritative status was again accorded to contemporary prophecy, pseudonymy could be dispensed with for Revelations, the Apocalypse of John.

118. On inner-biblical exegesis, see, e.g., Fishbane 1985; Koenig 1962.

119. Daniel 9:2. Applegate (1997, 107) notes that the number seventy is emphasized by location—the number is at the end of the sentence and in its own clause—and thereby identified as the main exegetical concern of Daniel's reading.

120. Daniel 9:21–22. Note that the verb employed by Gabriel, the Hiphil or causative conjugation of *śkl*, is at the root of the self-identity of the book of Daniel's authors, named *maśkilim* (wise teachers). This is applied to Daniel and his companions at 1:4 and to the sages of the end times at 11:32–35 and 12:3 and 10. On the second-century BCE *maśkilim*, see Portier-Young 2011, 229–234.

121. Daniel 9:24–27. On the figure of the *angelus interprens*, see Melvin 2013.

122. That is, the timeline runs from Nebuchadnezzar II's conquest of the west at the battle of Carchemish, in 605/604 BCE, until 165/164 BCE.

123. See Boccaccini 1991, 137; Barton 1986, 196–197.

124. On the *pesharim*, see, e.g., Charlesworth 2002, 68–83; Dimant 1993, 58–61; Blenkinsopp 1983, 256; Brownlee 1979, 23–36.

125. On such valorization of this exegetical activity, see, e.g., Tzoref 2011; Charlesworth 2002, 14–16; Funkenstein 1993, 74–76; Bickerman 1988, 60–61; Brownlee 1979, 30; Hengel 1974, 206; Gruenwald 1973, 70–75. On the theoretical distinction between past and present in exegesis, see Hughes 2003. Note that the manuscripts from Qumran allow us to observe a real shift in the perception of sacred writings. Lange (2004) has demonstrated a change in the

style of scriptural citation, from free-form allusions in the Ptolemaic period to anthological collections and explicit quotations, introduced by formulae such as "as it is written," under the Seleucids and Hasmoneans.

126. 1QpHab vii, 1–5, discussed by Charlesworth 2002, 83–88; Brownlee 1979, 107–113. For a broader discussion of the progressive and time-dependent nature of prophetic interpretation, see Bull 1999, 118–124.

127. Scholem 1971, 6–7. See also Gruenwald 1973.

128. Exodus 20:2.

129. See, e.g., Corley 2006, 201; Yerushalmi 1982, 5–26; Fensham 1981, 50–51; Pannenberg 1961, 91–95.

130. Yerushalmi 1982, 9. Derrida "would have liked to spend hours, in truth an eternity, meditating while trembling before this sentence" (1995, 50). See also Momigliano 1966, 19–20. On possible Near Eastern parallels or precedents for such *Heilsgeschichte*, see, e.g., Gnuse 1989.

131. Deuteronomy 26:5–9 and 32; Joshua 24:2–18; Nehemiah 9; and Psalms 105, 106, 135, and 136 have been identified as precedents. Note that these reviews, unlike those of the second century BCE, neither are extensive nor assume a profound separation from the present. For instance, Joshua son of Nun's farewell address (Joshua 24:2–18) incorporates a historical account from the call of Abraham to the conquest of the land; not only is its scope limited, but Joshua is directly implicated as a participant-actor in this history-in-the-making. Note that the closest of the precedents, the historical review of Nehemiah 9, is most likely a late (perhaps even Hasmonean-period) interpolation: in contrast to the generally pro-Persian tendency of Ezra-Nehemiah, this prayer considers the Persian kings as no less tyrannical than the Assyrians: "And so today we are slaves (*ʾᵃnaḥnû hayyôm ʿᵃḇāḏim*), slaves in the land that You have given to our ancestors to eat of its fruit and enjoy its bounty. But its abundant yield goes to the kings whom You have put over us because of our sins; they have power also over our bodies and our livestock according to their pleasure, and we are in great distress" (9:36–37). See, e.g., Carr 2011, 209; Bautch 2003, 111 and 136; Blenkinsopp 1988, 301–308; Fensham 1981; Kellermann 1967, 35–36.

132. Dillery 2005, 352. See also Yerushalmi 1982, 14: "By the second century BCE the corpus of biblical writings was already complete, and its subsequent impact upon Jewry was in its totality."

133. On the social setting of Ben Sira, see Tiller 2007, 244–250.

134. On this absence of history in Jewish wisdom literature, see, e.g., Olson 2013, 238, J. Collins 1997, 97–98; Siebeneck 1959, 411. On the Ptolemaic date of Qoheleth, see, e.g., Bundvad 2015, 5–9; Krüger 2000: 39. On its exploration of the temporal contradictions between the natural world and humanity, the inaccessibility of humanity's past or future, and so the overriding concern with the present moment "under the sun," see Bundvad 2015.

135. Enoch is missing in several ancient witnesses to the text (see J. Collins 1997, 100–101).

136. Included are: Enoch; Noah and the flood; Abraham and the covenant of circumcision; Isaac; Jacob and his twelve sons; Moses, the Sinai theophany, and the giving of the Law; Aaron and the establishment of the priestly

service and costume; the conspiracy of Dathan, Abiram, and Korach; the covenant with Phinehas; Joshua, Caleb, and the conquest of the land; the judges, as a collective; Samuel and the establishment of monarchy; Nathan, David, and the establishment of his dynasty; Solomon, his kingdom, and his wisdom; Jeroboam, Rehoboam, and the division of the state; Elijah; Elisha; Hezekiah, his tunnel, and Sennacherib's invasion; Isaiah; Josiah; Jeremiah; Ezekiel; the Twelve prophets, as a collective (see above); Zerubbabel and Joshua; and Nehemiah. A final *inclusio* reverses back to Enoch, Joseph, Shem, Seth, Enosh, and Adam (49:14–16), sealing off this section by returning to its point of origin. Note that to incorporate the high priest Simon (see below, n139), Mack (1985, 17) and others require that the 49:14–16 coda be an interpolation. Daniel is not present (his pseudonymous book had not yet been composed), Ezra is a more curious absence, and no women are included.

137. On the explicit temporal interrelations within the hymn, especially for the "prophetic" period after Moses, see Goshen-Gottstein 2002, 258–260.

138. In contrast to praising God or recounting Israel's obedience or disobedience. See, e.g., Di Lella 2006; Mack 1985.

139. Ben Sira 50, the chapter immediately following this historical review, is an encomium to the high priest Simon II, who had welcomed Antiochus III into Jerusalem. It is much debated whether or not this hymn in praise of Simon ought to be understood as belonging to, and so the culmination of, the Hymn in Honor of our Ancestors. See, e.g., Mulder 2003; Goshen-Gottstein 2002, 261–264; Di Lella 1987, 499; Lee 1986, 10–21; Mack 1985, 17, 35, 55–56, and 195–196; Smend 1906, 412. Pace Mack (1985, 55), in either scenario the praise of the high priest Simon is distinguished from the historical review, both as a separate, present-focused literary unit and as an ahistorical, realized eschatology. On Ben Sira's pro-priestly agenda more generally, see Wright 1997.

140. Polyb. 12.25a.3.

141. See, e.g., Arr. *Anab.* 2.10.2, and 3.9.5–8; Hdt. 8.83.1–2 and 9.17.4; Polyb. 3.108–109 and 111 and 15.10–11; Thuc. 2.11.1–9, 87.1–9, 89.1–11, 4.10.1–5, 92.1–7, 95.1–3, 126.1–6, 5.9.1–10, 69.1, 6.68.1–4, and 7.5.3–4 and 61–64; Xen. *An.* 3.2.2–39 and 4.8.14, *Cyr.* 3.3.34–45, and *Hell.* 2.4.13–17 and 7.1.30; Caes. *B Civ.* 3.90; Curt. 3.10.4–10; Liv. 36.17.2–16; Tac. *Ann.* 2.14–15. On battle exhortations in Greek historiography, see, e.g., M. Hansen 1993; Wooten 1974.

142. 1 Macc. 2:49–61. See Hieke 2007; Lee 1986, 35–36.

143. 1 Macc. 2:62–63.

144. 1 Macc. 1:1–10: "And it came to pass (καὶ ἐγένετο) after Alexander, son of Philip, the Macedonian, who came from the land of Kittim, had smote Darius, king of the Persians and the Medes, he ruled as king instead of him. . . . And Alexander reigned for twelve years and died. And his sons gained power, each in his own place. And after his death they all put on diadems, and their sons after them for, many years, and they filled the earth with evils. And from them came forth a sinful root (ῥίζα ἁμαρτωλός), Antiochus Epiphanes, son of Antiochus the king, who was a hostage in Rome; and he began to reign in the 137th year of the kingdom of the Greeks." Note

that the historical horizon is reinforced by the programmatic use of the Seleucid Era year date for the sprouting forth of Antiochus IV.

145. 2 Macc. 15:8-9. This is likely a compression of Jason of Cyrene's extended speech. Note that in 2 Macc. 8:23, before the first battle with Nicanor, Judas "declaimed from the holy scroll, and gave as watchword 'God's Help.'"

146. Possibly Antiochus IV, his Seleucid successors, including Antiochus VII and Demetrius III, or even Tigranes the Great (see, e.g., D. Patterson 2008; Baslez 2004; Delcor 1967). The original language was, in all likelihood, Greek (see Joosten 2007). On the genre of the Jewish novel, see, e.g., Wills 2011; Raup Johnson 2004.

147. On the book of Judith's ludic qualities, see Esler 2002; Gruen 2002, 158-170.

148. Judith 1:1.

149. Esler 2002, 117.

150. Judith 1:6-10.

151. Judith 1:13-16.

152. Diod. Sic. 31.19. See, e.g., Raup Johnson 2004, 27-28.

153. Judith 12:11-13:3 and 14:14-18. On the eunuchs Bagoas, see, e.g., Ruzicka 2012, 175-176; Badian 1958.

154. Judith 2:7. See also Hdt. 4.126, 127.4 and 132, 5.17.1, 18.1, and 73.2-3, 6.48.1-2, 49, and 94, 7.32, 132-133, 163.2, and 233.1; Corcella 1999, 83-87.

155. Jeremiah 32:1.

156. See Wills 2011, 154.

157. According to the Song of Deborah (Judges 4-5), Jael, wife of Heber the Kenite, hospitably welcomed Sisera, commander of the Canaanite army of king Jabin of Hazor, offered him milk, and then hammered a tent peg into his temple as he slept.

158. Judith 13:6 and 16:9. Priebatsch (1974) suggests that Greek *Persica* literature, perhaps Dinon (*FGrHist* 690), is behind such Achaemenid coloring.

159. Judith 8-15.

160. Judith 5:3-4. Note the significance, within the Seleucid empire, of the *apantēsis* ritual (see Kosmin 2014a, 151-157).

161. Judith 5:5-19. Note that the monarchic period, typically for Jewish fictions, goes unmentioned, suggesting both a deliberate focusing of Israelite history on the Jerusalem sanctuary (see Joosten 2007, 171) and a concern with the experience of foreign rule (see Raup Johnson 2004, 6n9).

162. Judith 5:21. On the character of Achior, see Roitman 1992.

163. Hdt. 7.101-104. For the comparison, see, e.g., Puerto 2006, 123; Raup Johnson 2004, 46-47; Caponigro 1992; Momigliano 1982, 227-228.

164. On Tarantino's *Inglourious Basterds*, for instance, see Suleiman 2014; Rieder 2011.

165. Phoenician auto-ethnographies or local histories seem to have been composed in the second century BCE (see Laitos, *BNJ* 784).

166. See Kosmin and Moyer forthcoming and Moyer 2011, 93-94 and 139-140.

167. Eratosthenes =*BNJ* 241 F1a = Clem. Al. *Strom.* 1.21.138.1-3 and F1d = Clem. Al. *Strom.* 1.138.4. On the diastematic method, see Clarke 2008, 64-70; Möller 2005; Fraser 1972, 456-457. On the method's application to his geography of the *oikoumenē*, see Kosmin 2017b.

168. Pace Feeney, who writes, "The death of Alexander is such a famous event that readers can construct their own links back to it" (2007, 14).

169. *BNJ* 532 F2.i.

170. The change occurs after entry xxxiii (see Higbie 2003, 174).

171. Note that the Lindian Chronicle (*BNJ* 532 F2.xv) also observes the "archaic stances" of a painted plaque ("πάντες ἀρχαϊκῶς ἔχοντες τοῖ<ς> σχήμασι"), dedicated by the island's presynoecism *phylai*. Additionally, the absence of Trojans and Romans from among this catalogue of famous devotees marks out a distinctly Greek chronotope.

172. The importance of this move is emphasized by Fasolt 2004, 12–15.

173. On past and present as correlative, nonoverlapping notions, see King 2000, 25–39.

174. See, e.g., Mosès 1994, 117; Collingwood 1946, 233.

175. See Fasolt 2004, 4–20.

176. See Bevernage and Lorenz 2013 (the icicle image is on page 10). Virno (2015, 41–46) speaks in terms of a reaction to historical hypertrophy.

177. For comparison, it has been argued by Raulff (1999, 13–49) concerning twentieth-century CE French historical writing that a preference for deep history and the *longue durée* was motivated by the political rejection of recent traumatic experiences: the French Revolution in conservative thinking, the Restoration in Marxist thinking, the lost war(s) and Vichy collaboration in both nationalist and liberal thought.

178. On "la terreur de l'histoire," see Eliade 1969, 158–182. See also Ruiz 2011, 4–34; Ankersmit 2005, 355–359.

179. See, e.g., Rappaport 1992, 29.

180. Fasolt 2004: xiv and 15.

181. See also Hammer 2011, 37; Chakrabarty 2000, 244; Rousso 2001, 8 and 28. De Certeau observes, "writing makes the dead so that the living can exist elsewhere" (1988, 101).

5. TOTAL HISTORY 2: PERIODIZATION AND APOCALYPSE

1. The extended intramural debate on terminology need not detain us here. We should only note that it is problematic to derive reified descriptions of either theological systems or bounded social groups from a category originally devised to capture a literary phenomenon. See, e.g., Bazzana 2015, 17; P. Davies 2011, 100–119.

2. For an overview of such debates, see, e.g., DiTommaso 2007.

3. Barton characterized such futurological interpretations as follows: "Though a scheme of historical prediction and fulfilment reaching back into the remote past, sometimes as far as the creation, provides the vehicle for the commentator or pseudepigraphist to convey his message, the substance of that message is concerned with the present and immediate future; the grand historical scale is little more than stage machinery" (1986, 214).

4. Chakrabarty 2000, 62; see also 47–71. See also Murthy 2015, 124–135.

5. Chakrabarty 2000, 63.

6. See especially Portier-Young 2011; Horsley 2010. On the difficulties of applying the modern colonial situation to the Hellenistic kingdoms, see, e.g., Manning 2010, 49–54; Bagnall 1997.

7. The transition occurs at Daniel 11:40 ("and in the time of the end [ûbᵉʿet qes]").

8. Cyrus is mentioned very briefly, in the closing phrase of the sixth chapter (6:29: "And the man Daniel prospered in the reign of Darius and in the reign of Cyrus the Persian").

9. Newsom (2014, 14) and Henze (2001, 20–21) contrast this theological program with the political context more typical of court tales, such as Esther and Aḥiqar.

10. The literary integrity of all parts of Daniel has been much investigated. For an overview of such arguments, see, e.g., J. Collins 1993, 12–13 and 24–38; Louis Hartman and DiLella 1978, 9–18. Van Deventer (2013) has suggested reversing this order of composition.

11. The Dead Sea Scrolls preserve the shift from Hebrew to Aramaic at 2:4 (1QDanielᵃ) and from Aramaic to Hebrew at the beginning of 8 (4QDanielᵃ⁻ᵇ); see Newsom 2014, 3–4; Portier-Young 2010; Arnold 1996, 9–14; J. Collins 1993, 3.

12. See, for sixteenth-century Europe alone, Johann Sleidan's *De quatuor summis imperiis* (1556), Philip Melanchthon's *Chronicon Carionis* (1532), David Gans's *Ṣemaḥ David* (1592), and Jean Bodin's famous critique in his *Methodus ad facilem historiarum cognitionem* (1566). On the reception of the four kingdoms, see, e.g., Pomian 1984, 107–115.

13. Daniel 1:1–2.

14. In the absence of a cult statue, the Temple's vessels were invested with a similar symbolic value; their exile and return epitomize and periodize the destiny of the Temple itself (see, e.g., Honigman 2014, 217; Ackroyd 1972).

15. On such "universalization" of Israel's *Heilsgeschichte*, see Koch 1961, 27–29; Pannenberg 1961, 96–97.

16. This was placed in the first year of Cyrus. On the paired dates of Jehoiakim and Cyrus, see Newsom 2014, 40.

17. See Knibb 1976, 255–256; Koch 1961, 28. Although the book of Daniel's account is limited to an exilic narrative setting and a postexilic visionary insight, new manuscript fragments uncovered at Qumran (4QPseudo-Danielᵃ⁻ᵇ [4Q243–244]) reveal an additional and more extensive history, which points toward the Enochic total histories discussed below (see DiTommaso 2005; J. Collins and Flint 1996; J. Collins 1996). Little more than isolated phrases survive in the scraps of this so-called Pseudo-Daniel text, but the tale seems to tell of an elderly Daniel, summoned before king Belshazzar and his assembled court and reciting from a scroll the total history of Israel and the world, *Urzeit* to *Endzeit*. Extant passages indicate that this historical review incorporated in a noncryptic, third-person narrative the primordial period, the flood, the tower of Babel, the conquest of the promised land, Nebuchadnezzar's destruction of Jerusalem, and the period of Macedonian rule, before concluding with a great eschatological battle and the establishment of "the holy kingdom." The hopelessly scrappy traces

render any interpretation of this lost work utterly speculative. Nonetheless, references to a heavenly scroll, the predicted regnal lengths of individual Hellenistic monarchs, and "seventy years" of, perhaps, oppression indicate both a deterministic historical model and an interest in the periodized chronography of the future. The episode, set in the court of king Belshazzar, recalls the "writing on the wall" fifth chapter from the biblical book of Daniel; with its shared interest in a fixed and countable historical order, perhaps this tale too takes place on the eve of the Median conquest, as the sun sets on the first world empire. And does the Judean hero, delivering a historical account before the king, parody and surpass the official court historians of the author's day?

18. Daniel is not a well-known biblical figure. A Daniel appears in Ezekiel as a non-Israelite but proverbially righteous and wise man of deep antiquity, paired with Noah and Job in 14:14 and used to taunt the king of Tyre's self-claimed wisdom in 28:3. A different Daniel is mentioned in 1 Chronicles 3:1, where Daniel is the alternative name for David's son by Abigail (Kileab in 2 Samuel 3:3). Danel is the name of one of the Watchers in 1 Enoch 6. The righteous, childless king in the Ugaritic Aqhat story is called *Dn'il*. On the possible identity of these Daniels or Danels, see Day 1980.

19. See Dinon's *Persica* (*FGrH* 690 F4 = Ath. 11.110.503f): "There is *potibazis*, which is baked barley- and wheat-bread (ἐστὶ δὲ ποτίβαζις ἄρτος κρίθινος καὶ πύρινος ὀπτός), a wreath made of cypress, and wine mixed in a golden *ōion*, which the king himself drinks."

20. This is an ancient observation (Jerome, *in Dan.* 1:3), seen as fulfilling the prophecy of Isaiah 39:7 that "some of your descendants, your own flesh and blood, will be taken into exile, and they will become eunuchs in the palace of the king of Babylon." See Grillo 2017; Braverman 1978, 53–66. As Newsom (2014, 57) notes, Daniel and his companions have no wives or children. Daniel appears as a eunuch in the "Monk's Tale" of Chaucer's *Canterbury Tales*. There is a remarkable fourteenth-century illumination of the castration of Daniel in the Hague (KB 71A23, fol. 222v). For Breed's contribution, see Newsom 2014, 58.

21. 1 Macc. 4:54. It approximates the various temporal predictions for this desolating abomination in the Daniel's apocalyptic visions: "a time, times, and half a time" (7:25); half a week of years (9:26); the precise, numerical calculations (8:14, 12:11, and 12:12); and perhaps also the added-up sum of "*m^{e}ne' m^{e}ne' t^{e}qel û)arsin*" (5:25).

22. Daniel 1:8–16.

23. 2 Kings 25:27; and Ezekiel 1:2, 40:1. See also Arnold 1996, 10.

24. Note that Louis Hartman and DiLella (1978, 139) propose, on the basis of certain intratextual contradictions, that verses 2:13–23 are secondary additions. See also the more extensive claims of Kratz (1991, 32–35 and 55–62), requiring drastic textual surgery.

25. Note that *ḥ^{a}sa)* (clay) does not have a Hebrew cognate and is rare in Aramaic; it appears to connote fired terracotta and not raw clay (see Reynolds 2011, 102).

26. Daniel 2:37–45.

27. Nebuchadnezzar's actions before Daniel—prostrating himself in worship and offering up sacrifice and incense—recall the actions associated with Hellenistic benefactor cult (see Mastin 1973).

28. Wills 1990.

29. The connection has been much observed (see, e.g., Henze 2012, 282–283; Boccaccini 1991, 138; G. Davies 1978, 24).

30. See, e.g., Hes. *Op.* 109–201. Serv. ad Verg. (*Ecl.* 4) indicates that the Cumaean Sibyl "divided the ages by metals and even declared who would rule in each age (*saecula per metalla divisit, dixit etiam quis quo saeculo imperaret*)."

31. Overlooked by, e.g., Horsley 2010, 38. It is discussed by Newsom (2014, 75).

32. See Newsom 2014, 75.

33. On this development and its internalization by the Seleucid monarchy, see Kosmin 2014a, 31–92.

34. Newsom 2014, 76.

35. Indeed, the positioning can usefully be compared with the statue *ecphraseis* of contemporary Hellenistic poetry, which similarly explored the subject experience of viewing a molded metal statue "down to the tips of its toes" (Milan Papyrus AB 63.2: "ἄκρους . . . εἰς ὄνυχας"). On the poetics of sculptural *ecphraseis*, see, e.g., Sens 2005; A. Stewart 2005; Kosmetatou 2004. Take the epigram on a famous statue of Alexander the Great (*Anth. Pal.* 16.120): "Both Alexander's boldness and his entire form has Lysippus molded; what great power does this bronze have (τίν᾽ ὁδὶ χαλκὸς ἔχει δύναμιν)! For this bronze figure looks as if it almost says, staring down Zeus, 'I am the master of the earth; Zeus, you keep Olympus (γᾶν ὑπ᾽ ἐμοὶ τίθεμαι, Ζεῦ, σὺ δ᾽ Ὄλυμπον ἔχε).'" Lysippus's bronze captures not only Alexander's nature but also the force (δύναμιν) of his rule; the employment of a medium of cult for the mortal monarch sacralizes the political, and kingship as statue displaces god.

36. On such temporal and ethical dualism in apocalyptic texts, see Gammie 1974, 357 and 377–383.

37. See Isaiah 17:13, 29:5, and 41:15; Jeremiah 51:33; Hosea 13:3; Psalms 1:4, 35:5. See also, e.g., Newsom 2014, 77; Silverman 2012, 150; Henze 2012, 284–285; Portier-Young 2011, 263.

38. See Isaiah 2:2. 11:9, Micah 4:1; see, e.g., Louis Hartman and DiLella 1978, 149–150.

39. See, especially, Ma 2013b, 45–63.

40. *OGIS* 219 = *SEG* 49 1752, lines 34–35.

41. See the similar episode of Alexander spotting ("κατανοήσας") a statue of Ariobarzanes, the former Persian satrap of Phrygia, fallen to the ground at the temple of Athena in Ilium (Diod. Sic. 17.17.6); this is considered among "other favorable omens (οἰωνῶν αἰσίων ἄλλων)."

42. Diog. Laert. 5.75–77; Plin. *HN* 34.12.27; Strabo 9.1.20; Plut. *Praecepta gerendae reipublicae* 820e–f; *SEG* 25 206. For a brilliant analysis of this episode, see Azoulay 2009.

43. See Polyb. 36.13, who uses a discussion of the toppling of statues of Callicrates to meditate on the peculiar ways of *Tychē* ("ἴδιον ἐπιτήδευμα τῆς τύχης").

44. Liv. 31.44.4–8. The passage derives from Polybius (see Briscoe 1973, 150–151). On the Athenian actions, see Habicht 1982, 145–150. Note that the Roman attack on Philip V's base at Chalcis included the casting down and breaking up of the king's statues (Liv. 31.23.10: "*statuis inde regis deictis truncatisque*").

45. Shear 1973, 165–168. It is interesting to note, for our understanding of the multimetallic imagery of Daniel 2, that the statue of Demetrius Poliorcetes was bronze, having been molded around a clay core using the lost-wax technique, and then entirely covered with gold foil.

46. Other examples include Hipparchus, son of Charmus (Lycurg, *Leocr.* 117–118), Philip II at Ephesus (Arr. *Anab.* 1.17.11), and Demades (Plut. *Praecepta gerendae reipublicae* 820f).

47. Daniel 2:20–21. The Greek translations of these crucial lines are "καὶ αὐτὸς ἀλλοιοῖ καιροὺς καὶ χρόνους, μεθιστῶν βασιλεῖς καὶ καθιστῶν" (Old Greek) and "καὶ αὐτὸς ἀλλοιοῖ καιροὺς καὶ χρόνους, καθιστᾷ βασιλεῖς καὶ μεθιστᾷ" (Theodotion).

48. On the dynamics of such parallelism, see Alter 2011, 1–28. On the rhetorical function of this doxology, see Kim 2011, 120–149.

49. In the doxology of the chastened Nebuchadnezzar in Old Greek Daniel 4:37, the same recognition is attributed to the Babylonian king: "I confess and praise, because He is God of gods, Lord of lords, and Lord of kings, because He does signs and wonders and changes times and seasons (ἀλλοιοῖ καιροὺς καὶ χρόνους), taking away the kingdom of kings and setting up others instead of them (ἀφαιρῶν βασιλείαν βασιλέων καὶ καθιστῶν ἑτέρους ἀντ᾽ αὐτῶν)."

50. Alternative versions of this tale are found at Qumran, the so-called Prayer of Nabonidus (4QPrNab [4Q242]), and in Eusebius, citing Abydenus *BNJ* 685 F6b = Euseb. *Praep. evang.* 9.41.1–9.

51. Daniel 4:25–27.

52. On the ideology of "king makes city," see Kosmin 2014a, 183–221.

53. Indeed, even though Daniel's interpretation develops only three of these four terms, the writing on the wall suggests the sequence and comparative evaluation of the four Near Eastern empires, much like the metallic statue of Nebuchadnezzar's dream in Daniel 2. Specifically, the last of the four terms, *parsin* or *pᵉres*, is interpreted as "your kingdom has been divided (*pᵉrisaṭ malḵûṭāḵ*)," an explicit echo of the "divided kingdom (*malḵû pᵉlîpâ*)" of the quadrimetallic statue's iron and clay feet (2:41). The repeated *mᵉne᾽* in the Masoretic Text is not present in the Greek versions (Old Greek 5:17; Theodotion 5:25) and has been regarded as secondary by several scholars (e.g., Charles 1929, 137; Montgomery 1927, 262), but the *lectio difficilior* should be retained. Moreover, note that *mᵉne᾽ mᵉne᾽ tᵉqel ûparsin* (1 mina + 1 mina + 1 sheqel + 2 half-minas) amounts to a little over 3 minas; with the equation of weight and time already established, and with the mina and sheqel being units of Babylonian water clocks (see Ben-Dov 2008, 157–159), this may be yet another allusive assertion of three and a bit years from the opening of the Antiochid persecution to the *eschaton*.

54. Most scholars accept the influence of 1 Enoch 14–15 on this throne vision.

55. For observations on each beast, see, e.g., J. Collins 1993, 295-299; Wittstruck 1978; Louis Hartman and DiLella 1978, 212-214. Note that this also actualizes biblical prophetic metaphors of Israel's enemy kingdoms and neighbors as animals (e.g., Hosea 13.7-8; Jeremiah 5:6; Ezekiel 34).

56. Kosmin 2014a, 1-3, and 2013a.

57. See Staub 1978.

58. Newsom 2014, 217 and 219; J. Collins 1993, 280-294; Gunkel 1895, 327-335.

59. Note that Horsley proposes, in general rather than specific terms, that the vision's imagery "is a symbolic 'turning of the tables' on imperial mythic ideology" (2010, 87).

60. The psalm's Maccabean dating was first proposed by Theodore of Mopsuestia. For the arguments in favor, see Donner 1973; for doubts, see B. Russell 2007, 125-127. The alternative proposed date is the Babylonian destruction of the Temple in 586 BCE. Tanner (in Declaissé-Walford, Jacobson, and Tanner 2014, 600), sees in the psalm a clear response to the *Enūma eliš* and a displacing of Marduk. Note that the personal pronoun "You" is emphatically repeated seven times in verses 12-17.

61. For discussion of the identifications, see, e.g., J. Collins 1993, 320-321; Caragounis 1987; Charles 1929, 172-173. It is almost certain that the horn list began with Alexander.

62. This is a proposal of VanderKam 1979 and 1981. See also Goldberg 2016, 183-197; H. Eshel 2008, 47-48; Sacchi 2007; Portier-Young 2011, 181-182 and 198-199; Tiller 1993, 107-108. However, note the objections, to my eyes not persuasive, of Bedenbender 2000, 168-173; P. Davies 1983. On the nature of the 364-day calendar, see Ben-Dov 2008, 21-62.

63. Both phrases employ *šny*, in the Haphel construction, with *zimnin* (see, e.g., Segal 2009, 145-149).

64. Noted by several scholars (see, e.g., Newsom 2014, 221; Louis Hartman and DiLella 1978, 211).

65. Gunkel 1895, 329: "ein Regiment der Schrecken."

66. See Boccaccini 1991, 154.

67. See also Kosmin 2016.

68. The identification is controversial and the debate fierce. See, e.g., Newsom 2014, 234-236; J. Collins 1993, 304-310. The grammatical construction "son (*bar*) of X" merely designates a member of a species group; here, the preposition *kᵉ-* (as in "*kᵉbar ᵃⁿāš*, "one like a son of man") establishes a symbolic association. There is little doubt that the imagery goes back to the Canaanite cloud rider Baal, appearing before El, the transcendent sky god of justice, as attested in the tablets from Ugarit (see Pope 1955, 32-35). Here, he has been transformed into a symbol to fit a monotheistic interpretation (see Boyarin 2012).

69. The earliest explicit attestation of this epithet is Plaut. *Mostell.* 775, in the early second century BCE.

70. See Newsom 2014, 265-268; Burgmann 1974, 544-545.

71. On this heptadic structure, see Berner 2006, 19-99.

72. As described in 2 Macc. 4:33-38.

73. On the notion that different angels were allotted to different nations, see Deuteronomy 32:8–9, Ben Sira 17:17, 1 Enoch 20. See also Newsom 2014, 332–333; J. Collins 1993, 374–375; Louis Hartman and DiLella 1978, 282–284; Charles 1929, 262 and 267.

74. Pace, e.g., Horsley 2010, 94; Redditt 1998, 471–473; J. Collins 1993, 377; Bickerman 1988, 122; Lebram 1975, 760–761.

75. Compare, for example, the highly detailed information recorded in the Babylonian Astronomical Diaries, including Antiochus IV's victories in Egypt during the Sixth Syrian War (AD -168 A Obv. 15). See also 4Q248, known as "Acts of a Greek King," from Qumran (see the commentary of Broshi and Eshel 1997). This parchment fragment, written in the Herodian period, narrates Antiochus IV's siege of Alexandria in the Sixth Syrian War. 4Q386ii–iii, or "Pseudo-Ezekiel," may even recount the same king's administrative appointments at Memphis, according to Dimant (1998, 523–526; though note the doubts of H. Eshel 2008, 151–161).

76. On the reworking of key texts from Isaiah, Habakkuk, and Ezekiel see, e.g., Newsom 2014, 337 and 342; Henze 2012, 291–295; Portier-Young 2011, 263–264. On the "enemy from the north," see, e.g., Childs 1959. On the identification of the early Seleucid foundation of Seleucia-in-Pieria with mount Ṣaphon (or Casius), see Kosmin 2016, 39.

77. Old Greek 11:5: "καὶ εἷς ἐκ τῶν δυναστῶν κατισχύσει αὐτὸν καὶ δυναστεύσει· δυναστεία μεγάλη ἡ δυναστεία αὐτοῦ." Theodotion 11:5: "καὶ εἷς τῶν ἀρχόντων αὐτοῦ ἐνισχύσει ἐπ᾽ αὐτὸν καὶ κυριεύσει κυριείαν πολλὴν ἐπ᾽ ἐξουσίας αὐτοῦ." Vulgate (Pouay-Rheims, Clementine) 11:5: "et de principibus eius praevalebit super eum, et dominabitur ditione: multa enim dominatio eius."

78. Scolnic (2014b) suggests a Ptolemaic source for this account, but the account is also hostile to the Ptolemaic monarchs.

79. BNJ 239 B16.

80. See, e.g., Newsom 2014, 327; Hall 1991, 89; Clifford 1975, 24.

81. Antiochus's seizure of the throne is described in mythopoeic language (see Scolnic 2014a).

82. Note that these events were also well-known in Babylon, where the author of AD -168 A Obv 14–15 reports: "That month, I heard as follows (al-te-e um-[ma]): king Antiochus marched victoriously through the cities of Meluḫḫa (ᶦan lugal ina uru^{meš} šá ^{kur}me-luḫ-ḫa šal-ṭa-niš gin.gin-ma)."

83. On the reuse of Canaanite mythic motifs for this third invasion and the final conflict in the holy land, see Clifford 1975, 25–26.

84. Note the employment of Seleucid Era dating in Egypt during Antiochus IV's second invasion (see Fischer-Bovet 2014, 227–228).

85. On this process of Fortschreibung, the continuous updating or prolonging of a text, see Henze 2012, 289–290.

86. On Daniel in the court of Cyrus, see Daniel 1:21, 6:29, and 10:1, as well as the apocryphal tale "Bel and the Dragon." See especially Bel et Draco 2 (Theodotion): "καὶ ἦν Δανιηλ συμβιωτὴς τοῦ βασιλέως (sc. Cyrus) καὶ ἔνδοξος ὑπὲρ πάντας τοὺς φίλους αὐτοῦ."

87. Daniel 12:13.

88. Of course, this was in part motivated by the limits of the human life span, but that in no way negates its programmatic function.

89. See Caragounis 1993, 390.

90. On deferred eschatology, see J. Collins 1990, 256–257.

91. Note that the Old Greek text and the Vulgate read "by the Ulai gate ("πρὸς τῇ πύλῃ Ωλαμ [or Ουλαμ or Αιλαμ]"; "super portam Ulai"), perhaps by metathesis from the Aramaic ᵃbul ("gate").

92. The most detailed modern discussion of Susa's hydrology is Potts 1999 (pace Le Rider 1965, 263–267).

93. Plin. HN 6.27.135.

94. Gen 15:18, Deut 1:7, and Josh 1:4 (see Newsom 2014, 330). This epithet also appears in Callim. Hymn 2.108 ("Ἀσσυρίου ποταμοῖο μέγας ῥόος").

95. Ezekiel 1:1.

96. 1 Enoch 13:7–8. Note that Janelle Peters (2009, 137–140) suggests the influence on Daniel 12:5–13 of Hellenistic literary and artistic depictions of river gods.

97. See, e.g., P. Jones 2005.

98. Pind. Nem. 1.1–2; Timaeus BNJ 566 F41b = Polyb. 12.4d.5–7, F41c = Strabo 6.2.4; Paus. 5.7.3; Ov. Met. 5.493–497.

99. See, e.g., Ferrari 2000.

100. Joseph. AJ 10.267. On this passage, see, e.g., Lars Hartman 1975–1976, 6–7.

101. For a discussion, see Horsley 2010, 82–83; Yarbro Collins 1996, 58–70 (I adapted the phrase on page 60); Wacholder 1975, 204–205.

102. See Taubes 2009, 32–33.

103. Koselleck 2004, 209.

104. See DiTommaso 2005, 124–125, especially 124n79; Helberg 1995.

105. De Certeau 1988, 7.

106. On this biblical material, see VanderKam 1984, 23–33. Enoch's sum of years, 365, is by far the lowest in the genealogy.

107. On the character of the Enochic corpus, see, e.g., Portier-Young 2011, 287–307; J. Collins 2011. It is unclear to what extent the corpus constituted a group polemic against the Mosaic Torah, as argued by Boccaccini (1998 and 2002) and others, or simply a parallel tradition with different emphases and based on an alternative, earlier authority (for which see, e.g., Nickelsburg 2007; VanderKam 2007, 16–20; Reed 2005b).

108. The myth of the Watchers' fall was incorporated into the Kebra Nagašt, the fourteenth-century legendary genealogy of Ethiopia's kings. On 1 Enoch's function in Ethiopian Christianity, see Nickelsburg 2001, 104–108.

109. For an introduction to the Aramaic fragments, see Stuckenbruck 2007. Note that there are later editorial additions, in particular 1 Enoch 106–108. Fragments of the Greek translation have been found at Oxyrhynchus (P.Oxy. 2069; see Chesnutt 2010). See also Reed 2003.

110. See, e.g., Reed 2005a, 62, 74, and 79–80.

111. The Geʿez script was borrowed by fourth-century CE missionaries from the Old South Arabian monumental epigraphy. The standard English-language grammar is Lambdin 2006, the simplest dictionary Leslau 2010. For commentaries, see below.

112. The combination of the two dream-visions in a single book is the product of later redaction (see Tiller 1993, 98–99).
113. On the date, see Gore-Jones 2015, 269; Bryan 2015, 161–163; Olson 2013, 5 and 84; Portier-Young 2011, 346–347. Contra Honigman (forthcoming), who prefers a later Hasmonean date, and Assefa (2007, 214–236), who considers the Judas passages an interpolation into an earlier text (see Portier-Young 2011, 349–352).
114. Nickelsburg 2001, 58.
115. On the significance of this hominization, see Bryan 2015, 47–50.
116. As laid out in Leviticus 11 and Deuteronomy 14; see Gore-Jones 2015, 281–282; Bryan 2015, 168–184; Olson 2013, 128–136; Nickelsburg 2001, 377; Tiller 1993, 28. The precise identifications develop both biblical metaphors and imperial self-identifications—e.g., the Philistines are under the sign of the dog, taking up Goliath's question to David, "Am I a dog, that you come to me with sticks?" (1 Samuel 17:43); the Midianites appear as wild asses, as their ancestor Ishmael "shall be a wild ass of a man" (Genesis 16:12).
117. Bryan 2015, 74–79; Nickelsburg 2001, 371; Bull 1999. 85–92.
118. Portier-Young 2011, 363–368; VanderKam 2004; Nickelsburg 2001, 380–381.
119. The myth is reported in the *Book of the Watchers* (1 Enoch 6–11); see Nickelsburg 2001, 165–228.
120. On the identification of this beast, see Tiller 1993, 242–243. While the elephant is absent from the Hebrew Bible, note that ivory is mentioned among the trade merchandise of Tyre (Ezekiel 27:15) and as an import of Solomon (1 Kings 10:22; 2 Chronicles: 9:21).
121. See the discussion of Bryan (2015, 88–91).
122. For parallel incidents of marked or symbolic anti-Seleucid elephant destruction, see Eleazar Avaran's attack on the "royal elephant" at the battle of Beth-Zechariah (1 Macc. 6:43–46; Joseph. *AJ* 12.373–374, and *BJ* 1.41–45), and the Roman legate Octavius's hamstringing of the Seleucid elephant force at Apamea (Polyb. 31.2.11; Cic. *Phil.* 9.4; Plin. *HN* 34.5.24). On a more general symbolic resistance to the Hellenistic kingdoms in the third-century *Book of the Watchers*, see Portier-Young 2014; and in the related *Book of the Giants*, see Angel 2014.
123. For the identifications of species with peoples, see the table in Olson 2013, 122, laying out the opinions of Tiller (1993), Nickelsburg (2001), and Olson (2013).
124. On the treatment of the exilic and postexilic periods as a unified era of history, to be ended only at the *eschaton*, see, e.g., Assefa 2007, 270–271; Knibb 1976, 256.
125. The major sources for this imagery of negligent or disobedient shepherds are Ezekiel 34 and perhaps Zechariah 11 (see, e.g., Nickelsburg 2001, 391).
126. On this condemnation of the Second Temple, see, e.g., Nickelsburg 2001, 394–395; Knibb 1976, 257.
127. See Olson 2013, 87.
128. This is a program of violent resistance, in contrast to the nonviolent resistance proposed by the book of Daniel (see Portier-Young 2011, 372–373; Venter 1997).

129. The plural of *sem'*, *'asmāt*, can also be translated as "reputations" and "evidence" (Leslau 2010, 67, s.v. *səm(ə)'*), perhaps implying something more discursive than a mere list of names.

130. 1 Enoch 89:72, 90:1, and 90:5. See Berner 2006, 215–216; Tiller 1993, 338.

131. Gore-Jones 2015, 276; Olson 2013, 99–100; Assefa 2007, 193–194.

132. Olson 2013, 100–101; Berner 2006, 216; Nickelsburg 2001, 392–393; Tiller 1993, 337–338.

133. As noted by Knibb 1976, 257.

134. Olson 2013, 31; Portier-Young 2011, 319; Kvanvig 2007, 144; Stuckenbruck 2007, 2 and 60–62; Berner 2006, 118–125; Nickelsburg 2001, 440–441. Note that Boccaccini (1998, 105–106) proposes a post-Maccabean date.

135. A sixth-to-seventh century CE fragmentary Coptic translation was found in the northern cemetery of Antinoe (see Stuckenbruck 2007, 17 and 52). The Ethiopic translation contains a manuscript dislocation for weeks 8–10.

136. For detailed exegetical and linguistic commentary, see Stuckenbruck 2007, 65–152; Berner 2006, 127–148; Nickelsburg 2001, 435–450; and Reese 1999, 54–69.

137. See Kvanvig 2007, 146; Stuckenbruck 2007, 2 and 54. Reference is made to the "seventh part" of weeks 1 (93:3) and 10 (91:15).

138. Bedenbender 2007, 74–75, and 2005; Nickelsburg 2007 and 2001, 446.

139. See Horsley 2010, 73–74.

140. Pace, e.g., Koch (1983), who suggests a duration of 490 years for each week.

141. Yarbro Collins 1996, 81; Licht 1965, 179.

142. See Berner 2006, 150–156 (with the diagram on 156) and 167–169; Henze 2005, 208–209; Boccaccini 1998, 108–109; Lars Hartman 1975–1976, 11–12; Licht 1965, 179.

143. For the former, see, e.g., Waerzeggers 2015, and, for the latter, see, e.g., J. Smith 1975, 136–137. Smith, discussing the recopying in 287/286 BCE of the late-third-millennium Sumerian literary composition *Lament over the Destruction of Sumer and Ur* notes that "the same text is, at one and the same time, a Sumerian 'original' religious expression and a Hellenistic Babylonian 'original' religious expression."

144. Letter of Sîn-šarru-iškun (CTMMA 44); discussed by Da Riva 2017, 80–81; De Breucker 2015, 76–77 and 87; and Lambert 2005, 203–207. Letter of Nabopolassar (BM 55467); discussed by Gerardi 1986. It is worth noting the remarkable similarity, in both the substance and order of accusations, between this cuneiform letter of Nabopolassar and Alexander the Great's epistle to Darius III, as reported in Arr. *Anab.* 2.14.4–9.

145. For a different contemporary resonance, see Frahm 2005. See also, more generally, R. Goldstein 2010.

146. This is persuasively argued by Waerzeggers 2015. However, note that van der Spek, responding to the proposal, requires the existence of a "proto-Nabonidus Chronicle" (2015, 453–461).

147. Dillery rightly observes: "Really for the first time, Berossus' *Babyloniaca* fundamentally 'periodized' the past of the ancient Near East, breaking it into temporal units defined by the transfer of dynastic power" (2015, 300). Indeed, as Dillery (ibid., 294) points out, the formula "βασιλείαν παραλαμβάνω"

used of the Persian conquest of Babylonia in Josephus's paraphrase "when Cyrus received the kingship of Asia (Κύρου τῆς Ἀσίας τὴν βασιλείαν παρειληφότος)" (*BNJ* 680 F9a = Joseph. *Ap.* 1.145) gives the sense of ordered, ordained succession.

148. BM 40623. Ed. pr. Grayson 1975, 24–37. See also the important re-edition in van der Spek 2003, 311–340.

149. On this genre, see Neujahr 2012, 4–99; van der Spek 2003, 324; Grayson 1975, 13–22.

150. Noted by Neujahr 2012, 7; van der Spek 2003, 324–325.

151. Van der Spek 2003, 318 and 323.

152. On late Babylonian injunctions to secrecy, see, e.g., Stevens 2013.

153. Edition and translation of van der Spek 2003, adapted.

154. For discussion of the term and its use, see Monerie 2014, 88–89; Kosmin 2014a, 91–92 (with references).

155. In the edition and translation of van der Spek 2003.

156. Van der Spek (2003, 322) has related it to the eschatological purification at Daniel 12:10.

157. Finn 2017, 195–197; Shayegan 2011, 56–57; Muccioli 2004, 133–134; Shahbazi 2003, 15–16; Sherwin-White 1987, 10–14; Marasco 1985, 532–533; Lambert 1978, 12; Grayson 1975, 26–27.

158. Several of these problems have been discussed by van der Spek (2003, 326–332), who honorably confesses to *aporia* in the face of such difficulties. For "the overthrow of the Hanaean army" at 5.17 he is obliged to assume a scribal error, replacing *Ha-ni-i* with *Gu-ti-i*; the end of column 5 would, therefore, refer to Alexander's conquest of the Persian empire. This argument was developed by Shayegan (2011, 137–140), who suggests that the end of column 5 was redacted during the reign of the Arsacid king Phraates II to represent the initial successes and ultimate failure of Antiochus VII Sidetes's *anabasis* of 131–129 BCE. However, it remains unclear how this would explain the regime change at the end of column 6 or accommodate the centuries of Macedonian rule between Darius III and the Parthian conquest.

159. This is argued in brief by Del Monte (2001, 146) and Geller (1990, 5–6).

160. *BCHP* 11 Obv., lines 6 and 11 and Rev., lines 7 and 13 (the Babylonian Chronicle describing the Third Syrian War) characterizes the army of Ptolemy III, "who do not fear the gods," as Hanaeans.

161. Note that this Seleucid reconstruction of this passage is criticized by van der Spek (2003, 330–331) on two grounds: the supposedly inappropriate characterization of Seleucus's enemies as Hanaean and the downplaying of Alexander the Great. As we have seen in the previous note, Hanaean was used to refer to the Seleucids' Macedonian enemies at the approximate date of the Dynastic Prophecy's composition. The condensing of Alexander's own conquest is found in other eschatological total histories (e.g., 1 Enoch's Apocalypse of Animals, discussed above) and reproduces the Seleucid dynasty's own memory traditions (see Chapter 3). Moreover, the *arkanu* ("later") of 5.13 represents a chronographic gap. Finally, the Dynastic Prophecy is concerned with the succession of the imperial dynasties ruling Babylon, for which Alexander was not responsible.

162. Van der Spek 2003, 311–312; Tadmor 1981, 338; Lambert 1978, 13. Contra Shayegan 2011, 50n41.

163. As observed by van der Spek (2003, 325).

164. See, e.g., 1 Macc. 7:41–42; 2 Macc. 15:22–24; 1QM 1.2; discussed by H. Eshel 2008, 170–171; Yadin 1962, 22–26. For Callimachus, see Catull. 66.12.

165. Silverman 2012, 154–167; Sundermann 2011; Boyce and Grenet 1991, 382–386; Boyce 1989 and 1984, 66–75; Hultgård 1992, 1991, and 1989; Kippenberg 1978; Eddy 1961, 14–32; Benveniste 1932, 372; Cumont 1931, 67 and 77–78; Meyer 1921–1923, 2:190–191.

166. *Zand ī Wahman Yasn* 1: 1–11. Trans. Cereti 1995, 149, modified by Vevaina (2011, 246–247).

167. 4QFour Kingdom (4Q553).

168. To argue for the direct influence of the book of Daniel on the *Zand ī Wahman Yasn*, Gignoux (1986, 58) suggested that the final tree branch was made of "iron mixed with dust *(āhan ī *xāk gumēxt)*," an emendation accepted by Boyce (1984, 71–72). Boyce and Grenet (1991, 386), who reverse the direction of influence, suggest Zoroastrian influence on Daniel. However, Vevaina (2011, 249–251) has shown, from an examination of the surviving manuscripts, that there is no support for such an emendation.

169. On the Kayanians, see, e.g., Wiesehöfer 1994, 16–17.

170. E.g., Cereti 2015, 271, and 1995, 12–14 and 26–27; Gignoux 1999; Duchesne-Guillemin 1982.

171. Duchesne-Guillemin 1982, 760.

172. See, e.g., Kellens 2015, 47; Skjærvø 1999, 4–9.

173. *Yasna* 26:10, 30:7–9; *Yašt* 19:10–12, 13:129. Attempts have even been made to identify behind these passages a common Indo-European eschatology (see O'Brien 1976; Lincoln 1977). On this Avestan theology, see Agostini 2016, 497–498, Humbach 2015, 42–43; Silverman 2012, 94; Vevaina 2009, 220; Boyce and Grenet 1991, 364–367; Boyce 1984, 57–59; Widengren 1983, 77–84; Kippenberg 1978, 53–58.

174. See the articles collected in Lincoln 2012.

175. *FGrH* 765 F32 = Diog. Laert. 1.2. On this passage, see Gnoli 2000, 46; Kingsley 1995, 182 and 192–194.

176. Eudoxus F342 (Lasserre) = Plin. *HN* 30.2.3; Aristotle, περὶ φιλοσοφίας F6 (Untersteiner). Discussed by Horky 2009, 83–84; Kingsley 1995, 195.

177. *BNJ* 115 F65 = Plut. *De Is. et Os.* 369d–370c. See the commentaries of de Jong 1997, 158–202: Widengren 1983, 127–133; Kippenberg 1978, 53; Hopfner 1940–1941, 201–211; Griffiths 1970, 470–482.

178. Especially, *Greater Bundahišn* 1 and 34. See, e.g., Panaino 2015, 238–241; Vevaina 2011, 253–254; Silverman 2012, 58–59.

179. Justin, *Apol.* 1.20.1; Clement, *Strom.* 6.5.1; Lactant. *Div inst.* 7.15.19, 7.18.2, and possibly 7.17.9–11. For commentary and analysis, see Agostini 2016, 499–501, Sundermann 2012; Widengren 1983, 119–127; Kippenberg 1978, 70–75; Hinnells 1973; Cumont 1931, 64–93; Windisch 1929. Flusser (1982) identifies *Hystaspes* as a Jewish apocalypse based on a Zoroastrian source.

180. Hinnells 1973, 135–136; Eddy 1961, 18; Benveniste 1932, 374–375; Cumont 1931, 77–78; Windisch 1929, 37–39.

181. Lact. *Div. inst.* 7.15.9.
182. This was first proposed by Benveniste (1932, 377–378) and adopted by, among others, Boyce and Grenet (1991, 378).
183. Sundermann 2012; Colpe 1983, 831; Kippenberg 1978, 70; Eddy 1961, 33–34.
184. Silverman 2012, 155–156; Hultgård 1991, 123–127.
185. *Zand ī Wahman Yasn* 3:26, discussed by Cereti 1995, 184–185.
186. *Zand ī Wahman Yasn* 6:5.
187. App. *Syr.* 56, discussed by Kosmin 2014a, 212–214.
188. Cereti 1995, 200.
189. Silverman 2012, 165–166.
190. *Seder 'Olam* 28 (trans. Guggenheimer 1998, 240).
191. Scholem 1971, 10. See also Bultmann 1957, 30–31.
192. See, e.g., Agostini 2016; Silverman 2015; Dobroruka 2012; Wiesehöfer 2003; West 1997, 312–319; Koenen 1994, 13; Boyce and Grenet 1991, 374–375, 384–386, and 412–413; Boyce 1984, 70–72; Burkert 1989, 244–250; Kvanvig 1988, 484–491; Gignoux 1986; Duchesne-Guillemin 1982; Momigliano 1980; D. Russell 1964, 228–229; Swain 1940.
193. The supposed precedents include Hesiod's myth of five ages (Hes. *Op.* 109–201; although these represented types of humans, not political regimes; only the iron belongs to history proper; and the schema is itself likely a borrowing), imperial sequencing in Greek historiography (Hdt. 1.95, 130; Ctesias *FGrHist* 688 F1; although these have less clarity and coherence than has often been stated); the periodizing concerns of Zoroastrian texts (discussed above); and certain inscriptions of the Neo-Babylonian monarch Nabonidus, as read by Haubold (2013, 78–97). These intimations or preliminary sketches of apocalyptic periodization have been overplayed: they are recognized as such only after their Hellenistic formulations are already known, much as, for Hegel, human anatomy contained the key to the anatomy of the ape.
194. The relevant Egyptian texts—the "Demotic Chronicle," the "Oracle of the Lamb," the "Oracle of the Potter," and the "Dream of Nectanebo"—have been republished with commentary in Blasius and Schipper 2002a. For a summary of their common features, and the derivation of these from earlier Egyptian ideas, see Moore 2015, 160–187; Blasius and Schipper 2002b, 282–302. Although these Ptolemaic texts share some important motifs with the eschatological total histories discussed in this chapter—among them pseudonymy, the condemnation of a current crisis, and the predicted arrival of a savior king—they do not give a sequential narrative of the land's history, offer schematic periodizations of that history, appear to oppose the ruling Hellenistic regime as such (but rather the Achaemenids, the Seleucids, the city of Alexandria, and the Jews of Egypt; see, e.g., Moyer 2011, 128–134; Blasius and Schipper 2002b, 294–298; Felber 2002, 106–110; Koenen 2002, 170–172 and 183–187; Thissen 2002, 123–125; Johnson 1984), or culminate in the end of historical time. Indeed, for our purposes this last may be the most significant difference, and perhaps diagnostic of the radical novelty of the Seleucid temporal order: these Egyptian prophecies and oracles culminate in a restoration of order and the reign of a good pharaoh within a continuing

historical time (a form and content that goes back to the "Prophecy of Neferti" at the very opening of the second millennium BCE), whereas, as we have seen in this chapter, the Judean and Iranian total histories conclude with the end of historical time as such.

195. Gasche 2010; Boiy 2004, 9; Haerinck 1973; Schmidt 1941, 820–822.
196. See Kosmin 2014a, 193–195, with Figure 6.
197. Martinez-Sève 2015. A Greek funerary stele was uncovered in its vicinity (see Martinez-Sève 2002, 50).
198. Berlin and Herbert 2003, 376–377.
199. In this most disputed of regions it is at least tempting to identify a symbolic act of annihilation and irreversible conquest in the destruction and treading underfoot of the adversary's materials of administration. On smashing pottery to symbolize irreversibility, see Kosmin 2015, 126–134.
200. For instance, Hellenistic Lydia contained the layered, Achaemenid-Seleucid population of "Macedonian-Hyrcanians" (Strabo 13.4.13; IGR 4.1354; see also G. Cohen 1995, 209–212) and Phrygia, the colony of "Docimeum of the Macedonians" beside a mountain named "Persis" (see Sergueenkova and Rojas 2017, 286–287; G. Cohen 1995, 295–299).
201. See, e.g., Jameson 2003, 699; Koselleck 2002, 102–103.
202. Boym 2001, 136 (developing a point in Benjamin 2009, 177–178).
203. M. Bloch 1992, 29.
204. On the pedagogical function of the maśkilim, see Portier-Young 2011, 254–276.
205. Hayek and Caldwell 2010, 67.
206. The term derives from Pomian 1984, 160–162. See also the similar claims of Davis 2008, 116; Maier 1999.
207. See, e.g., Hölscher 2014; Blumenberg 1983, 457–472.
208. On the aesthetics of miniaturization, see, e.g., S. Stewart 1984, 37–69.
209. On the periodizing function of ritual, see, e.g., Rappaport 1992, 9–12.
210. On teleology and the "sense of an ending," see Kermode 2000, 190–193. See also Simmel 1977, 150–153.

6. *ALTNEULAND*: RESISTANCE AND THE RESURRECTED STATE

1. See, e.g., Chrubasik 2016; Ehling 2008, 111–277.
2. On Commagene, see Diod. Sic. 31.19a on the revolt of governor Ptolemaeus, sometime before 163/162 BCE, discussed by Facella 2006, 199–205. On Sophene, see Strabo 11.14.15 on kings Zariadres and Artanes, discussed by Marciak 2012. On Osrhoene, according to the Syriac *Chronicle of Pseudo-Dionysius of Tell-Maḥre*, in 132/131 BCE (180 SE) Antioch-on-the-Callirhoe (Edessa) gained its first independent monarch, the Arab king Orhai ('rhy), son of Ḥewyā (or, if Orhai is fictive, 'Abdū bar Maṣ'ūr). See Tubach 2009, 295–311; Luther 1999, 437–440; Duval 1892: 30–39. On Adiabene, in the first half of the second century BCE a bearded king Abdissares (probably "Servant of Ištar") appears on very rare coins wearing the Iranian satrapal tiara tied with a diadem; see Marciak and Wójcikowski 2016, 80–88; Luther 2015, 276–280; and de Callataÿ 1996. The rupestral relief at Batas-Herir in northern

Iraq may depict the newly invested king; see Grabowski 2011, 132–134. On Media Atropatene, see, e.g., Boyce and Grenet 1991, 69–86. On the Nabateans, see, e.g., Schwentzel 2005; Sartre 2001, 411–424. On the Phoenicians, see, e.g., Duyrat 2005; Millar 1983.

3. This has been taken as evidence of both the late vitality of Babylonian cultural traditions and the openness of the Seleucids to Near Eastern monarchic forms (see, e.g., Haubold 2013, 127–177; Sherwin-White and Kuhrt 1993, 149–161.

4. App. *Syr.* 58, discussed by Kosmin 2014a, 212–214.

5. Following Clancier and Monerie 2014. The so-called Lehmann Tablet (MMA 86.11.299), dating to 173/172 BCE, records land donations by Laodice, wife of Antiochus II, to the inhabitants of Babylon, Borsippa, and Kutha in 236 BCE. It has been proposed that this was an attempt to confirm Babylonian ownership in the face of the reallocation of these lands to the new Seleucid colony in Babylon (see, e.g., Clancier and Monerie 2014, 192 and 219–220; van der Spek 2009; 112–113; Boiy 2004, 151, 160, and 207).

6. *BCHP* 14 Obv., lines 2–6.

7. Kosmin 2014b.

8. *BCHP* 6 Obv., line 6; *BCHP* 9 Rev., lines 6–7; *BCHP* 11 Rev., line 2; and *AD* -168 A Obv. 15 attested for Seleucus I or Antiochus I, the invading Ptolemy III, and the reign of Antiochus IV (see van der Spek's commentary on *BCHP* 6 Obv., line 9).

9. A tablet of 308/307 BCE (Porter 1822 II, pl. 77g; van der Spek 1995, #9) attests to an attempted Seleucid appropriation from the temple of Šamaš in Sippar or Larsa by the head of the royal treasury (see Capdetrey 2007, 182; Joannès 2006, 112–115; and van der Spek 1995, 238–241, and 1993, 65). *AD* -168 A Rev. 12–14 describes the appointment by royal decree of a *zazakku*, or financial officer, over Babylon's Esagil temple (see, e.g., Pirngruber 2017, 153, 175, and 188; Geller 1991: 2–4).

10. Hdt. 3.150–160; Beaulieu 2014.

11. Waerzeggers 2003–2004; Horowitz 1995, 65–67.

12. On fighting, see *AD* -261 B Obv. 1–3, C Rev. 9–12; -255 A Rev. 15; -237 Obv. 13; -234 Obv. 13; -229 A Rev. 5–6; -162 Rev. 11–17; -161 A_1A_2 Obv. 23–24 and Rev. 12; -156 A Rev. 19–20. On theft, see *AD* -277 C Obv. 14–15; -254 Obv. 13 and Lower Edge; -240 Obv. 6–8; -175 B Rev. and Left Edge; -168 A Rev. 12–20.

13. Eddy 1961, 157–162 (quotation from page 159).

14. The multiethnicity of Babylonia was a recognized characteristic: Herodotus 3.159.2 reported that the Babylonians of his day were descended from the women of all the surrounding nations ("περιοίκοισι ἔθνεσι"), who had been settled there by Darius I after the Babylonian revolt. Berossus (*BNJ* 680 F1b = Syncellus, *Chron.* 49, 19 (3) (Dindorf)) retrojected this situation to prehistoric times, reporting that the region was settled from the first by a mingled population ("πολὺ πλῆθος ἀνθρώπων γενέσθαι ἀλλοεθνῶν").

15. See, e.g., Clancier and Monerie 2014, 186.

16. See, e.g., Joseph. *AJ* 18.9.8–9; Tac. *Ann.* 6.42. On the necessity of strong, communal structures for a sustained, coordinated engagement with imperial agency, see, e.g., Boone 2003: 20–23.

17. Bar-Kochva 1976, 52. On Jews fighting for the Seleucids in Babylonia, see 2 Macc. 8:20, discussed by Bar-Kochva 1976, 500–507.

18. On the local case of Adad-nādin-aḫḫē, see below.

19. See, e.g., Kuhrt 1987, 56. Note, however, that it is possible to detect a spirit of competition in each of the witnesses to this supposed Seleucid prototype. Megasthenes's Nebuchadnezzar is esteemed among the Chaldeans even more than Heracles ("Ἡρακλέους μᾶλλον"), the Macedonian hero-ancestor par excellence and the stock figure of comparison for Alexander (Megasthenes BNJ 715 F11a = Strabo 15.1.6). I have argued elsewhere that Berossus's accession narrative for Nebuchadnezzar was deliberately contrived so as to contrast with the chaos of Antiochus I's, as a smooth old Babylonian success to a contemporary Seleucid failure (Kosmin 2013c, 209–210). In Chapter 2 I suggested that Antiochus I's reconstruction of the Ezida temple of Borsippa in 43 SE, on the eve of 44 SE, marked a deliberate surpassing of Nebuchadnezzar II's famed forty-three-year reign. Even the Babylonian gift of Nebuchadnezzar II's royal robe to Antiochus III in 187 BCE (AD -187 A Rev. 10–12) had an ambiguous valence—offered on the four hundredth anniversary of the Neo-Babylonian conquest of Jerusalem to a Seleucid king devastated by the disaster at Magnesia, the humiliations of the Apamea peace treaty, the loss of Asia Minor, the dispatching to Rome of his son as a hostage, the disbanding of his elephant corps, the independence of Armenia, and the rebellions in western Iran. Did the Babylonians join with the smug Romans in gloating, "There once was a great king" (App. Syr. 37)?

20. As Haubold (2016) observed.

21. See, e.g., Beaulieu 2013; Schaudig 2003.

22. Clancier and Monerie 2014, 181; Beaulieu 2006, 201–206.

23. See, e.g., Nielsen 2015.

24. See, e.g., Ristvet 2015. 203–204; Clancier 2011, 756.

25. I am following D. Brown 2008, 85–88. See also Pirngruber 2017, 41–42; van de Mieroop 2016, 212.

26. The latest cuneiform tablet so far discovered is an almanac from Uruk dated to 79/80 CE (see Hunger and de Jong 2014).

27. Taylor 2011, 16.

28. Ristvet (2015, 184–185) notes that at Uruk the courtyard of the private house belonging to the Ekur-zākir āšipu priests housed a tablet kiln. For archaeological details, see Hoh 1979, 29–30.

29. Ristvet (2015, 203) cites Clancier (2005) on the juridical document OECT 9, 24 (228 BCE), indicating that the cuneiform tablet copied a contract in parchment composed a month earlier.

30. Wallenfels 1996, 120–121.

31. See, e.g., Hicks 2016, 72 and 76; Robson 2011, 557–558.

32. MacRae 2016, 28.

33. Or his grandfather (see Chapter 4).

34. TU 45 (AO 6472) Obv. 22–25, trans. Linssen. On the second-person, casuistic form of these ritual texts, see Watts 2003, 92–94.

35. TU 45 (AO 6472) Rev. 1, trans. Linssen.

36. *RAcc.*, 127–154, 415–421, trans. Linssen.

37. See, e.g., Linssen 2004, 102; Mayer 1988, 150–154.

38. They are not like the anachronisms of the Mishnah, which can imagine a past of Temple worship in the image of the rabbinical present (see, e.g., Cohn 2009). Rather, these "catachronistic" ritual texts imagine the Seleucid present in the manner of the lost past of involved and present kings. On the textualization of ritual in general, see, e.g., Bell 1988.

39. Here, I virtualize the "nationalist" reading of the *akītu* rites by J. Z. Smith, who assumes the physical enactment of the king's striking (1976, 2–11, and 1978, 72–74). Noted already by Zimmern (1922, 192): "Ich möchte darum auch annehmen, daß dieses Neujahrsfestritual, wenigstens in seinem jetzigen Wortlaute, nicht etwa, wie mancher sonstige späte Text, eine sklavische Kopie aus älterer Zeit ist, sondern vielmehr auf der Grundlage älterer Vorlagen eine freie Konzeption des Neujahrsfestrituals für die Priesterschaft des Esagil-Tempels in Babylon aus der spätesten, seleukidisch-parthischen, Periode darstellt."

40. Waerzeggers 2010, 115–116; Linssen 2004, 132–138.

41. *TU* 38, Rev. 43–50; trans. Krul 2014: 74, adapted.

42. Hdt. 3.150–160; Waerzeggers 2003–2004, 162.

43. Though it drew on a very ancient past. See, e.g., Krul 2014, 19–20 and 56–64; Baker 2014, 188–198, and 2013, 56–58; Kose 2013b, 324–326.

44. Strikingly, it is the only temple name attested in Akkadian alone, without a Sumerian equivalent (see Krul 2014, 64; Röllig 1991, 127).

45. Da Riva 2017, 78–79; van de Mieroop 2016, 29; Waerzeggers 2010, 115–118; Beaulieu 2004, 315–316, and 1993, 47–50. Perhaps Kidin-Anu was responsible for this ritual encyclopedia.

46. Rassam Cylinder 6.107–124 (Streck 1916, 58–59), discussed by Scurlock 2006, 455–456.

47. Ezra 1:7–11, 5:14–15, and 6:5; 2 Chronicles 36; 1 Esdras 2:10–15 and 8:55–57, discussed by Honigman 2014, 217–218; Ackroyd 1972.

48. E.g., *OGIS* 54, lines 20–22, and 56, lines 10–11, discussed by Briant 2003, 175–180.

49. Paus. 1.8.5, 1.16.3; Val. Max. 2.10 ext.1; Gell. 7.17.2 (concerning Athens's public library).

50. This accords with the report of *ABC* 2 lines 16–17, that "Nabopolassar returned to Susa the gods of Susa, which the Assyrians had carried off and settled in Uruk."

51. See, e.g., Kose 2013a; Schaudig 2010, 142–143.

52. See Kose 2013b, 326–327, with the excavation photograph (figure 57.6) at 324. Pace Westh-Hansen 2017, 159–160.

53. On these urbanistic developments, see Baker 2014, 200–203 (updating Bergamini 1987, 204–212).

54. *AD* -140 C Obv. 35–44.

55. *AD* -132 B Rev. 18–20.

56. *AD* -124 A Rev. 4–5.

57. See the Astronomical Diaries for these years as well as *BCHP* 18, with commentary.

58. Noted by Pirngruber (2013, 202–204). These births recall the protases of the older Babylonian omen collection *Šumma izbu*.

59. *AD* -132 B Rev. 25–36, Lower edge, Left edge, Upper edge, C Obv. 26–34. This event has been discussed by van der Spek (2014a), Shayegan (2011, 83–85), Nissinen (2002), and Del Monte (1997, 123–127), but it remains poorly understood on account of tablet breaks and Akkadian ambiguities.

60. Note that, in the alternative and shorter account, the boatman sets up multiple daises ("bár^meš") (*AD* -132 C Obv. 26).

61. I am following van der Spek (2014a: 14–16) and Del Monte (1997, 126).

62. I am following Nissinen (2002, 66n20), who derives this noun from the verb *šabābu* ("to roast" or "to burn").

63. *AD* -132 C Obv. 26–34.

64. This does not seem to be an absence associated with the mid-year *akītu* (pace Nissinen 2002, 68), for popular rumors of Nanaya's return had already begun in Ulūlu 133 BCE, as reported in *AD* -132 B Obv. 29, and the episode falls outside the regular rhythms of the ritual year.

65. *māḫiṣu* is the G participle of the verb *maḫāṣu* ("to strike"), and substantivized forms can mean "weaver," "plowman," "hunter," or "military scout." As van der Spek (2014a, 14) observes, this cannot be the goddess Nanaya, as the grammatical gender is masculine (contra Shayegan 2011, 84; Nissinen 2002, 69). Note that Nissinen (2002, 69–70) reads this as "the strong, striking god, your god," treating the plural dingir^meš-*ku-nu, ilīkunu*, as plural in form and singular in meaning, like the Hebrew *ʾelōheḳem*.

66. Proposals have included the Greek thundering Zeus, the Zoroastrian wicked Ahriman, the Neo-Assyrian bowman god, and the jealous God of Israel (see van der Spek, 2014a, 16–17; Nissinen 2002, 70).

67. The Second Dynasty of Lagash was not incorporated into the Sumerian King List, and the chronology, sequence, and regnal lengths of its rulers remain obscure (see, e.g., Suter 2000, 15; Steinkeller 1988). In all likelihood, the capital had already been moved from Lagash (modern al-Hiba) to Girsu in the Early Dynastic period (the Lagash I dynasty; see D. Hansen 1992).

68. Suter 2012, 61; Edzard 1997, 26.

69. The palace's construction can be confidently dated to the mid- to late second century on paleographic, ceramic, and numismatic grounds (see, e.g., Kose 2000, 407). On the layout of the Tello site, see Rey 2016, 5–12 and 15–16.

70. Note that Rey (2016, 20–31) confirms that Tell A belonged to Girsu's Uruku, or holy quarter, but locates the Early Dynastic Eninnu on Tell K.

71. For a description of the palace, see Wood forthcoming; Caubet 2009; Kose 2000; Parrot 1948, 151–156; Heuzey 1888 and 1884–1912, 395–449; de Sarzac 1884–1912, 33 and 44–45.

72. *IEOG* 129a. The line break occurs in the same position in both the Greek and Aramaic. Ristvet (2015, 200) notes that the employment of these languages in Mesopotamian building inscriptions or monuments was paralleled by Anu-uballiṭ-Nicarchus and Anu-uballiṭ-Cephalon at third-century BCE Uruk.

73. Kose 2000, 398 and 407 (following the records of the excavator G. Cros). However, it is possible that this was the location of a well. Glazed ceramic lamps were also found.

74. These were also combined with the bricks of Gudea (see Huh 2008, 247; Ghirshman 1936).

75. Winckler (1898, 80) suggested that Adad-nādin-aḫḫē was identical with Sagdodonacus, who was the father of Hyspaosines (the Seleucid sea-land satrap and first king of Characene). This has been thoroughly refuted (see, e.g., Parrot 1948, 312–313).

76. Parrot 1948, 310–313; de Genouillac 1930, 183, and 1936, 53 and 56.

77. In corridor N (see Kose 2000, 417–418).

78. See the description of Heimpel (1996, 17).

79. A suggestion of Radner (2005, 234).

80. For an attempt to reconstruct the site's stratigraphy, see Huh 2008, 23–48. On the methodological backwardness of the excavation, see Liverani 2016, 65–66.

81. Note, though, that the reign of Šulgi, of the Ur III dynasty and a near contemporary of Gudea, was still written about in mid-third-century BCE Uruk (see Cavigneaux 2005).

82. Kose (2000, 424) believes so.

83. Note that one of the roughly contemporary Uruk ritual texts for temple building prescribed the deliberate veneration and reuse of ancient temple bricks: "[The builder of the new temple] shall take an ax of lead, remove the former brick, and put it in a restricted place. You set up an offering table in front of the brick for the god of foundations, and you offer sacrifices" (*TU* 46 (O.174) Obv. 14–17). Ellis interprets this as preserving the continuity of worship: "The single brick embodied the essence of the god's home and bridged the gap between the destruction of the old building and the founding of the new" (1968, 28–29).

84. Beaulieu 2013, 125. Eddy described it as "a safe form of nonpolitical activity" (1961, 152).

85. See, e.g., Radner 2005, 234 and Winter 1992, 26–30.

86. See, e.g., Wood forthcoming. Note the contrast with the neighboring or suzerain kingdom of Characene, based in a Hellenistic colony on soil that had not yet been deposited in the days of Gudea, whose early rulers fashioned themselves in the Seleucid manner, appearing on their coins with shaven faces, Hellenistic profiles, and diademed heads, displaying beneath the Seleucid Era count of years.

87. Ankersmit 2005. Note the important critique of Icke (2012, 106–151). See also Roshwald 2006, 21 and 48–61.

88. For references, see G. Cohen 2013, 185–187, 189, and 199.

89. Callieri 2004; Wiesehöfer 1994, 68, and 93–95.

90. Polyaenus *Strat.* 7.39. Bar-Kochva, noting that Antiochus III's Thracian forces at Raphia fought under the same commander as the Persians, suggested that they should be identified with these settlers in Persis (1976, 50).

91. Polyb. 5.40.7 (Alexander, brother of the rebel Molon). See also Polyb. 5.54.13.

92. Polyb. 5.79.6; Liv. 37.40; App. *Syr.* 32, discussed by Bar-Kochva 1976, 49–52.

93. Plin. *HN* 6.32.152. For the date of this expedition and its Gulf context, see Kosmin 2013b.

94. 2 Macc. 9:1–2. See also Kosmin 2016, 40–46.

95. Polyaenus *Strat.* 7.39.
96. Polyaenus *Strat.* 7.40.
97. Following Houghton, Lorber, and Hoover (2008, 213–215), the sequence is: Ardaxšīr I, Vahbarz, Baydād, Vadfradad I, and Vadfradad II. Müseler (2018) proposes an interruption and restoration of Baydād's rule.
98. For the high date, see Müseler 2018; Engels 2017, 247–306, and 2013; Chrubasik 2016, 34–35; Curtis 2010, 383–387; Houghton, Lorber, and Hoover 2008, 213–215; Klose and Müseler 2008, 15–46; Müseler 2005–2006, 75–77; Klose 2005; Eddy 1961, 75–80. For the low date, see Strootman 2017, 194–197; Plischke 2014, 298–312; Wiesehöfer 2013a, 2011, and 1994, 101–136; Rezakhani 2013, 768 and 775; Shayegan 2011, 168–178; Haerinck and Overlaet 2008, 208; Callieri 2007, 115–145; Alram 1986, 162–171; Altheim 1959, 375–379. The argument for the high date is based on three strictly numismatic grounds: (i) that no coins of a Seleucid king later than Seleucus I have been found in the few excavated hoards from Persis that contain *fratarakā* issues; (ii) that some *fratarakā* coins are overstrikes of Seleucus I types; and (iii) that there is a stylistic—and thus a chronological—gap between the coins of the first *frataraka* rulers down to Vadfradad I and those of Vadfradad II and his successors. To advocates of the low date, these factors are not compelling: (i) Persis appears to have suffered a severe shortage of precious metal in the third and second centuries BCE and did not participate deeply in interregional trade, as suggested by the *frataraka* overstrikes, the absence of *frataraka* coins outside Persis, and the scarcity of post-Seleucus I coins within Persis; (ii) the issues of Alexander and Seleucus I, struck from Achaemenid bullion, dominated this region; (iii) the *frataraka* coins were likely more of a prestige strike than functioning currency (Callieri 2007, 113 and 128); (iv) the original undertypes of most overstruck *frataraka* coins cannot be identified, and the evidence is so scarce that one new coin may completely challenge the high-date reconstruction (Engels 2017, 256–257, and 2013, 37)—note that in Judea, Seleucid coins survived to be overstruck by the Jewish forces of the First and Second Revolts (Miller 2011; Mildenberg 1984, 89n242); (v) the presence in hoards of Seleucus I coins alongside those of the later *frataraka* rulers demonstrates the long circulation of these coins, and post-Seleucus I coins are absent despite the certain and attested presence in Persis of later Seleucid administrators, armies, and kings; and (vi) the coins show a stylistic continuity until the reign of Darev II.
99. Potts 2007, 275. The same headdress was adopted on the coins of Arsaces I and Arsaces II (Martinez-Sève 2014, 133–134). Baydād's earflaps have been tucked in, which has been taken as an iconographic marker of either superior or inferior status (see, e.g., Engels 2017, 296, and 2013, 73–74; Wiesehöfer 1994, 103). Note the similarly tucked headdress of the contemporary Abdissares, likely first king of Adiabene, on his coins and in the Batas-Herir relief (see Marciak and Wójcikowski 2016, 80–88; Grabowski 2011).
100. For the Achaemenid prototypes, see, e.g., Garrison 2013, 580–581; Root 1979, 162–181.
101. See, e.g., Potts 2007, 278–296. For doubts about this identification, see Callieri 2007, 122; Haerinck and Overlaet 2008.

102. See de Jong 2003; Nylander 1983. Haerinck and Overlaet (2008, 208) note that a nearly identical image—of a ruler figure, building, and standard—is known from an unprovenanced clay *bulla*, suggesting a wider employment of this typology.

103. E.g., Darius I's Behistun relief, the Çan Sarcophagus, and Graeco-Persian gemstones (see Klose and Müseler 2008, 27–30; Ma 2008; Müseler 2005–2006, 93–94; Klose 2005, 97–98; Boardman 2000, 160, figures 5.6 and 5.7, and 1970, plate 881).

104. As suggested by, e.g., Plischke (2014, 302) and Wiesehöfer (2011, 113–114), though note the doubts of Engels (2013, 62–69).

105. For the Achaemenid prototypes, see, e.g., Garrison 2013, 578–580; Root 1979, 232–240. Note that the Achaemenid palace on the site of Palace H, the likely *frataraka* residence, was decorated with at least one bas-relief showing the king on his throne (see Tilia 1977, 77).

106. Noted by Rezakhani (2013, 777).

107. Ito 1976.

108. The *fratarakā* adopted the royal title *mlk'* either just prior to or immediately after the Parthian conquest of Persis.

109. Klose and Müseler 2008, 18–19; Panaino 2003.

110. I am following Skjærvø 1997, 102. Contra Altheim (1959, 378), who reads the legend as an abbreviation of *"pārs bīrtā"* ("citadel of Persis")—i.e., Persepolis.

111. See, e.g., Rung 2015; Martinez-Sève 2014, 132; Olbrycht 2013, 65–68.

112. Diod. Sic. 17.70; Curt. 5.6.4–8.

113. In an assessment unique to Persis, Diod. Sic. 17.71.3 reports that Alexander "did not trust the inhabitants and felt bitter enmity toward them."

114. Arr. *Anab.* 3.18.11–12; Diod. Sic. 17.72; Plut. *Alex.* 38; Curt. 5.7.1–9. For interpretations of this act, see Mousavi 2012, 60–71; Sancisi-Weerdenburg 1993; Briant 2002, 850–851; Hammond 1992; Borza 1972; Badian 1967, 185–190.

115. Note that when Peucestas, satrap of Persis, hosted a grand feast for the successor Eumenes at Persepolis in 316 BCE, a tent city had to be erected at the base of the rubble-strewn terrace (Diod. Sic. 19.22.1–3, discussed by Mousavi 2012, 73; Henkelman 2011, 116–118).

116. Mousavi 2012, 74; Callieri 2007, 134; Boucharlat 2006, 451–452; Wiesehöfer 1994, 68–70; Tilia 1977 and 1972, 253–316; Schmidt 1953, 279–282. Such post-Achaemenid reuse was first proposed by Flandin (1851, 2:184–185).

117. Haerinck and Overlaet 2008, 210–212; Tilia 1977, 77, 1972, 315n1, and 1969.

118. Istakhr would replace Persepolis at some point in the Parthian period (see Callieri 2007, 23–24 and 136–137; Wiesehöfer 1994, 64). Note that, at about the same time or a little earlier, the northeastern part of the so-called Fratadara temple, about 300 meters to the northwest of the terrace, was constructed, also partly out of toppled Achaemenid-period blocks. Two reliefs on the doorjambs closely resemble the adoration pose from the *frataraka* coinage (see, e.g., Mousavi 2012, 76–77).

119. The metaphorical interpretation of erect columns was part of Achaemenid visual vocabulary. For instance, a hundred spear-holding guards were de-

picted on the northern doorjambs of the Persepolis terrace's Hall of a Hundred Columns (see, e.g., Gopnik 2010, 204-205).

120. See also Thonemann 2015, 88–91. On the role of beards in scholarly discussions of anti-Hellenism, see Baumgarten 2010, 264-267.

121. On "anti-languages," see Halliday 1976.

122. *BNJ* 188, and perhaps also Metrodorus of Scepsis's history of Tigranes the Great (*BNJ* 184 F1 = schol. Ap. Rhod. *Argon.* 4.131).

123. On the use of Moses of Khoren, see the optimistic assessment of Tirats'yan (2003, 139-149) and the more cautious approach of Thomson (1978). Undoubtedly, many of Moses's sources on Artaxias are invented—including, probably, Polycrates, Euagoras, Scamandrus, Phlegonius, and, certainly, the "Chaldean Books" of Nineveh, translated at the request of Alexander the Great, and brought by Mar Abas Catina in a Syriac translation to Nisibis where the Armenian king Valarshak secured them in his palace. Furthermore, Moses's narrative was strongly influenced, down to specific lexical formulae, by 1 Maccabees. On Moses's sources, see Thomson 1978, 10-22. On the Artaxiad epic, see Redgate 1998, 68; J. Russell 1986-1987. Note that at least some elements of the epic—the expansion into the Caspian regions, for example—are supported by classical sources (Moses of Khoren 2.53; see also Strabo 11.14.5).

124. See, e.g., A. Smith 2003, 156-183.

125. On this period of Armenian history, see Khatchadourian 2016, 81–152. On the near invisibility of the region's Urartian past in the Achaemenid period, see Zimansky 1995.

126. See, e.g., Bosworth 1980, 315-316.

127. Strabo 11.14.9, discussed by Bernard 1999, 37–38 and 46–52.

128. Armenia did not appear in the settlements of Babylon or Triparadeisus, although Orontes appears as a close friend of Peucestas, governor of Persis (Diod. Sic. 19.23.3). Neoptolemus was campaigning in the region in 321 BCE (Plut. *Eum.* 4.1, 5.1; Diod. Sic. 18.29.2).

129. *Kartlis Tskhovreba* 1.22–23 (Gamq'relidze); see Braund 1994, 144. On the account of Parnavaz's rise, see, e.g., Rapp 2009, 668-674.

130. A remarkable Greek funerary epigram from Armavir, the traditional satrapal center, has a certain Numenius, perhaps a Seleucid official and the later governor of the Arab-Persian Gulf, report Orontes IV's death to his sister-wife in dactylic pentameter (Armavir #7 = *SEG* 44 1297, with Mahé 1994, 577 and 584-585.

131. Strabo 11.14.5, 15. On the date of this transition, see Bengtson 1944, 2:157-158.

132. Strabo 11.14.5, 15. See also L. Patterson 2001. According to Polyb. 25.2.12, Artaxias was included in the Pharnaces' peace treaty with Eumenes and Ariarathes. Note that the Georgian *Kartlis Tskhovreba* 1.26 (Gamq'relidze) correlates the rise of Artaxias with the end of Seleucid authority over Iberia in the reign of "Antiokoz of Babylon."

133. Diod. Sic. 31.17a; App. *Syr.* 45, 66; *AD* –164 B Obv. 15, C Obv. 13–14. The war was waged against "the fortresses of the city / land of Ḫabigalbat, which they call the land of Armenia (⌜e⌝.bàd^{meš} šá ^{uru}ḫa-bi-gal-bat šá ^{kur}ar-mi-ni$_5$ mu-šú sa$_4$-ú)." Ḫabigalbat was an Akkadian archaism, employed in the second millen-

nium BCE for the Hurrian federation of Mittani (see Clancier 2014, 421–431; Gera and Horowitz 1997, 243–244). On the determinative ᵘʳᵘ (possibly meaning "land" in late contexts), see van der Spek 2003, 306. In his commentary on the book of Daniel, the late antique scholar Porphyry identified Artaxias I as one of the three horns of the fourth beast of Daniel 7 pushed out by Antiochus IV Epiphanes (Porph. *BNJ* 260 F38 = Jerome, *in Dan.* 7:8). Jerome rightly dismissed this identification on the ground that Artaxias remained in possession of his kingdom as before *(sed illum in regno pristino permansisse)*.

134. Diod. Sic. 31.27a.

135. Diod. Sic. 31.17a coordinates these events: "Ἀρτάξης ὁ τῆς Ἀρμενίας βασιλεὺς ἀποστὰς Ἀντιόχου πόλιν ἔκτισεν ἐπώνυμον ἑαυτοῦ."

136. Traina 2007. The attribution of the city's plan to the famed general Hannibal—"the Armenian Carthage!" (Strabo 11.14.6; Plut. *Lucullus* 31.3-4 and 32.3)—likely derives from the self-promotion of its Roman conqueror, Lucullus.

137. Khatchadourian 2007, 60; Khachatrian 1998, 99–103; Tonikian 1992, 162 and 172–185. I am grateful to Mkrtich Zardarian and his team, of the Armenian Institute of Archaeology and Ethnography, for their generous hospitality and guided tour.

138. Moses of Khoren 2.49.

139. Sargsyan, cited in Tirats'yan 2003, 142. Perhaps it was modeled on Antiochus III's recent introduction of a Seleucid dynastic cult.

140. Tirats'yan 2003, 128.

141. A. Smith 2003, 166–168.

142. For Epiphania-on-the-Tigris, see Steph. Byz. s.v. *Epiphaneia* 4, discussed by G. Cohen 2013, 47–48; Chaumont 1993, 433. The *Chronicle of Pseudo-Dionysius of Tell-Maḥre* lists Amida as a foundation of Seleucus I (see G. Cohen 2013, 56–57).

143. Strabo 11.14.12.

144. Berossus *BNJ* 680 F4b = Syncellus, *Chron.* 53, 19 (16) (Dindorf) and F4c = Joseph. *AJ* 1.93.

145. I am following the interpretation of Périkhanian (1971, 170–173). Unfortunately, the reading of Dupont-Sommer (1946–1948), proposing a record of Artaxias's fishing for carp, cannot be sustained.

146. Tirats'yan 2003, 121; Redgate 1998, 81; Naveh 1971, 45; Périkhanian 1971, 171.

147. Note that the faked letter of the Achaemenid satrap Orontes, produced by Eumenes, was written in Aramaic (Diod. Sic. 19.23.3).

148. Armavir #5 = *SEG* 44 1295. This was presumably for didactic purposes (see Habicht 1953, 255–256; J. and L. Robert 1952).

149. See Khatchadourian 2016, 11–16.

150. Strabo 11.14.5.

151. Moses of Khoren 2.56: "After all these noble deeds and acts of wise government, Artašēs ordered the boundaries of villages and estates to be distinguished. . . . And he established markers for the borders in the following way: he ordered four-sided stones to be hewn, their centers to be hollowed

out like plates, and that they be buried in the earth. Over them he had fitted four-sided obelisks, a little higher than the ground."

152. Moses of Khoren 2.59.

153. The suffix -kn on 'rwndkn seems to belong to the Iranian adjectival morpheme -akān (see Naveh 1971, 45n21). For the Orontid-Achaemenid relation, see Xen. Anab. 2.4.8 and 3.4.13; Plut. Artax. 27.4, discussed by Tirats'yan 2003, 117 and 123; J. Russell 1987, 46.

154. See, e.g., Virno 2015, 7-19.

155. Kovacs 2016, 9-10; Vardanyan and Vardanyan 2008.

156. See, e.g., Y. Zerubavel 1995, 22-26 and 218-221; Don-Yehiya 1992.

157. See, e.g., Doran 2012, 14-17; D. Schwartz 2008, 11-15.

158. On the significance for dating of 1 Maccabees' final verses, mentioning John and his deeds, see, e.g., S. Schwartz 1991, 36-37; Bar-Kochva 1989, 162-164; Momigliano 1976.

159. The bibliography is enormous (see, e.g., Eckhardt 2016; Chrubasik 2016, 123-145 and 179-184.

160. Ben Sira 36:1-3 and 12. Note that this prayer is considered a Hasmonean-period interpolation by J. Collins (1997, 109-111).

161. See Cotton and Wörrle 2007, with the improved readings of C. Jones 2009 and the new version from Byblus (Yon 2015, 92-101); Aperghis 2004, 166-171; Honigman 2014, 321-361.

162. 2 Macc. 4 and, for the intentions of Lysias and Antiochus V, 11:1-3. See also, e.g., Monson 2016, 28-35; Rooke 2000, 273-276.

163. 2 Macc. 4:9.

164. This ordering of events—"it was not the revolt which came as a response to the persecution, but the persecution which came as a response to the revolt" (Tcherikover 1961, 191)—was first proposed by Tcherikover (ibid., 183-203) and has gained a certain acceptance.

165. 2 Macc. 11:24.

166. Not including 4 Maccabees (a rewriting of 2 Maccabees). The words πάτριος (which appears in Sir. Prologue 8 and Isaiah 8:21 [a corrupt passage]), πατρῷος and προγονικός do not appear in this sense elsewhere.

167. For πάτριος, see 2 Macc. 6:1, 7:2, 8, 21, 24, 27, 37, 12:37, and 15:29. For πατρῷος, see 2 Macc. 4:15, 5:10, 6:1, 6, and 12:39. For προγονικός, see 2 Macc. 8:17 and 14:7. On such language, see, e.g., Nongbri 2005; Himmelfarb 1998, 27-28.

168. 2 Macc. 4:11 and 7:31. Jason's "ἐκαίνιζεν" may be a deliberate pun on Judas's "ἐγκαινίσαι" ("dedicate") and "ἐγκαινισμός" ("dedication") of the new altar (1 Macc. 4:36; 2 Macc. 2:19).

169. 1 Macc. 2:19-20.

170. 1 Macc. 1:56.

171. 1 Macc. 1:44-50 (A).

172. 1 Macc. 8:22; 10:3, 17, 25, and 51; 11:29; 12:5, 7, and 19; 13:35; 14:20; 15:1 and 15.

173. Typically translated, only here, as "letters" (1 Macc. 1:56 and 57; 3:48; 12:9; 14:23 [of Spartan public archives in a cited letter]; and 16:24).

174. See, e.g., Malachi 1:1 (the last of the Twelve minor prophets): "λῆμμα λόγου κυρίου ἐπὶ τὸν Ισραηλ ἐν χειρὶ ἀγγέλου αὐτοῦ."

175. Noted by Doran 1996, 39. See, e.g., Exodus 17:1, Leviticus 17:2, Numbers 30:2, and Deuteronomy 13:1.
176. The account of Antiochus IV's persecution in 2 Macc. 6:1–7 is less directly crafted to the biblical model, yet it also depicts a kind of Seleucid super-sessionism. The monarch dispatches a certain Geron of Athens "to compel the Jews to change from their ancestral laws" and in their place to rededicate the temple to Zeus Olympius, honor Dionysus and the king's birthday, and violate the Sabbath and the festivals. Nothing more is known of Geron, except that this lawgiver or philosopher from Athens was to oust the shepherd-prince of Egypt. Similarly, the book of Daniel characterizes Antiochus IV's persecution as the directing of disruptive, arrogant, slippery speech against God and those who honored His covenant (Daniel 7:8 and 11:32 and 36).
177. There are considerable inconsistencies with the parallel account in 2 Macc. 10–12 (see, e.g., Berthelot 2014; Brutti 2010, 157–158; S. Schwartz 1991, 25–26).
178. 1 Macc. 5:1, 3, 65, 66, and 68. See also S. Schwartz 1991; Bickerman 1979, 18–20.
179. 1 Macc. 5:4–5. See also Joseph. *AJ* 12.328.
180. On Jericho, see Joshua 6:15–21. On Amalek, see 1 Samuel 15:2–33; 1 Chronicles 4:42–43. See also, e.g., Berthelot 2014, 81; Batsch 2005, 408–439.
181. 1 Macc. 5:28, 35, 51, and 43–44 (interpreted as the *ḥērem* by Doran [1996, 21]); Joseph. *AJ* 12.344.
182. 1 Macc. 5:68
183. 1 Macc. 10:83–85. Cf. Joseph. *AJ* 13.100. Famously, the strong-man judge Samson destroyed the Dagon temple at Gaza (Judges 16:23–31), and the Philistine captivity of the Ark of the Covenant within Azotus's Dagon temple caused the cult statue to lose its hands and head (1 Samuel 5:1–5).
184. See Exodus 34:13; Deuteronomy 7:5 and 12:3; Joshua 6:24; 1 Samuel 15:8, 15, and 33; 2 Kings 18:4 and 23:4–20. See also Tilly 2015, 155; Doran 1996, 82.
185. 1 Macc. 5:51. Cf. Joseph. *AJ* 12.347. 2 Maccabees does not share this hostility to gentiles (see, e.g., D. Schwartz 1998; Doran 1981, 110–111).
186. 1 Macc. 9:73. Note that king Saul had settled in Michmash (1 Samuel 13:2). On the allusion to the book of Judges, see Rooke 2000, 284–285.
187. The topographic reliability of the campaigns has been asserted by Bar-Kochva (1989, 153–162). Also see below on 1 Enoch's Apocalypse of Animals.
188. 1 Enoch 90:9, 13, and 31.
189. 1 Enoch 89:12, 14, 42, 45, 47, and 48.
190. On such a re-creation of a heroic or mythic past in Greek ritual, see, e.g., Sinos 1998, 84–86; Nagy 1990, 345–349.
191. E.g., the video of the Islamic State's destruction of antique statuary in Iraq's Mosul Museum in February 2015 was accompanied by Quranic quotations recounting Ibrahim's foundational iconoclasm (see Flood 2016). An outraged and fearful Iraqi Christian, in flight from the Islamic State's brutality, asked "What century are we in?" (quoted in Freedland 2014). We might also compare how the Greek War of Independence was made to evoke the Persian Wars of antiquity, from the readoption of classical names to the reuse of sanctified landscapes. For instance, the first issue of the *Ephimeris ton Athinon*

(July 6, 1824) reported, in an archaizing mode that comes close to that of 1 Maccabees, that "General Gouras camped a few days ago at Marathon, very close to the cave of Pan, now called Nenoi, with around three hundred (!) soldiers . . . [and] carried out a war plan against the enemy" (quoted in Hamilakis 2007, 78).

192. 1 Macc. 3:35–36.

193. 1 Macc. 3:46.

194. Judges 10:17–18 and 11:11.

195. Judges 19–21. Note that this latter grim conflict, to avenge the gang rape and murder of a Levite's concubine, was set during the priesthood of Pinḥas (Judges 20:28), the zealot-priest claimed by the Hasmoneans as their ancestor (see 1 Macc. 2:54; see also Babota 2014, 269–284).

196. 1 Samuel 7:2–14.

197. 1 Samuel 10:17–25.

198. 1 Enoch 89:41. See Olson 2013, 182.

199. Jeremiah 40–41; 2 Kings 25:22–26. The Jeremiah account places extraordinary emphasis on Mizpah, with several redundant references to the site. On the events, see, e.g., Blenkinsopp 2013, 30–45; Weinberg 2007.

200. Deuteronomy 20:5–9.

201. Exodus 18:21 and 25; Deuteronomy 1:15.

202. 2 Macc. 8:12–29; Joseph. AJ 12.300–303, with reference to the Jews' supplication "in their ancestral manner (τῷ πατρίῳ νόμῳ)" and the troops being drawn up "according to the ancient and ancestral mode (τὸν ἀρχαῖον . . . τρόπον καὶ πάτριον)." For differences between the 1 and 2 Maccabees' accounts, see D. Schwartz 2008, 323–325; Bar-Kochva 1989, 219–274 and 494–499.

203. 1 Macc. 3:43. See also 1 Macc. 2:7, 12. On Maccabean slogans, see Nongbri 2005, 105.

204. 1 Macc. 4:36–61; 2 Macc. 1:8; Joseph. AJ 12.317. Note that the Seleucid-appointed high priest Alcimus, rival of the Hasmoneans, was held responsible for tearing down a wall of the Temple courtyard, "destroying the work of the prophets" (1 Macc. 9:54; Joseph. AJ 12.413). There is the evident influence of Jewish lamentation literature on this picture (see, e.g., Lamentations 5).

205. 1 Macc. 4:47. See also 2 Macc. 10:3.

206. 1 Macc. 9:62; Joseph. AJ. 13.26. See also 2 Chron. 26:10; Meshel 2000.

207. 2 Macc. 1:18.

208. Bergren 1997.

209. 1 Macc. 10:21; 2 Macc. 1:9, 18, and 10:6–7. On the character and identity of this new festival, see, e.g., Honigman 2014, 101–102, 113, and 122–136; Regev 2013, 36–57, and 2008, 93–97; Goldstein 1976, 273–284.

210. 1 Kings 8:65; 2 Chron. 5–6.

211. Ezra 3:1–6; 1 Esdras 5:48–51.

212. Nehemiah 6:15.

213. Nehemiah 8:13–18.

214. See, e.g., Berthelot 2014, 74–75, S. Schwartz 1991, 30–31; Bickerman 1979, 17–21; Goldstein 1976, 21. The big difference from the older scriptural histories

is that God does not intervene in 1 Maccabees (see, e.g., Dąbrowa 2010, 139–140).

215. Note, however, that many in the Maccabean movement did use Greek names (e.g., Eupolemus, Dositheus, and Sosipater). See, e.g., Hengel 1974, 63–65.

216. Joseph. *AJ* 12.239.

217. See, especially, Portier-Young 2010, contra S. Schwartz 1995, 11–12.

218. Jubilees 12:26: "I began to speak with him in Hebrew, the language of the creation." See also, e.g., Schniedewind 2013, 170.

219. 2 Macc. 12:37.

220. 2 Macc. 15:29.

221. 2 Macc. 15:33.

222. D. Schwartz 2008, 509.

223. 2 Macc. 7:8.

224. 2 Macc. 7:10.

225. 2 Macc. 7:21 and 27.

226. 2 Macc. 7:24.

227. On such resistance discourse, see the classic study of Scott 1990.

228. Regev 2013, 175–221; Dąbrowa 2010, 136–138; Meshorer 1982, 1:39–62.

229. Schniedewind 2013, 139–184; S. Schwartz 1995, 25–26; Meshorer and Qedar 1991, 10–11; Barag 1986–1987, 18–19.

230. Zissu and Abadi 2014; Schniedewind 2013, 180–182; Tov 1988, 13–14 and 19; Siegel 1971.

231. Schniedewind 2013, 160–161.

232. See, e.g., Honigman 2014; Ma 2013a and 2012, 70–84.

233. See, e.g., Honigman: "In historical terms, the idea that Antiochos prohibited circumcision is totally implausible" (2014, 246), and "no one forced the Judeans to eat forbidden foods" (ibid., 249).

234. See, e.g., Honigman 2014, 229–258; Weitzman 2004. On SpTU 3 58, "The Crimes and Sacrileges of Nabû-šuma-iškun," see also Finn 2017, 117–125.

235. For similar criticism, see, e.g., J. Collins 2016, 199–201.

236. Diod. Sic. 34/35.1.4. The passage is discussed in Chapter 3. See, similarly, Tac. *Hist.* 5.8.2: "*rex Antiochus demere superstitionem et mores Graecorum dare adnisus.*"

237. See, above all, Rajak 1996.

238. It may be helpful to compare contemporary populations, like Chasidic Jews or the Amish, who refuse to adopt the temporal structures and horizons of modernity. See, e.g., Rosa 2003, 16 and 22–23; Kraybill 1994; Olshan 1994, 233–240.

239. Droysen 1877–1878, 3:597 and 605–606. See the discussions of Briant (2017, 297–301), Moyer (2011, 11–15), and Momigliano (1970).

240. Eddy 1961.

241. The best study of this genre remains Ma 2002.

242. Memnon *BNJ* 434 F1 6.2 = Photius, *Bibliotheca* 224.222b9–239b43.

243. Memnon *BNJ* 434 F1 7.1 = Photius, *Bibliotheca* 224.222b9–239b43.

244. See, e.g., Alexander 2002, 243–248.

245. Theophr. *Char.* 16.8.

246. Herter 1950, 118–119.

247. Sophron F59 (Hordern).
248. Tober forthcoming; Hordern 2004, 11.
249. Polyb. 18.52.
250. Liv. 33.38.7. Note, by contrast, that the king spoke coaxingly through his ambassadors: *"nec vi tantum terrebat, sed per legatos leniter adloquendo castigandoque temeritatem"* (Liv. 33.38.5).
251. Summers 2006. See also Mbembe 2001, 37–38 and 103–104.
252. See, e.g., Fields 2009, 30–47 and 93–109.
253. Kosmin 2014a, 230–242. On the possible revolt of Aradus, reported by Porphyry (*BNJ* 260 F56 = Jerome, *in Daniel* 11:44–45: *"regrediens capiet Aradios resistentes et omnem in litore Phoenicis vastabit provinciam"*). But see Duyrat 2005, 251–252; J. Schwartz 1982; Will 1966–1967, 2:286; Mørkholm 1966, 122–124.
254. Williamson 2013. Note, however, the caution of Piras (2010, 225–229 and 232–233).
255. See, e.g., Momigliano 1987; Löwith 1949, 6–9; Collingwood 1946, 33–36.
256. See Kosmin 2016.
257. Benjamin writes, "It is precisely the modern which always conjures up prehistory" (1997, 171). And Boym claims that "the sense of historicity and discreteness of the past is a new nineteenth-century sensibility" (2001, 15).
258. See, e.g., Chowers 2012, 13–15 and 48–62.
259. Boym 2001, 41, see also 41–51.
260. On nonsynchronism, allochrony, and *die Gleichzeitigkeit des Ungleichzeitigen*, see Harootunian 2007, 472–479; Banerjee 2006, 3–8; Koselleck 2004, 93–104; E. Bloch 1977.

CONCLUSION

1. The midrash is discussed by Neusner (2001, 131–132 and 180) and Kugel (1995, 212–216).
2. See, e.g., Koselleck 2004, 12–17; Neusner 1988.

BIBLIOGRAPHY

Aaboe, A., J. Britton, J. Henderson, O. Neugebauer, and A. Sachs. 1991. *Saros Cycle Dates and Related Babylonian Astronomical Texts*. Philadelphia.

Ackroyd, P. 1972. "The Temple Vessels—A Continuity Theme." In *Studies in the Religion of Ancient Israel*, 166–181. Leiden.

Adiego, Ignacio-Javier, Pierre Debord, and Ender Varinlioğlu. 2005. "La stèle caro-grecque d'Hyllarima (Carie)." *Revue des études anciennes* 107:601–653.

Agamben, Giorgio. 1998. *Homo Sacer: Sovereign Power and Bare Life*. Translated by Daniel Heller-Roazen. Stanford.

Agostini, Domenico. 2016. "On Iranian and Jewish Apocalyptics, Again." *JAOS* 136:495–505.

Albertz, Rainer. 1994. *A History of Israelite Religion in the Old Testament Period*. Translated by John Bowden. Louisville.

Alcock, Susan. 1994. "Breaking Up the Hellenistic World: Survey and Society." In *Classical Greece: Ancient Histories and Modern Archaeologies*, edited by Ian Morris, 171–190. Cambridge.

Alcock, Susan, Jennifer Gates, and Jane Rempel. 2003. "Reading the Landscape: Survey Archaeology and the Hellenistic *Oikoumene*." In *A Companion to the Hellenistic World*, edited by Andrew Erskine, 354–372. Oxford.

Alexander, Loveday. 2002. "'Foolishness to the Greeks': Jews and Christians in the Public Life of the Empire." In *Philosophy and Power in the Graeco-Roman World. Essays in Honour of Miriam Griffin*, edited by Gillian Clark and Tessa Rajak, 229–249. Oxford.

Allen, Danielle. 1996. "A Schedule of Boundaries: An Exploration, Launched from the Water-Clock, of Athenian Time." *Greece & Rome* 43:157–168.

Allen, R. 1983. *The Attalid Kingdom: A Constitutional History*. Oxford.

Allen, Thomas. 2008. *A Republic in Time: Temporality and Social Imagination in Nineteenth-Century America*. Chapel Hill.

Alram, Michael. 1986. *Iranisches Personennamenbuch IV. Nomina Propria Iranica in Nummis*. Vienna.

Alter, Robert. 2011. *The Art of Biblical Poetry*. New York.

Altheim, Franz. 1959. *Geschichte der Hunnen I: Von den Anfängen bis zum Einbruch in Europa*. Berlin.

Andronicos, Manolis. 1984. *Vergina: The Royal Tombs and the Ancient City*. Athens.

Angel, Joseph. 2014. "Reading the *Book of Giants* in Literary and Historical Contexts." *Dead Sea Discoveries* 21:313–346.

Ankersmit, Frank. 2005. *Sublime Historical Experience.* Stanford.

Anson, Edward. 2005. "Idumaean Ostraca and Early Hellenistic Chronology." *JAOS* 125:263–266.

Aperghis, G. 2004. *The Seleukid Royal Economy: The Finances and Financial Administration of the Seleukid Empire.* Cambridge.

Appadurai, Arjun. 1996. *Modernity at Large: Cultural Dimensions of Globalization.* Minneapolis.

———. 2003. "Archive and Aspiration." In *Information Is Alive,* edited by J. Brouwer and A. Mulder, 14–25. Rotterdam.

Applegate, John. 1997. "Jeremiah and the Seventy Years in the Hebrew Bible: Inner-Biblical Reflections on the Prophet and his Prophecy." In *The Book of Jeremiah and Its Reception / Le livre de Jérémie et sa réception,* edited by A. Curtis and T. Römer, 91–110. Leuven.

Arendt, Hannah. 1958. *The Human Condition.* New York.

Ariel, D., I. Sharon, J. Gunneweg, and I. Perlman. 1985. "A Group of Stamped Hellenistic Storage-Jar Handles from Dor." *IEJ* 35:135–152.

Arnold, B. 1996. "The Use of Aramaic in the Hebrew Bible: Another Look at Bilingualism in Ezra and Daniel." *Journal of Northwest Semitic Languages* 22:1–16.

Assar, G. 2003. "Parthian Calendars at Babylon and Seleucia on the Tigris." *Iran* 41:171–191.

Assefa, Daniel. 2007. *L'Apocalypse des animaux (1 Hen 85-90): Une propagande militaire? Approches narrative, historico-critique, perspectives théologiques.* Leiden.

Azoulay, Vincent. 2009. "La gloire et l'outrage: Heurs et malheurs des statues honorifiques de Démétrios de Phalère." *Annales* 64:303–340.

Babelon, Ernest. 1890. *Catalogue des monnaies grecques de la Bibliothèque nationale. Les rois de Syrie, d'Arménie et de Commagène.* Paris.

Babota, Vasile. 2014. *The Institution of the Hasmonean High Priesthood.* Leiden.

Bach, Johannes. 2013. "Berossos, Antiochos und die Babyloniaka." *Ancient West and East* 12:157–180.

Badian, Ernst. 1958. "The Eunuch Bagoas." *CQ* 8:144–157.

———. 1967. "Agis III." *Hermes* 95:170–192.

Bagnall, Roger. 1997. "Decolonizing Ptolemaic Egypt." In *Hellenistic Constructs: Essays in Culture, History, and Historiography,* edited by Paul Cartledge, Peter Garnsey, and Erich Gruen, 225–241. Berkeley.

Baker, Heather. 2013. "The Image of the City in Hellenistic Babylonia." In *Shifting Social Imaginaries in the Hellenistic Period: Narrations, Practices, and Images,* edited by Eftychia Stavrianopoulou, 51–65. Leiden.

——. 2014. "Temple and City in Hellenistic Uruk: Sacred Space and the Transformation of Late Babylonian Society." In *Redefining the Sacred: Religious Architecture and Text in the Near East and Egypt 1000 BC–AD 300*, edited by Elizabeth Frood and Rubina Raja, 183–208. Turnhout.

Banerjee, Prathama. 2006. *Politics of Time: "Primitives" and History-Writing in Colonial Society*. Oxford.

Barag, Dan. 1986–1987. "A Silver Coin of Yoḥanan the High Priest and the Coinage of Judea in the Fourth Century BC." *Israel Numismatic Journal* 9:4–21.

Bar-Kochva, Bezalel. 1976. *The Seleucid Army: Organization and Tactics in the Great Campaigns*. Cambridge.

——. 1989. *Judas Maccabaeus: The Jewish Struggle against the Seleucids*. Cambridge.

——. 2010. *The Image of the Jews in Greek Literature*. Berkeley.

Barnes, T. 1994. "Scholarship or Propaganda? Pophyry *Against the Christians* and Its Historical Setting." *Bulletin of the Institute of Classical Studies* 39:53–65.

Baron, Christopher. 2013. *Timaeus of Tauromenium and Hellenistic Historiography*. Cambridge.

Barrows, Adam. 2011. *The Cosmic Time of Empire: Modern Britain and World Literature*. Berkeley.

Barton, John. 1986. *Oracles of God: Perceptions of Ancient Prophecy in Israel after the Exile*. Oxford.

Baslez, Marie-Françoise. 2004. "Polémologie et histoire dans le livre de Judith." *Revue biblique* 111:362–376.

Baslez, Marie-Françoise, and F. Briquel-Chatonnet. 1991. "Un exemple d'intégration phénicienne au monde grec: Les Sidoniens au Pirée à la fin du IVe siècle." *Atti del II congresso internazionale di studi fenici e punici* 1: 229–240. Rome.

Batsch, Christophe. 2005. *La guerre et les rites de guerre dans le judaïsme du deuxième Temple*. Leiden.

Baumgarten, Albert. 2010. *Elias Bickerman as a Historian of the Jews*. Tübingen.

Bautch, Richard. 2003. *Developments in Genre between Post-Exilic Penitential Prayers and the Psalms of Communal Lament*. Leiden.

Bazzana, Giovanni. 2015. *Kingdom of Bureaucracy: The Political Theology of Village Scribes in the Sayings Gospel Q*. Leuven.

Bearzot, Cinzia. 1984. "Il santuario di Apollo Didimeo e la spedizione di Seleuco I a Babulonia (312 a.C.)." In *I santuari e la guerra nel mondo classico*, edited by Marta Sordi, 51–81. Milan.

Beaulieu, Paul-Alain. 1989. *The Reign of Nabonidus, King of Babylon, 556–539 B.C.* New Haven.

——. 1993. "The Historical Background of the Uruk Prophecy." In *The Tablet and the Scroll: Near Eastern Studies in Honor of William W. Hallo*, edited by M. Cohen, D. Snell, and D. Weisberg, 41–52. Bethesda.

———. 2000. "The Descendants of Sîn-lēqi-unninni." In *Assyriologica et Semitica: Festschrift für Joachim Oelsner*, edited by Joachim Marzahn and Hans Neumann, 1-16. Münster.

———. 2004. "New into Old: Religious Reforms under Nabonidus and in Late Babylonian Uruk." In *Mythen der Anderen: Mythopoetik und Interkulturalität*, edited by Manfried Dietrich, Wilhelm Dupré, Ansgar Häußling, Konrad Meising, and Annemarie Mertens, 305-319. Münster.

———. 2006. "Official and Vernacular Languages: The Shifting Sands of Imperial and Cultural Identities in First-Millennium B.C. Mesopotamia." In *Margins of Writing, Origins of Cultures*, edited by Seth Sanders, 187-216. Chicago.

———. 2013. "Mesopotamian Antiquarianism from Sumer to Babylon." In *World Antiquarianism: Comparative Perspectives*, edited by Alain Schnapp, 121-139. Los Angeles.

———. 2014. "An Episode in the Reign of the Babylonian Pretender Nebuchadnezzar IV." In *Extraction and Control: Studies in Honor of Matthew W. Stolper*, edited by Michael Kozuh, Wouter Henkelman, Charles Jones, and Christopher Woods, 17-26. Chicago.

Beaulieu, Paul-Alain, and Francesca Rochberg. 1996. "The Horoscope of Anu-bēlšunu." *JCS* 48:89-94.

Becking, Bob. 2004. *Between Fear and Freedom: Essays on the Interpretation of Jeremiah 30-31*. Leiden.

Bedenbender, Andreas. 2000. *Der Gott der Welt tritt auf den Sinai: Entstehung, Entwicklung und Funktionsweise der frühjüdischen Apokalyptik*. Berlin.

———. 2005. "Reflection on Ideology and Date of the Apocalypse of Weeks." In *Enoch and Qumran Origins: New Light on a Forgotten Connection*, edited by Gabriele Boccaccini, 200-203. Grand Rapids.

———. 2007. "The Place of the Torah in the Early Enochic Literature." In *The Early Enochic Literature*, edited by Gabriele Boccaccini and John Collins, 65-79. Leiden.

Bell, Catherine. 1988. "Ritualization of Texts and Textualization of Ritual in the Codification of Taoist Liturgy." *History of Religions* 27:366-392.

Bencivenni, Alice. 2010. "Il re scrive, la città iscrive. La pubblicazione su pietra della epistole regie nell'Asia ellenistica." *Studi ellenistici* 24:149-178.

———. 2011. "'Massima considerazione.' Forma dell'ordine e immagini del potere nella corrispondenza di Seleuco IV." *ZPE* 176:139-153.

———. 2014. "The King's Words: Hellenistic Royal Letters in Inscriptions." In *State Correspondence in the Ancient World from New Kingdom Egypt to the Roman Empire*, edited by Karen Radner, 141-171. Oxford.

Ben-Dov, Jonathan. 2008. *Head of All Years: Astronomy and Calendars at Qumran in Their Ancient Context*. Leiden.

Bengtson, Hermann. 1944. *Die Strategie in der hellenistischen Zeit. Ein Beitrag zum antiken Staatsrecht II.* 3 vols. Munich.

Benjamin, Walter. 1997. *Charles Baudelaire: A Lyric Poet in the Era of High Capitalism.* Translated by Harry Zohn. London.

———. 2009. *The Origin of German Tragic Drama.* Translated by John Osborne. London.

Bennett, Chris. 2011. *Alexandria and the Moon: An Investigation into the Lunar Macedonian Calendar of Ptolemaic Egypt.* Leuven.

Benveniste, Émile. 1932. "Une apocalypse pehlevie: Le Žamāsp-Nāmak." *Revue de l'histoire des religions* 106:337–380.

Bergamini, Giovanni. 1987. "Parthian Fortifications in Mesopotamia." *Mesopotamia* 22:195–214.

———. 2011. "Babylon in the Achaemenid and Hellenistic Period: The Changing Landscape of a Myth." *Mesopotamia* 46:23–34.

Berger, P.-R. 1970. "Das Neujahrfest nach den Königsinschriften des ausgehenden babylonischen Reiches." In *Actes de la XVIIᵉ Rencontre Assyriologique Internationale,* edited by André Finet, 155–159. Ham-Sur-Heure.

Bergren, Theodore. 1997. "Nehemiah in 2 Maccabees 1:10–2:18." *JSJ* 28:249–270.

Berlin, Andrea. 2014. "Household Judaism." In *Galilee in the Late Second Temple and Mishnaic Periods 1. Life, Culture, and Society,* edited by David Fiensy and James Strange, 208–215. Minneapolis.

Berlin, Andrea, and Sharon Herbert. 2003. "Tel Kedesh." In *The Oxford Encyclopedia of the Bible and Archaeology,* edited by Daniel Master, 2.373–381. Oxford.

Bernard, Paul. 1985. *Fouilles d'Aï Khanoum IV. Les monnaies hors trésors. Questions d'histoire gréco-bactrienne.* Paris.

———. 1999. "Alexandre, Ménon et les mines d'or d'Arménie." In *Travaux de numismatique grecque offerts à Georges Le Rider,* edited by Michel Amandry and Silvia Hurter, 37–64. London.

Berner, Christoph. 2006. *Jahre, Jahrwochen und Jubiläen. Heptadische Geschichtskonzeptionen im antiken Judentum.* Berlin.

Bertelli, Lucio. 2001. "Hecataeus: From Genealogy to Historiography." In *The Historian's Craft in the Age of Herodotus,* edited by Nino Luraghi, 67–94. Oxford.

Berthelot, Katell. 2014. "Judas Maccabeus' Wars against Judaea's Neighbours in 1 Maccabees 5: A Reassessment of the Evidence." *Electrum* 21:73–85.

Bertrand, J. 1982. "Sur l'inscription d'Hefzibah." *ZPE* 46:167–174.

———. 2006. "Réflexions sur les modalités de la correspondance dans les administrations hellénistiques: La réponse donnée par Antiochos IV Épiphane à une requête des Samaritains (Flavius Josèphe, *Antiquités juives,* 12.258–264)." In *La circulation de l'information dans les états antiques,* edited by Laurent Capdetrey and Jocelyne Nelis-Clément, 89–104. Bordeaux.

Bevernage, Berber, and Chris Lorenz. 2013. Introduction to *Breaking Up Time: Negotiating the Borders between Present, Past and Future*, edited by Chris Lorenz and Berber Bevernage, 7-35. Göttingen.

Bickerman, Elias. 1937. "Un document relatif à la persécution d'Antiochos IV Épiphane." *Revue de l'histoire des religions* 115:188-223.

———. 1944. "Notes on Seleucid and Parthian Chronology." *Berytus* 8:73-83.

———. 1944-1945. "L'européanisation de l'Orient classique: A propos du livre de Michel Rostovtzeff." *Renaissance* 2-3:381-392.

———. 1968. *Chronology of the Ancient World*. New York.

———. 1979. *The God of the Maccabees: Studies on the Meaning and Origin of the Maccabean Revolt*. Translated by Horst Moehring. Leiden.

———. 1983. "Time-Reckoning." In *The Cambridge History of Iran*, vol. 3: *The Seleucid, Parthian and Sasanid Periods, Part 2*, edited by Ehsan Yarshater, 778-791. Cambridge.

———. 1988. *The Jews in the Greek Age*. Cambridge, MA.

Bidmead, Julye. 2004. *The Akītu Festival: Religious Continuity and Royal Legitimation in Mesopotamia*. Piscataway, NJ.

Bilfinger, G. 1886. *Die Zeitmesser der antiken Völker*. Stuttgart.

Bivar, David. 1969. "The Achaemenids and the Macedonians: Stability and Turbulence." In *Central Asia*, edited by Gavin Hambly, 19-34. London.

———. 1983. "The Political History of Iran under the Arsacids." In *The Cambridge History of Iran*, vol. 3: *The Seleucid, Parthian and Sasanid Periods, Part 1*, edited by Ehsan Yarshater, 21-99. Cambridge.

Black, J. 1981. "The New Year Ceremonies in Ancient Babylon: 'Taking Bel by the Hand' and a Cultic Picnic." *Religion* 11:39-59.

Blasius, A., and B. Schipper, eds. 2002a. *Apokalyptik und Ägypten. Eine kritische Analyse der relevanten Texte aus dem griechisch-römischen Ägypten*. Leuven.

———. 2002b. "Apokalyptik und Ägypten? Erkenntnisse und Perspektiven." In *Apokalyptik und Ägypten. Eine kritische Analyse der relevanten Texte aus dem griechisch-römsichen Ägypten*, edited by A. Blasius and B. Schipper, 277-302. Leuven.

Blenkinsopp, Joseph. 1983. *A History of Prophecy in Israel from the Settlement of the Land to the Hellenistic Period*. Philadelphia.

———. 1988. *Ezra-Nehemiah: A Commentary*. London.

———. 2013. *David Remembered: Kingship and National Identity in Ancient Israel*. Grand Rapids.

Bloch, Ernst. 1977. "Nonsynchronism and the Obligation to its Dialectics." Translated by Mark Ritter. *New German Critique* 11:22-38.

Bloch, Marc. 1992. *The Historian's Craft*. Translated by Peter Putnam. Manchester.

Blumenberg, Hans. 1983. *The Legitimacy of the Modern Age.* Translated by Robert Wallace. Cambridge, MA.

Boardman, John. 1970. *Greek Gems and Finger Rings: Early Bronze Age to Late Classical.* London.

———. 2000. *Persia and the West: An Archaeological Investigation of the Genesis of Achaemenid Art.* London.

Boccaccini, Gabriele. 1991. *Middle Judaism: Jewish Thought, 300 BCE to 200 CE.* Minneapolis.

———. 1998. *Beyond the Essene Hypothesis: The Parting of the Ways between Qumran and Enochic Judaism.* Grand Rapids.

———. 2002. *Roots of Rabbinic Judaism: An Intellectual History, from Ezekiel to Daniel.* Grand Rapids.

Boffo, Laura. 2013. "La 'presenza' dei re negli archivi delle *poleis* ellenistiche." In *Archives and Archival Documents in Ancient Societies,* edited by M. Faraguna, 201–244. Trieste.

Boiy, Tom. 1998. "Dating in Early Hellenistic Babylonia: Evidence on the Basis of CT 49 13, 1982.A.1853 and HSM 1893.5.6." *NABU* §134.

———. 2000. "Dating Methods during the Early Hellenistic Period." *JCS* 52:115–121.

———. 2001. "Dating Problems in Cuneiform Tablets concerning the Reign of Antigonus Monophthalmus." *JAOS* 121:645–649.

———. 2002a. "The 'Accession Year' in the Late Achaemenid and Early Hellenistic Period." In *Mining the Archives: Festschrift for Christopher Walker on the Occasion of His 60th Birthday, 4th October 2002,* edited by Cornelia Wunsch, 25–34. Dresden.

———. 2002b. "Early Hellenistic Chronography in Cuneiform Tradition." *ZPE* 138:249–255.

———. 2004. *Late Achaemenid and Hellenistic Babylon.* Leuven.

———. 2007. *Between High and Low: A Chronology of the Early Hellenistic Period.* Frankfurt.

———. 2009. "Date Formulas in Cuneiform Tablets and Antigonus Monophthalmus, Again." *JAOS* 129:467–476.

———. 2010. "Local and Imperial Dates at the Beginning of the Hellenistic Period." *Electrum* 18:9–22.

———. 2011. "The Reigns of the Seleucid Kings According to the Babylon King List." *JNES* 70:1–12.

Boncquet, B. 1987. *Diodorus Siculus (II, 1–34) over Mesopotamië. Een historische commentaar.* Brussels.

Boone, Catherine. 2003. *Political Topographies of the African State: Territorial Authority and Institutional Choice.* Cambridge.

Bopearachchi, Osmund. 2007. "Some Observations on the Chronology of the Early Kushans." *Res orientales* 17:41–53.

———. 2008. "Les premiers souverains kouchans: Chronologie et iconographie monétaire." *Journal des savants*, 3–56.

Borza, E. 1972. "Fire from Heaven: Alexander at Persepolis." *CP* 67:233–245.

Bosworth, A. B. 1980. *A Historical Commentary on Arrian's History of Alexander I. Commentary on Books I–III*. Oxford.

———. 2002. *The Legacy of Alexander: Politics, Warfare, and Propaganda under the Successors*. Oxford.

Botha, Phil. 2007. "The Relevance of the Book of Daniel for Fourth-Century Christianity according to the Commentary Ascribed to Ephrem the Syrian." In *Die Geschichte der Daniel-Auslegung in Judentum, Christentum und Islam*, edited by Katharina Bracht and David du Toit, 99–122. Berlin.

Boucharlat, Rémy. 2006. "Le destin des résidences et sites d'Iran dans la seconde moitié du IVᵉ siècle avant J.-C." In *La transition entre l'empire achéménide et les royaumes hellénistiques*, edited by Pierre Briant and Francis Joannès, 443–470. Paris.

Boussac, Marie-Françoise. 1982. "À propos de quelques sceaux déliens." *BCH* 106:427–446.

———. 1992. *Les sceaux de Délos 1: Sceaux public, Apollon, Hélios, Artémis, Hécate*. Paris.

Boyarin, Daniel. 2012. "Daniel 7, Intertextuality, and the History of Israel's Cult." *Harvard Theological Review* 105:139–162.

Boyce, Mary. 1979. *Zoroastrians: Their Religious Beliefs and Practices*. London.

———. 1984. "On the Antiquity of Zoroastrian Apocalyptic." *BSOAS* 47:57–75.

———. 1989. "The Poems of the Persian Sibyl and the *Zand ī Vahman Yašt*." In *Études irano-aryennes offertes à Gilbert Lazard*, edited by C.-H. Fouchécour and P. Gignoux, 59–77. Paris.

Boyce, Mary, and Frantz Grenet. 1991. *A History of Zoroastrianism III: Zoroastrianism under Macedonian and Roman Rule*. Leiden.

Boym, Svetlana. 2001. *The Future of Nostalgia*. New York.

Braund, David. 1994. *Georgia in Antiquity: A History of Colchis and Transcaucasian Iberia 550 BC—AD 562*. Oxford.

Braverman, Jay. 1978. *Jerome's* Commentary on Daniel: *A Study of Comparative Jewish and Christian Interpretations of the Hebrew Bible*. Washington, DC.

Briant, Pierre. 1990. "The Seleucid Kingdom, the Achaemenid Empire and the History of the Near East in the First Millennium BC." In *Religion and Religious Practice in the Seleucid Kingdom*, edited by Per Bilde, Troels Engberg-Pedersen, Lise Hannestad, and Jan Zahle, 40–65. Aarhus.

———. 2002. *From Cyrus to Alexander: A History of the Persian Empire*. Translated by Peter Daniels. Winona Lake.

———. 2003. "Quand les rois écrivent l'histoire: La domination achéménide vue à travers les inscriptions officielles lagides." In *Événement, récit, histoire officielle. L'écriture de l'histoire dans les monarchies antiques*, edited by N. Grimal and M. Baud, 171–183. Paris.

———. 2017. *The First European: A History of Alexander in the Age of Empire*. Translated by Nicholas Elliott. Cambridge, MA.

Brijder, Herman. 2014a. "The East Terrace." In *Nemrud Dağı: Recent Archaeological Research and Conservation Activities in the Tomb Sanctuary on Mount Nemrud*, edited by Herman Brijder, 314–356. Göttingen.

———. 2014b. "The West Terrace." In *Nemrud Dağı: Recent Archaeological Research and Conservation Activities in the Tomb Sanctuary on Mount Nemrud*, edited by Herman Brijder, 357–381. Göttingen.

Brinkman, J. 1979. "Babylonia under the Assyrian Empire 745–627 BC." In *Power and Propaganda: A Symposium on Ancient Empires*, edited by M. Larsen, 223–250. Copenhagen.

Briscoe, John. 1973. *A Commentary on Livy Books XXXI–XXXIII*. Oxford.

Britton, John. 2007. "Studies in Babylonian Lunar Theory: Part 1. Empirical Elements for Modeling Lunar and Solar Anomalies." *Archive for History of Exact Sciences* 61:83–145.

———. 2008. "Remarks on Strassmeier Cambyses 400." In *From the Banks of the Euphrates: Studies in Honor of Alice Louise Slotsky*, edited by Micah Ross, 7–33. Winona Lake.

Brodersen, Kai. 1986. "The Date of the Secession of Parthia from the Seleucid Kingdom." *Historia* 35:378–381.

Broshi, Magen, and Esther Eshel. 1997. "The Greek King Is Antiochus IV (4QHistorical Text = 4Q248)." *JJS* 48:120–129.

Brown, David. 2000. *Mesopotamian Planetary Astronomy-Astrology*. Groningen.

———. 2008. "Increasingly Redundant: The Growing Obsolescence of the Cuneiform Script in Babylonia from 539 BC." In *The Disappearance of Writing Systems: Perspectives on Literacy and Communication*, edited by John Baines, John Bennet, and Stephen Houston, 73–101. London.

Brown, F., and C. Welles. 1944. "Inscriptions." In *The Excavations at Dura-Europos. Preliminary Report of the Ninth Season of Work 1935–1936. Part I, The Agora and Bazaar*, edited by M. Rostovtzeff, A. Bellinger, F. Brown, and C. Welles, 168–185. London.

Brownlee, William. 1979. *The Midrash Pesher of Habakkuk*. Ann Arbor.

Brutti, Maria. 2010. "War in 1 Maccabees." *Deuterocanonical and Cognate Literature Yearbook*, 147–172.

Bryan, David. 2015. *Cosmos, Chaos, and the Kosher Mentality*. London.

Bull, Malcolm. 1999. *Seeing Things Hidden: Apocalypse, Vision and Totality*. London.

Bultmann, Rudolf. 1957. *The Presence of Eternity: History and Eschatology.* New York.

Bundvad, Mette. 2015. *Time in the Book of Ecclesiastes.* Oxford.

Buraselis, Kostas. 2010. "Eponyme Magistrate und hellenistischer Herrscherkult." In *Symposion 2009. Vorträge zur griechischen und hellenistischen Rechtsgeschichte (Seggau, 25-30, August 2009)*, edited by Gerhard Thür, 419-434. Vienna.

Burgmann, Hans. 1974. "Die vier Endzeittermine in Danielbuch." *ZAW* 86:543-550.

Burkert, Walter. 1989. "Apokalyptik im frühen Griechentum: Impulse und Transformationen." In *Apocalypticism in the Mediterranean World and the Near East: Proceedings of the International Colloquium on Apocalypticism, Uppsala, August 12-17, 1979*, edited by David Hellholm, 235-254. Tübingen.

———. 1995. "Lydia between East and West, or How to Date the Trojan War: A Study in Herodotus." In *The Ages of Homer: A Tribute to Emily Townsend Vermeule*, edited by Jane Carter and Sarah Morris, 139-148. Austin.

Burman, Rickie. 1981. "Time and Socioeconomic Change on Simbo, Solomon Islands." *Man* 16:251-267.

Burstein, Stanley. 1978. *The Babyloniaca of Berossus.* Malibu.

Calame, Claude. 1998. "Mémoire collective et temporalités en contact: Somare et Hérodote." *Revue de l'histoire des religions* 215:341-367.

Callieri, Pierfrancesco. 2004. "Again on the Chronology of the Tall-e Takht at Pasargadae." *Parthica* 6:95-100.

———. 2007. *L'archéologie du Fārs à l'époque hellénistique.* Paris.

Capdetrey, Laurent. 2006. "Pouvoir et écrit: Production, reproduction et circulation des documents dans l'administration séleucide." In *La circulation de l'information dans les états antiques*, edited by Laurent Capdetrey and Jocelyne Nelis-Clément, 105-125. Bordeaux.

———. 2007. *Le pouvoir séleucide: Territoire, administration, finances d'un royaume hellénistique (312-129).* Rennes.

Caponigro, Mark. 1992. "Judith, Holding the Tale of Herodotus." In *"No One Spoke Ill of Her": Essays on Judith*, edited by James VanderKam, 47-59. Atlanta.

Caragounis, Chrys. 1987. "The Interpretation of the Ten Horns in Daniel 7." *Ephemerides Theologicae Lovanienses* 63:106-113.

———. 1993. "History and Supra-History: Daniel and the Four Empires." In *The Book of Daniel in the Light of New Findings*, edited by A. S. van der Woude, 387-397. Leuven.

Carr, David. 1996. "Canonization in the Context of Community: An Outline of the Formation of the Tanakh and the Christian Bible." In *A Gift of God in Due Season: Essays on Scripture and Community in Honor of James A. Sanders*, edited by Richard Weis and David Carr, 22-64. Bath.

———. 2011. *The Formation of the Hebrew Bible: A Reconstruction.* Oxford.

Caubet, Annie. 2009. "The Historical Context of the Sumerian Discoveries." *Museum International* 61:74-80.

Cavigneaux, Antoine. 2005. "Shulgi, Nabonide, et les Grecs." In *An Experienced Scribe Who Neglects Nothing: Ancient Near Eastern Studies in Honor of Jacob Klein*, edited by Y. Sefati, P. Artzi, C. Cohen, B. Eichler, and V. Hurowitz, 63-72. Bethesda, MD.

Ceccarelli, Paola. 2013. *Ancient Greek Letter Writing: A Cultural History (600 BC—150 BC)*. Oxford.

Cereti, Carlo. 1995. *The Zand ī Wahman Yasn: A Zoroastrian Apocalypse*. Rome.

———. 2015. "Myths, Legends, Eschatologies." In *The Wiley-Blackwell Companion to Zoroastrianism*, edited by Michael Stausberg and Yuhan Vevaina, 259-272. Oxford.

Chakrabarty, Dipesh. 2000. *Provincializing Europe: Postcolonial Thought and Historical Difference*. Princeton.

Chaniotis, Angelos. 2010. "Epigraphic Bulletin for Greek Religion 2007." *Kernos* 23:271-327.

———. 2011. "The Ithyphallic Hymn for Demetrios Poliorketes and Hellenistic Religious Mentality." In *More than Men, Less than Gods: Studies on Royal Cult and Imperial Worship*, edited by Panagiotis Iossif, Andrzej Chankowski, and Catharine Lorber, 157-195. Leuven.

Charles, Robert. 1929. *A Critical and Exegetical Commentary on the Book of Daniel*. London.

Charlesworth, James. 2002. *The Pesharim and Qumran History*. Grand Rapids.

Charpin, Dominique, and Nele Ziegler. 2013. "Masters of Time: Old Babylonian Kings and Calendars." In *Time and History in the Ancient Near East: Proceedings of the 56th Rencontre Assyriologique Internationale at Barcelona 26-30 July 2010*, edited by L. Feliu, J. Llop, A. Millet Albà, and J. Sanmartín, 57-68. Winona Lake.

Chaumont, Marie-Louise. 1993. "Fondations séleucides en Arménie méridionale." *Syria* 70:431-441.

Chesnutt, Randall. 2010. "*Oxyrhynchus Papyrus* 2069 and the Compositional History of 1 *Enoch*." *JBL* 129:485-505.

Childs, Brevard. 1959. "The Enemy from the North and the Chaos Tradition." *JBL* 78:187-198.

Chowers, Eyal. 2012. *The Political Philosophy of Zionism: Trading Jewish Words for a Hebraic Land*. Cambridge.

Christesen, Paul. 2007. *Olympic Victor Lists and Ancient Greek History*. Cambridge.

Chrubasik, Boris. 2013. "The Attalids and the Seleukid Kings, 281-175 BC." In *Attalid Asia Minor: Money, International Relations, and the State*, edited by Peter Thonemann, 83-119. Oxford.

———. 2016. *Kings and Usurpers in the Seleukid Empire: The Men Who Would Be King.* Oxford.

Clancier, Philippe. 2005. "Les scribes sur parchemin du temple d'Anu." *Revue d'assyriologie et d'archéologie orientale* 99:85–104.

———. 2007. "La Babylone hellénistique. Aperçu d'histoire politique et culturelle." *Topoi* 15:21–74.

———. 2011. "Cuneiform Culture's Last Guardians: The Old Urban Notability of Hellenistic Uruk." In *The Oxford Handbook of Cuneiform Culture,* edited by Karen Radner and Eleanor Robson, 752–773. Oxford.

———. 2014. "Antiochos IV dans les sources babyloniennes." In *Le projet politique d'Antiochos IV,* edited by Christophe Feyel and Laetitia Graslin-Thomé, 415–437. Nancy.

Clancier, Philippe, and Julien Monerie. 2014. "Les sanctuaires babyloniens à l'époque hellénistique: Evolution d'un relais de pouvoir." *Topoi* 19:181–237.

Clarke, Katherine. 2008. *Making Time for the Past: Local History and the* Polis. Oxford.

Clarysse, Willy, and Dorothy Thompson. 2007. "Two Greek Texts on Skin from Hellenistic Bactria." *ZPE* 159:273–279.

Clermont-Ganneau, M. 1887. "Mene, Tekel, Peres, and the Feast of Belshazzar." *Hebraica* 3:87–102.

Clifford, Richard. 1975. "History and Myth in Daniel 10–12." *BASOR* 220:23–26.

Cohen, Getzel. 1995. *The Hellenistic Settlements in Europe, the Islands, and Asia Minor.* Berkeley.

———. 2006. *The Hellenistic Settlements in Syria, the Red Sea Basin, and North Africa.* Berkeley.

———. 2013. *The Hellenistic Settlements in the East from Armenia and Mesopotamia to Bactria and India.* Berkeley.

Cohen, Shaye. 2006. *From the Maccabees to the Mishnah.* Louisville.

Cohn, Naftali. 2009. "Rabbis as Jurists: On the Representation of Past and Present Legal Institutions in the Mishnah." *JJS* 60:245–263.

Collingwood, R. G. 1946. *The Idea of History.* Oxford.

Collins, John. 1977. "Pseudonymity, Historical Reviews and the Genre of the Revelation of John." *Catholic Biblical Quarterly* 39:329–343.

———. 1990. "Nebuchadnezzar and the Kingdom of God: Deferred Eschatology in the Jewish Diaspora." In *Loyalitätskonflikte in der Religionsgeschichte,* edited by Christoph Elsas and Hans Kippenberg, 252–257. Würzburg.

———. 1993. *Daniel: A Commentary on the Book of Daniel.* Minneapolis.

———. 1995. *The Scepter and the Star: The Messiahs of the Dead Sea Scrolls and Other Ancient Literature.* New York.

———. 1996. "Pseudo-Daniel Revisited." *Revue de Qumran* 17:111–135.

———. 1997. *Jewish Wisdom in the Hellenistic Age.* Louisville.

———. 2011. "Enochic Judaism: An Assessment." In *The Dead Sea Scrolls and Contemporary Culture. Proceedings of the International Conference Held at the Israel Museum, Jerusalem (July 6–8, 2008),* edited by A. Roitman, L. Schiffman, and S. Tzoref, 219–234. Leiden.

———. 2016. "Temple or Taxes? What Sparked the Maccabean Revolt?" In *Revolt and Resistance in the Ancient Classical World and the Near East,* edited by John Collins and Joseph Manning, 189–201. Leiden.

Collins, John, and P. Flint. 1996. "243–245. 4QPseudo-Daniel^a–c ar." In George Brooke, John Collins, Torleif Elgvin, Peter Flint, Jonas Greenfield, Erik Larson, Carol Newsom, et al. *Qumran Cave 4.XVII. Parabiblical Texts. Part 3,* 95–164. Oxford.

Collins, N. 1997. "The Various Fathers of Ptolemy I." *Mnemosyne* 50:436–476.

Coloru, Omar. 2009. *Da Alessandro a Menandro: Il regno greco di Battriana.* Pisa.

Colpe, Carsten. 1983. "Development of Religious Thought." In *The Cambridge History of Iran,* vol. 3: *The Seleucid, Parthian and Sasanid Periods, Part 2,* edited by Ehsan Yarshater, 819–865. Cambridge.

Confino, Alon. 2014. *A World without Jews: The Nazi Imagination from Persecution to Genocide.* New Haven.

Cook, L. Stephen. 2011. *On the Question of the "Cessation of Prophecy" in Ancient Judaism.* Tübingen.

Coqueugniot, Gaëlle. 2011. "The Oriental Agora: The Case of Seleucid Europos-Dura, Syria." In *The Agora in the Mediterranean from Homeric to Roman Times,* edited by Angelike Giannikoure, 295–309. Athens.

———. 2012. "Le *chreophylakeion* et l'agora d'Europos-Doura: Bilan des recherches, 2004–2008." In *Europos-Doura Varia,* edited by Pierre Leriche, Gaëlle Coqueugniot, and Ségolène de Pontbriand, 1:93–110. Beirut.

———. 2013. *Archives et bibliothèques dans le monde grec: Edifices et organisation Ve siècle avant notre ère—IIe siècle de notre ère.* Oxford.

———. 2014. "The Agora of Europos-Doura (Syria): An Archival and Archaeological Reappraisal." *Pegasus* 57:27–35.

Corcella, Aldo. 1999. "Giuditta e i Persiani." *Scritti in ricordo di Giacomo Bona (Annali della Facoltà di Lettere e Filosofia dell'Università degli studi della Basilicata 9):* 73–90.

Corley, Jeremy. 2006. "The Review of History in Eleazar's Prayer in 3 Macc. 6:1–15." *Deuterocanonical and Cognate Literature Yearbook,* 201–229.

Cotton, Hannah, and Michael Wörrle. 2007. "Seleukos IV to Heliodoros. A New Dossier of Royal Correspondence from Israel." *ZPE* 159:191–205.

Courtray, Régis. 2011. "Porphyre et le livre de Daniel au travers du *Commentaire sur Daniel de Jérôme.*" In *Le traité de Porphyre Contre les Chrétiens,* edited by Sébastien Morlet, 329–356. Paris.

Cribb, Joe. 2005. "The Greek Kingdom of Bactria, Its Coinage and Its Collapse." In *Afghanistan: Ancien carrefour entre l'est et l'ouest*, edited by Osmund Bopearachchi and Marie-Françoise Boussac, 207–225. Turnhout.

Cumont, Franz. 1931. "La fin du monde selon les mages occidentaux." *Revue de l'histoire des religions* 103:29–96.

Curtis, Vesta. 2010. "The Frataraka Coins of Persis: Bridging the Gap between Achaemenid and Sasanian Persia." In *The World of Achaemenid Persia: History, Art and Society in Iran and the Ancient Near East*, edited by John Curtis and St John Simpson, 379–394. London.

Da Riva, Rocío. 2008. *The Neo-Babylonian Royal Inscriptions: An Introduction.* Münster.

———. 2017. "The Figure of Nabopolassar in Late Achaemenid and Hellenistic Historiographic Tradition: BM 34793 and CUA 90." *JNES* 76:75–92.

Dąbrowa, Edward. 2010. *The Hasmoneans and Their State: A Study in History, Ideology, and the Institutions.* Krakow.

Dalley, Stephanie. 1989. *Myths from Mesopotamia: Creation, The Flood, Gilgamesh and Others.* Oxford.

Darshan, Guy. 2016. "The Calendrical Framework of the Priestly Flood Story in Light of a New Akkadian Text from Ugarit (RS 94.2953)." *JAOS* 136:507–514.

Davies, Graham. 1978. "Apocalyptic and Historiography." *Journal for the Study of the Old Testament* 5:15–28.

Davies, Philip. 1983. "Calendrical Change and Qumran Origins: An Assessment of VanderKam's Theory." *Catholic Biblical Quarterly* 45:80–89.

———. 2000. "'Pen of Iron, Point of Diamond' (Jer. 17:1): Prophecy as Writing." In *Writings and Speech in Israelite and Ancient Near Eastern Prophecy*, edited by Ehud Ben Zvi and Michael Floyd, 65–81. Atlanta.

———. 2011. *On the Origins of Judaism.* London.

Davis, Kathleen. 2008. *Periodization and Sovereignty: How Ideas of Feudalism and Secularization Govern the Politics of Time.* Philadelphia.

Day, John. 1980. "The Daniel of Ugarit and Ezekiel and the Hero of the Book of Daniel." *VT* 30:174–184.

———. 2011. "The Flood and the Ten Antediluvian Figures in Berossus and in the Priestly Source in Genesis." In *On Stone and Scroll: Essays in Honour of Graham Ivor Davies*, edited by James Aitken, Katharine Dell, and Brian Mastin, 211–223. Berlin.

De Breucker, Geert. 2015. "Heroes and Sinners: Babylonian Kings in Cuneiform Historiography of the Persian and Hellenistic Periods." In *Political Memory in and after the Persian Empire*, edited by Jason Silverman and Caroline Waerzeggers, 75–94. Atlanta.

De Callataÿ, F. 1996. "Abdissarès l'Adiabénien." *Iraq* 58:135–145.

De Certeau, Michel. 1988. *The Writing of History*. Translated by Tom Conley. New York.

De Genouillac, Henri. 1930. "Rapport sur les travaux de la Mission de Tello, IIᵉ campagne: 1929–1930." *Revue d'assyriologie et d'archéologie orientale* 27:169–186.

———. 1936. *Fouilles de Telloh*. Vol. 2, *Époques d'Ur IIIᵉ dynastie et de Larsa*. Paris.

De Giorgi, Andrea. 2016. *Ancient Antioch: From the Seleucid Era to the Islamic Conquest*. Cambridge.

De Jong, Albert. 1997. *Traditions of the Magi: Zoroastrianism in Greek and Latin Literature*. Leiden.

———. 2003. "Vexillologica Sacra: Searching the Cultic Banner." In *Religious Themes and Texts of Pre-Islamic Iran and Central Asia: Studies in Honour of Professor Gherardo Gnoli on the Occasion of his 65th Birthday on 6th December 2002*, edited by Carlo Cereti, Mauro Maggi, and Elio Provasi, 191–202. Wiesbaden.

De Sarzec, Ernest. 1884–1912. *Découvertes en Chaldée*. 2 vols. Paris.

Declaissé-Walford, Nancy, Rolf Jacobson, and Beth Tanner. 2014. *The Book of Psalms*. Grand Rapids.

Del Monte, Giuseppe. 1997. *Testi dalla Babilonia ellenistica*. Pisa.

———. 2001. "Da 'barbari' a 're di Babilonia': I Greci in Mesopotamia." In *I Greci: Storia cultura arte società III. I Greci oltre la Grecia*, edited by Salvatore Settis, 137–166. Turin.

Delcor, Mathias. 1951. "Les allusions à Alexandre le Grand dans Zach ix 1–8." *VT* 1:110–124.

———. 1967. "Le livre de Judith et l'époque grecque." *Klio* 49:151–179.

Depuydt, Leo. 1995. "'More Valuable than All Gold': Ptolemy's Royal Canon and Babylonian Chronology." *JCS* 47:97–117.

Derrida, Jacques. 1995. "Archive Fever: A Freudian Impression." Translated by Eric Prenowitz. *Diacritics* 25:9–63.

Di Lella, Alexander. 1987. *The Wisdom of Ben Sira. A New Translation with Notes by Patrick Skehan, Introduction and Commentary by Alexander A. Di Lella*. New York.

———. 2006. "Ben Sira's Praise of the Ancestors of Old (Sir 44–49): The History of Israel as Parenetic Apologetics." *Deuterocanonical and Cognate Literature Yearbook*, 151–170.

Dillery, John. 2015. *Clio's Other Sons: Berossus and Manetho*. Ann Arbor.

Dimant, Devorah. 1993. "The Seventy Weeks Chronology (Dan 9,24–27) in the Light of New Qumranic Texts." In *The Book of Daniel in the Light of New Findings*, edited by A. S. van der Woude, 57–76. Leuven.

———. 1998. "4Q386 ii–iii: A Prophecy on Hellenistic Kingdoms?" *Revue de Qumrân* 18:511–529.

Dirven, Lucinda. 1997. "The Exaltation of Nabû: A Revision of the Relief Depicting the Battle against Tiamat from the Temple of Bel in Palmyra." *Welt des Orients* 28:96–116.

DiTommaso, Lorenzo. 2005. "4QPseudo-Daniel[A-B] (4Q243-244) and the Book of Daniel." *Dead Sea Discoveries* 12:101–133.

———. 2007. "Apocalypses and Apocalypticism in Antiquity." *Currents in Biblical Research* 5:235–286 and 367–432.

Dobroruka, Vicente. 2012. "Hesiodic Reminiscences in Zoroastrian-Hellenistic Apocalypses." *BSOAS* 75:275–295.

Donner, Herbert. 1973. "Argumente zur Datierung des 74. Psalms." In *Wort, Lied und Gottesspruch: Beiträge zu Psalmen und Propheten: Festschrift für Joseph Ziegler*, edited by J. Schreiner, 41–50. Würzburg.

Don-Yehiya, Eliezer. 1992. "Hanukkah and the Myth of the Maccabees in Zionist Ideology and in Israeli Society." *Jewish Journal of Sociology* 34:5–23.

Doran, Robert. 1981. *Temple Propaganda: The Purpose and Character of 2 Maccabees.* Washington, DC.

———. 1996. "The First Book of Maccabees: Introduction, Commentary, and Reflections." In *The New Interpreter's Bible*, 4:3–178. Nashville.

———. 2012. *2 Maccabees: A Critical Commentary.* Minneapolis.

Dörner, F. 1996. "Epigraphy Analysis." In *Nemrud Dağı. The Hierothesion of Antiochus I of Commagene. Volume 1: Text*, edited by Donald Sanders, 361–377. Winona Lake.

Downey, R. 1935. "References to Inscriptions in the Chronicle of Malalas." *TAPA* 66:55–72.

———. 1961. *A History of Antioch in Syria: From Seleucus to the Arab Conquest.* Princeton.

Doyen, Charles. 2014. "Le système monétaire et pondéral d'Antiochos IV." In *Le projet politique d'Antiochos IV*, edited by Christophe Feyel and Laetitia Graslin-Thomé, 261–299. Nancy.

Droysen, Johann Gustav. 1877–1878. *Geschichte des Hellenismus.* 3 vols. Gotha.

Duchesne-Guillemin, Jacques. 1982. "Apocalypse juive et apocalypse iranienne." In *La soteriologia dei culti orientali nell'Impero Romano. Atti del colloquio internazionale su la soteriologia dei culti orientali nell'Impero Romano, Roma 24–28 Settembre 1979*, edited by Ugo Bianchi and Maarten Vermaseren, 753–761. Leiden.

Dunn, Francis. 2007. *Present Shock in Late Fifth-Century Greece.* Ann Arbor.

Dupont-Sommer, A. 1946–1948. "Deux inscriptions araméennes trouvées près du lac Sevan (Arménie)." *Syria* 25:53–66.

Duval, Rubens. 1892. *Histoire politique, religieuse et littéraire d'Edesse jusqu'à la première croisade.* Paris.

Duyrat, Frédérique. 2005. *Arados hellénistique: Etude historique et monétaire.* Beirut.

Eck, Bernard. 2003. *Diodore de Sicile, bibliothèque historique Tome II Livre II.* Paris.

Eckhardt, Benedikt. 2016. "The Hasmoneans and Their Rivals in Seleucid and Post-Seleucid Judea." *JSJ* 47:55–70.

———. 2017. "Memories of Persian Rule: Constructing History and Ideology in Hasmonean Judea." In *Perisanism in Antiquity*, edited by Rolf Strootman and Miguel Versluys, 249–265. Stuttgart.

Eco, Umberto. 2009. *The Infinity of Lists: From Homer to James Joyce*. Translated by Alastair McEwen. London.

Eddy, Samuel. 1961. *The King Is Dead: Studies in Near Eastern Resistance to Hellenism 334–31 B.C.* Lincoln, NE.

Edensor, Tim. 2006. "Reconsidering National Temporalities: Institutional Times, Everyday Routines, Serial Spaces and Synchronicities." *European Journal of Social Theory* 9:525–545.

Edzard, Dietz. 1997. *Gudea and His Dynasty*. Toronto.

Ehling, Kay. 2008. *Untersuchungen zur Geschichte der späten Seleukiden (164–63 v. Chr.): Vom Tode des Antiochos IV. bis zur Einrichtung der Provinz Syria unter Pompeius*. Stuttgart.

Eliade, Mircea. 2005. *The Myth of the Eternal Return: Cosmos and History*. Translated by William Trask. Princeton, NJ.

Ellis, Richard. 1968. *Foundation Deposits in Ancient Mesopotamia*. New Haven.

Engels, David. 2011. "Posidonius of Apameia and Strabo of Amasia on the Decline of the Seleucid Kingdom." In *Seleucid Dissolution: The Sinking of the Anchor*, edited by Kyle Erickson and Gillian Ramsey, 181–194. Wiesbaden.

———. 2013. "A New Frataraka Chronology." *Latomus* 72:28–80.

———. 2017. *Benefactors, Kings, Rulers. Studies on the Seleukid Empire between East and West*. Leuven.

Eph'al, Israel, and Joseph Naveh. 1996. *Aramaic Ostraca of the Fourth Century BC from Idumaea*. Jerusalem.

Errington, Elizabeth, and Vesta Curtis. 2007. *From Persepolis to the Punjab: Exploring Ancient Iran, Afghanistan and Pakistan*. London.

Errington, R. 1986. "Antiochos III, Zeuxis und Euromos." *Epigraphica Anatolica* 8:1–7.

Eshel, Esther. 2010. "Inscriptions in Hebrew, Aramaic and Phoenician Script." In Amos Kloner, Esther Eshel, Hava Korzakova, and Gerald Finkielsztejn, *Maresha Excavations Final Report III. Epigraphic Finds from the 1989–2000 Seasons*, 35–88. Jerusalem.

Eshel, Hanan. 2008. *The Dead Sea Scrolls and the Hasmonean State*. Grand Rapids.

Esler, Philip. 2002. "Ludic History in the Book of Judith: The Reinvention of Israelite Identity?" *Biblical Interpretation* 10:107–143.

Evans, James. 2016. "Images of Time and Cosmic Connection." In *Time and Cosmos in Greco-Roman Antiquity*, edited by Alexander Jones, 143–169. New York.

Evans-Pritchard, Edward. 1948. *The Divine Kingship of the Shilluk of the Nilotic Sudan.* Cambridge.

Fabian, Johannes. 1983. *Time and the Other: How Anthropology Makes Its Object.* New York.

Facella, Margherita. 2006. *La dinastia degli Orontidi nella Commagene ellenistico-romana.* Pisa.

Falk, Harry, and Chris Bennett. 2009. "Macedonian Intercalary Months and the Era of Azes." *Acta Orientalia* 70:197–216.

Falkenstein, A. 1959. "Akiti-Fest und akiti-Festhaus." In *Festschrift Johannes Friedrich,* edited by R. von Kienle, A. Moortgat, H. Otten, E. von Schuler, and W. Zaumseil, 147–182. Heidelberg.

Fantalkin, Alexander, and Oren Tal. 2012. "The Canonization of the Pentateuch: When and Why? (Part I)." *ZAW* 124:1–18.

Fasolt, Constantin. 2004. *The Limits of History.* Chicago.

Feeley-Harnik, Gillian. 1985. "Issues in Divine Kingship." *Annual Review of Anthropology* 14:273–313.

Feeney, Denis. 2007. *Caesar's Calendar: Ancient Time and the Beginnings of History.* Berkeley.

Felber, Heinz. 2002. "Die demotische Chronik." In *Apokalyptik und Ägypten. Eine kritische Analyse der relevanten Texte aus dem griechisch-römsichen Ägypten,* edited by A. Blasius and B. Schipper, 65–111. Leuven.

Fensham, F. 1981. "Neh. 9 and Pss. 105, 106, 135 and 136. Post-Exilic Historical Traditions in Poetic Form." *Journal of Northwest Semitic Languages* 9:35–51.

Ferrari, Gloria. 2000. "The Geography of Time: The Nile Mosaic and the Library at Praeneste." *Ostraka* 8:359–386.

Fields, Dana. 2009. "The Rhetoric of *Parrhēsia* in Roman Greece." PhD diss. Princeton University.

Finkielsztejn, Gerald. 1998a. "More Evidence on John Hyrcanus I's Conquests: Lead Weights and Rhodian Amphora Stamps." *Bulletin of the Anglo-Israel Archaeological Society* 16:33–63.

———. 1998b. "Timbres amphoriques du Levant d'époque hellénistique." *Transeuphratène* 15:83–121.

———. 1999. "A Standard of Volumes for Liquids from Hellenistic Marisa." *'Atiqot* 38:51–64.

———. 2001. "Administration and Trade in the Hellenistic Period: Dating and Standards of Measure." In *Measuring and Weighing in Ancient Times,* edited by Yisrael Ronen, 15–19. Haifa.

———. 2003. "Administration du Levant sud sous les Séleucides. Remarques préliminaires." *Topoi,* supplement 4, 465–484.

———. 2006. "An Incised Bag-Shaped Hellenistic Jar Handle from the Kaplan In-terchange (Tel Aviv)." *IEJ* 56:51–56.

———. 2007. "Poids de plomb inscrits du Levant: Une réforme d'Antiochos IV?" *Topoi*, supplement 8, 35–60.

———. 2010a. "The Maresha Scale Weights: Metrology, Administration, and His-tory." In Amos Kloner, Esther Eshel, Hava Korzakova, and Gerald Finkielsz-tejn, *Maresha Excavations Final Report III. Epigraphic Finds from the 1989–2000 Seasons*, 175–192. Jerusalem.

———. 2010b. "The *Sekoma*: A Volume Standard for Liquids." In Amos Kloner, Es-ther Eshel, Hava Korzakova, and Gerald Finkielsztejn, *Maresha Excavations Final Report III. Epigraphic Finds from the 1989–2000 Seasons*, 193–203. Jerusalem.

———. 2014. "The Weight Standards of the Hellenistic Levant, Part One—The Evi-dence of the Syrian Scale Weights." *Israel Numismatic Research* 9:61–94.

Finn, Jennifer. 2017. *Much Ado about Marduk: Questioning Discourses of Royalty in First Millennium Mesopotamian Literature*. Berlin.

Fischer, Thomas. 1979. "Zur Seleukideninschrift von Hefzibah." *ZPE* 33:131–138.

———. 1992. "Tryphons verfehlter Sieg von Dor?" *ZPE* 93:29–30.

Fischer-Bovet, Christelle. 2014. "Est-il facile de conquérir l'Égypte? L'invasion d'Antiochos IV et ses conséquences." In *Le projet politique d'Antiochos IV*, edited by Christophe Feyel and Laetitia Graslin-Thomé, 209–259. Nancy.

Fishbane, Michael. 1985. *Biblical Interpretation in Ancient Israel*. Oxford.

Flandin, Eugène. 1851. *Voyage en Perse de mm. Eugène Flandin, peintre, et Pascal Coste, architecte: Enterpris par ordre de m. le ministre des affaires étrangères, d'après les instructions dressés par l'Institut. Publié sous les auspices de m. le ministre de l'intérieur*. 2 vols. Paris.

Flood, Finbarr. 2016. "Idol-Breaking as Image-Making in the 'Islamic State.'" *Reli-gion and Society* 7:116–126.

Flusser, David. 1982. "Hystaspes and John of Patmos." In *Irano-Judaica: Studies Re-lating to Jewish Contacts with Persian Culture throughout the Ages*, edited by Shaul Shaked, 12–75. Jerusalem.

Fortes, Meyer. 1967. "Of Installation Ceremonies." *Proceedings of the Royal Anthro-pological Institute of Great Britain and Ireland*, 5–20.

Frahm, Eckart. 2005. "On Some Recently Published Late Babylonian Copies of Royal Letters." *NABU* §43.

———. 2010. "Counter-Texts, Commentaries, and Adaptations: Politically-Motivated Responses to the Babylonian Epic of Creation in Mesopotamia, the Biblical World, and Elsewhere." *Orient* 45:3–33.

Fraser, Peter. 1972. *Ptolemaic Alexandria*. Oxford.

———. 1996. *Cities of Alexander the Great*. Oxford.

Frazer, James. 1890. *The Golden Bough: A Study in Comparative Religion*. London.

Freedland, Jonathan. 2014. "This Islamic State Nightmare Is Not a Holy War but an Unholy Mess." *Guardian*, August 8.

Fröhlich, Ida. 1996. *'Times and Times and Half a Time': Historical Consciousness in the Jewish Literature of the Persian and Hellenistic Eras*. Sheffield.

Froidefond, Christian. 1971. *Le mirage égyptien dans la littérature grecque d'Homère à Aristote*. Paris.

Frost, S. 1952. *Old Testament Apocalyptic*. London.

Funkenstein, Amos. 1993. *Perceptions of Jewish History*. Berkeley.

Fussman, Gérard. 1985. "Nouvelles inscriptions Śaka (III)." *Bulletin de l'École française d'Extrême-Orient* 74:35–42.

Gabrielsen, Vincent. 2008. "Provincial Challenges to the Imperial Centre in Achaemenid and Seleucid Asia Minor." In *The Province Strikes Back: Imperial Dynamics in the Eastern Mediterranean*, edited by Björn Forsén and Giovanni Salmeri, 15–44. Helsinki.

Gadd, C. 1958. "The Harran Inscriptions of Nabonidus." *Anatolian Studies* 8:35–92.

Gafni, Isaiah. 1996. "Concepts of Periodization and Causality in Talmudic Literature." *Jewish History* 10:21–38.

Gammie, John. 1974. "Spatial and Ethical Dualism in Jewish Wisdom and Apocalyptic Literature." *JBL* 93:356–385.

Gardiner-Garden, John. 1987. *Apollodorus of Artemita and the Central Asian Skythians*. Bloomington, IN.

Garlan, Yvon. 1993. "À qui étaient destinés les timbres amphoriques grecs?" *Comptes rendus de l'Académie des inscriptions et belles-lettres* 137:181–190.

Garrison, Mark. 2013. "Royal Achaemenid Iconography." In *The Oxford Handbook of Ancient Iran*, edited by Daniel Potts, 566–595. Oxford.

Gasche, Hermann. 2010. "Les palais perses achéménides de Babylone." In *Le palais de Darius à Suse: Une résidence royale sur la route de Persépolis à Babylone*, edited by Jean Perrot, 446–463. Paris.

Geertz, Clifford. 1973. *The Interpretation of Cultures*. New York.

Geller, M. 1990. "Babylonian Astronomical Diaries and Corrections of Diodorus." *BSOAS* 53:1–7.

———. 1991. "New Information on Antiochus IV from Babylonian Astronomical Diaries." *BSOAS* 54:1–4.

Genovese, Eugene. 1976. *Roll, Jordan, Roll: The World the Slaves Made*. New York.

George, Andrew. 1985–1986. "The Topography of Babylon Reconsidered." *Sumer* 44:7–24.

———. 1992. *Babylonian Topographical Texts*. Leuven.

———. 1993. *House Most High: The Temples of Ancient Mesopotamia*. Winona Lake.

——. 1996. "Studies in Cultic Topography and Ideology." *Bibliotecha Orientalis* 53:363–395.

——. 2000. "Four Temple Rituals from Babylon." In *Wisdom, Gods, and Literature. Studies in Assyriology in Honour of W. G. Lambert*, edited by A. R. George and I. Finkel, 259–299. Winona Lake.

——. 2003. *The Babylonian Gilgamesh Epic: Introduction, Critical Edition and Cuneiform Texts*. Vol. 1. Oxford.

Gera, Dov. 1985. "Tryphon's Sling Bullet from Dor." *IEJ* 35:153–163.

——. 2009. "Olympiodoros, Heliodoros and the Temples of Koilê Syria and Phoinikê." *ZPE* 169:125–155.

Gera, Dov, and Wayne Horowitz. 1997. "Antiochus IV in Life and Death: Evidence from the Babylonian Astronomical Diaries." *JAOS* 117:240–252.

Gerardi, Pamela. 1986. "Declaring War in Mesopotamia." *Archiv für Orientsforschung* 33:30–38.

Ghirshman, Roman. 1936. "Appendice sur les fouilles de Medaïn." In *Fouilles de Telloh*, vol. 2: *Époques d'Ur IIIᵉ dynastie et de Larsa*, edited by Henri de Genouillac, 139–150. Paris.

Gibson, McGuire. 1994. "Parthian Seal Style: A Contribution from Nippur." *Mesopotamia* 29:89–105.

Giddens, Anthony. 1990. *The Consequences of Modernity*. Cambridge.

Gignoux, Philippe. 1986. "Sur l'inexistence d'un Bahman Yasht avestique." *Journal of Asian and African Studies* 32:53–64.

——. 1999. "L'apocalyptique iranienne est-elle vraiment ancienne? Notes critiques." *Revue de l'histoire des religions* 216:213–227.

Ginzburg, Carlo. 2010. "Mircea Eliade's Ambivalent Legacy." In *Hermeneutics, Politics, and the History of Religions: The Contested Legacies of Joachim Wach and Mircea Eliade*, edited by Christian Wedemeyer and Wendy Doniger, 307–323. Oxford.

Gnoli, Gherardo. 2000. *Zoroaster in History*. New York.

Gnuse, Robert. 1989. *Heilsgeschichte as a Model for Biblical Theology: The Debate Concerning the Uniqueness and Significance of Israel's Worldview*. Lanham, MD.

Goldberg, Sylvie. 2000. "Questions of Times: Conflicting Time Scales in Historical Perspective." *Jewish History* 14:267–286.

——. 2016. *Clepsydra: Essay on the Plurality of Time in Judaism*. Translated by Benjamin Ivry. Stanford.

Goldstein, Jonathan. 1976. *1 Maccabees: A New Translation with Introduction and Commentary*. New Haven.

——. 1987. "How the Authors of 1 and 2 Maccabees Treated the 'Messianic' Promises." In *Judaisms and Their Messiahs at the Turn of the Christian Era*, edited by Jacob Neusner, William Green, and Ernest Frerichs, 69–96. Cambridge.

Goldstein, Ronnie. 2010. "Late Babylonian Letters on Collecting Tablets and Their Hellenistic Background—A Suggestion." *JNES* 69:199–207.

Gonzalez, Hervé. 2013. "Zechariah 9–14 and the Continuation of Zechariah during the Ptolemaic Period." *Journal of Hebrew Scriptures* 13:1–43.

Goodman, Martin. 1982. "The First Jewish Revolt: Social Conflict and the Problem of Debt." *JJS* 33:417–427.

Gopnik, Hilary. 2010. "Why Columned Halls?" In *The World of Achaemenid Persia: History, Art and Society in Iran and the Ancient Near East,* edited by John Curtis and St John Simpson, 195–206. London.

Gore-Jones, Lydia. 2015. "Animals, Humans, Angels and God: Animal Symbolism in the Historiography of the 'Animal Apocalypse' of 1 Enoch." *Journal for the Study of the Pseudepigrapha* 24:268–287.

Gorre, Gilles. 2009. "'Nectanébo-le-faucon' et le dynastie lagide." *Ancient Society* 39:55–69.

Goshen-Gottstein, Alon. 2002. "Ben Sira's Praise of the Fathers: A Canon-Conscious Reading." In *Ben Sira's God: Proceedings of the International Ben Sira Conference, Durham—Ushaw College, 2001,* edited by Renate Egger-Wenzel, 235–267. Berlin.

Grabowski, Maciej. 2011. "Abdissares of Adiabene and the Batas-Herir Relief." *Światowit* 9:117–140.

Grayson, A. 1975. *Babylonian Historical-Literary Texts.* Toronto.

Greenspahn, Frederick. 1989. "Why Prophecy Ceased." *JBL* 108:37–49.

Griffiths, J. Gwyn. 1970. *Plutarch's De Iside et Osiride, Edited with an Introduction, Translation, and Commentary.* Cambridge.

Grillo, Jennie. 2017. "'From a Far Country': Daniel in Isaiah's Babylon." *JBL* 136:363–380.

Gruen, Erich. 2002. *Diaspora: Jews amidst Greeks and Romans.* Cambridge, MA.

———. Forthcoming. "The Maccabean Model: Resistance or Adjustment?" In *The Maccabean Moment,* edited by Paul Kosmin and Ian Moyer.

Gruenwald, Ithamar. 1973. "Knowledge and Vision: Towards a Clarification of Two 'Gnostic' Concepts in the Light of Their Alleged Origins." *Israel Oriental Studies* 3:63–107.

Grzybek, Erhard. 1992. "Zu einer babylonischen Königsliste aus der hellenistischen Zeit (Keilschrifttafel BM 35603)." *Historia* 41:190–204.

Guggenheimer, Heinrich. 1998. *Seder 'olam: The Rabbinic View of Biblical Chronology, Translated with Commentary.* Northvale, NJ.

Gunkel, Hermann. 1895. *Schöpfung und Chaos in Urzeit und Enzeit. Eine religionsgeschichtliche Untersuchung über Gen 1 und Ap Joh 12.* Göttingen.

Guyer, Jane. 2007. "Prophecy and the Near Future: Thoughts on Macroeconomic, Evangelical, and Punctuated Time." *American Ethnologist* 34:409–421.

Habicht, Christian. 1953. "Über eine armenische Inschrift mit Versen des Euripides." *Hermes* 81:251–256.

———. 1982. *Studien zur Geschichte Athens in hellenisticher Zeit.* Göttingen.

Haerinck, Ernie. 1973. "Le palais achéménide de Babylone." *Iranica Antiqua* 10:108–132.

Haerinck, Ernie, and Bruno Overlaet. 2008. "Altar Shrines and Fire Altars? Architectural Representations on *Frataraka* Coinage." *Iranica Antiqua* 43:207–233.

Haldar, Alfred. 1950. *The Notion of the Desert in Sumero-Accadian and West-Semitic Religions.* Uppsala.

Hall, Robert. 1991. *Revealed Histories: Techniques for Ancient Jewish and Christian Historiography.* Sheffield.

Halliday, M. 1976. "Anti-Languages." *American Anthropologist* 78:570–584.

Hallo, William. 1963. "On the Antiquity of Sumerian Literature." *JAOS* 83:167–176.

———. 1984–1985. "The Concept of Eras from Nabonassar to Seleucus." *JANES* 16–17:143–151.

———. 1988. "The Nabonassar Era and Other Epochs in Mesopotamian Chronology and Chronography." In *A Scientific Humanist: Studies in Memory of Abraham Sachs,* edited by Erle Leichty, Maria Ellis, and Pamela Gerardi, 175–190. Philadelphia.

Hamilakis, Yannis. 2007. *The Nation and Its Ruins: Antiquity, Archaeology, and National Imagination in Greece.* Oxford.

Hammer, Espen. 2011. *Philosophy and Temporality from Kant to Critical Theory.* Cambridge.

Hammond, N. 1992. "The Archaeological and Literary Evidence for the Burning of the Persepolis Palace." *CQ* 42:358–364.

Handy, Lowell. 1994. *Among the Host of Heaven: The Syro-Palestinian Pantheon as Bureaucracy.* Winona Lake.

Hannah, Robert. 2009. *Time in Antiquity.* Abingdon.

Hansen, Donald. 1992. "Royal Building Activity at Sumerian Lagash in the Early Dynastic Period." *Biblical Archaeologist* 55:206–211.

Hansen, Mogens. 1993. "The Battle Exhortation in Ancient Historiography: Fact or Fiction?" *Historia* 42:161–180.

Harootunian, Harry. 2007. "Remembering the Historical Present." *Critical Inquiry* 33:471–494.

Hartal, Moshe. 2002. "Excavations at Khirbet Zemel, Northern Golan: An Ituraean Settlement Site." In *Eretz Zafon: Studies in Galilean Archaeology,* edited by Zvi Gal, 74–117. Jerusalem.

Hartman, Lars. 1975–1976. "The Functions of Some So-Called Apocalyptic Timetables." *New Testament Studies* 22:1–14.

Hartman, Louis, and Alexander DiLella. 1978. *The Book of Daniel.* New York.

Hartog, François. 2014. "L'apocalypse, une philosophie d'histoire." *Esprit* 6:22–32.

———. 2015. *Regimes of Historicity: Presentism and Experiences of Time.* Translated by Saskia Brown. New York.

Hatzopoulos, Miltiade. 2007. "Bulletin épigraphique #358 (Démétrias)." *REG* 120:694–695.

Hauben, Hans. 1973. "On the Chronology of the Years 313–311 BC." *AJP* 94:256–267.

Haubold, Johannes. 2013. *Greece and Mesopotamia: Dialogues in Literature.* Cambridge.

———. 2016. "History and Historiography in the Early Parthian Diaries: A Close Reading." Paper presented at the Keeping Watch in Babylon: From Evidence to Text in the Astronomical Diaries conference, Durham, June 25.

Haubold, Johannes, Giovanni Lanfranchi, Robert Rollinger, and John Steele, eds. 2013. *The World of Berossos.* Wiesbaden.

Hayek, Friedrich, and Bruce Caldwell. 2010. *Studies on the Abuse and Decline of Reason: Text and Documents.* Chicago.

Hazzard, R. 2000. *Imagination of a Monarchy: Studies in Ptolemaic Propaganda.* Toronto.

Heidel, William. 1935. "Hecataeus and the Egyptian Priests in Herodotus, Book II." *Memoirs of the American Academy of Arts and Sciences* 18:53–134.

Heimpel, Wolfgang. 1996. "The Gates of the Eninnu." *JCS* 48:17–29.

Helberg, J. 1995. "The Determination of History According to the Book of Daniel." *ZAW* 107:273–287.

Hengel, Martin. 1974. *Judaism and Hellenism: Studies in Their Encounter in Palestine during the Early Hellenistic Period.* Translated by John Bowden. London.

Henkelman, Wouter. 2011. "Parnakka's Feast: Šip in Pārsa and Elam." In *Elam and Persia,* edited by Javier Álvarez-Mon and Mark Garrison, 89–166. Winona Lake.

Henze, Matthias. 2001. "The Narrative Frame of Daniel: A Literary Assessment." *JSJ* 32:5–24.

———. 2005. "The Apocalypse of Weeks and the Architecture of the End Time." In *Enoch and Qumran Origins: New Light on a Forgotten Connection,* edited by Gabriele Boccaccini, 207–209. Grand Rapids.

———. 2012. "The Use of Scripture in the Book of Daniel." In *A Companion to Biblical Interpretation in Early Judaism,* edited by Matthias Henze, 279–307. Grand Rapids.

Herbert, Sharon, and Andrea Berlin. 2003. "A New Administrative Center for Persian and Hellenistic Galilee: Preliminary Report of the University of Michigan / University of Minnesota Excavations at Kedesh." *BASOR* 329:13–59.

Herter, Hans. 1950. "Böse Dämonen im frühgriechischen Volksglauben." *Rheinisches Jahrbuch für Volkskunde* 1:112–143.

Heuzey, Léon. 1884–1912. "Les constructions de Tello d'après les fouilles d'Ernest de Sarzec, notes complémentaires." In Ernest de Sarzec, *Découvertes en Chaldée*, 1:395–449. Paris.

———. 1888. *Un palais chaldéen d'après les découvertes de M. de Sarzec.* Paris.

Hicks, Jennifer. 2016. "Hollow Archives: Bullae as a Source for Understanding Administrative Structures in the Seleukid Empire." PhD diss., University College London.

Hieke, Thomas. 2007. "The Role of 'Scripture' in the Last Words of Mattathias (1 Macc 2:49–70)." In *The Books of the Maccabees: History, Theology, Ideology. Papers of the Second International Conference on the Deuterocanonical Books, Pápa, Hungary, 9–11 June, 2005*, edited by Géza Xeravits and József Zsengellér, 61–74. Leiden.

Higbie, Carolyn. 2003. *The Lindian Chronicle and the Greek Creation of Their Past.* Oxford.

Himmelfarb, Martha. 1998. "Judaism and Hellenism in 2 Maccabees." *Poetics Today* 19:19–40.

Hinnells, John. 1973. "The Zoroastrian Doctrine of Salvation in the Roman World: A Study of the Oracle of Hystaspes." In *Man and His Salvation: Studies in Memory of S. G. F. Brandon*, edited by Eric Sharpe and John Hinnells, 125–148. Manchester.

Hintzen-Bohlen, Brigitte. 1990. "Die Familiengruppe—Ein Mittel zur Selbstdarstellung hellenistischer Herrscher." *Jahrbuch des deutschen archäologischen Instituts* 105:129–154.

Hocart, A. M. 1927. *Kingship.* Oxford.

Hoh, Manfred. 1979. "Die Grabung in Ue XVIII 1." In *XXIX. und XXX. Vorläufiger Bericht über die von dem Deutschen Archäologischen Institut aus Mitteln der Deutschen Forschungsgemeinschaft unternommenen Ausgrabungen in Uruk-Warka 1970/1 und 1971/2*, edited by Jürgen Schmidt, 28–35. Berlin.

Hölscher, Lucian. 2014. "Time Gardens: Historical Concepts in Modern Historiography." *History and Theory* 53:577–591.

Holt, Frank. 1984. "The So-Called 'Pedigree Coins' of the Bactrian Greeks." In *Ancient Coins of the Graeco-Roman World: The Nickle Numismatic Papers*, edited by W. Heckel and R. Sullivan, 69–85. Waterloo.

Honeyman, A. 1941. "Observations on a Phoenician Inscription of Ptolemaic Date." *JEA* 26:57–67.

Honigman, Sylvie. 2014. *Tales of High Priests and Taxes: The Books of the Maccabees and the Judean Rebellion against Antiochus IV.* Berkeley.

———. Forthcoming. "Diverging Memories, Not Resistance Literature: The Maccabean Crisis in the Animal Apocalypse and 1 and 2 Maccabees." In *The Maccabean Moment*, edited by Paul Kosmin and Ian Moyer.

Hooker, Mischa. Forthcoming. "Polychronius of Apamea and Daniel 11: Seleucid History through the Eyes of an Antiochene Biblical Interpreter." In *Seleukeia: Studies in Seleucid History, Archaeology, and Numismatics in Honor of Getzel M. Cohen*, edited by Roland Oetjen and Francis Ryan. Berlin.

Hoover, Oliver. 2009. *Handbook of Syrian Coins: Royal and Civic Issues. Fourth to First Centuries BC*. London.

Hopfner, Theodor. 1940–1941. *Plutarch, Über Isis und Osiris II. Die Deutungen der Sage. Übersetzung und Kommentar*. Darmstadt.

Hopkins, Clark. 1934. "The Agora and the Shops." In *The Excavations at Dura-Europos. Preliminary Report of Fifth Season of Work October 1931–March 1932*, edited by M. Rostovtzeff, 73–95. New Haven.

———. 1972. *Topography and Architecture of Seleucia on the Tigris*. Ann Arbor.

Hordern, J. 2004. *Sophron's Mimes: Text, Translation, and Commentary*. Oxford.

Horky, Philip. 2009. "Persian Cosmos and Greek Philosophy: Plato's Associates and the Zoroastrian *Magoi*." *Oxford Studies in Ancient Philosophy* 38:47–102. Oxford.

Horowitz, Wayne. 1995. "An Astronomical Fragment from Columbia University and the Babylonian Revolts against Xerxes." *JANES* 23:61–67.

Horsley, Richard. 2010. *Revolt of the Scribes: Resistance and Apocalyptic Origins*. Minneapolis.

Houghton, Arthur, and Catharine Lorber. 2002. *Seleucid Coins: A Comprehensive Catalogue. Part 1. Seleucus I through Antiochus III*. London.

Houghton, Arthur, Catharine Lorber, and Oliver Hoover. 2008. *Seleucid Coins: A Comprehensive Catalogue. Part 2. Seleucus IV through Antiochus XIII*. London.

Hughes, Aaron. 2003. "Presenting the Past: The Genre of Commentary in Theoretical Perspective." *Method and Theory in the Study of Religion* 15:148–168.

Huh, Su Kyung. 2008. *Studien zur Region Lagaš: Von der Ubaid- bis zu altbabylonischen Zeit*. Münster.

Hultgård, Anders. 1989. "Forms and Origins of Iranian Apocalyptic." In *Apocalypticism in the Mediterranean World and the Near East: Proceedings of the International Colloquium on Apocalypticism, Uppsala 12–17, 1979*, edited by David Hellholm, 387–411. Tübingen.

———. 1991. "*Bahman Yasht*: A Persian Apocalypse." In *Mysteries and Revelations: Apocalyptic Studies since the Uppsala Colloquium*, edited by John Collins and James Charlesworth, 114–134. Sheffield.

———. 1992. "Zoroastrian Myth in Bahman Yasht." In *The Middle East Viewed from the North: Papers from the First Nordic Conference in Middle Eastern Studies, Uppsala 26–29 January 1989*, edited by Bo Utas and Knut Vikør, 15–27. Bergen.

Humbach, Helmut. 2015. "Interpretations of Zarathustra and the *Gāthās*: The *Gāthās*." In *The Wiley-Blackwell Companion to Zoroastrianism*, edited by Michael Stausberg and Yuhan Vevaina, 39–43. Oxford.

Hunger, Hermann, and Teije de Jong. 2014. "Almanac W22340a from Uruk: The Latest Datable Cuneiform Tablet." *ZA* 103:182–194.

Hunger, Hermann, Abraham Sachs, and John Steele. 2001. *Astronomical Diaries and Related Texts from Babylonia V: Lunar and Planetary Texts*. Vienna.

Huß, Werner. 1977. "Der 'König der Könige' und der 'Herr der Könige.'" *ZDPV* 93:131–140.

Icke, Peter. 2012. *Frank Ankersmit's Lost Historical Cause: A Journey from Language to Experience*. New York.

Invernizzi, Antonio. 1996. "Gli archivi pubblici di Seleucia sul Tigri." In *Archives et sceaux du monde hellénistique*, edited by Marie-Françoise Boussac and Antonio Invernizzi, 131–143. Athens.

Isaac, Benjamin. 1991. "A Seleucid Inscription from Jamnia-on-the-Sea: Antiochus V Eupator and the Sidonians." *IEJ* 41:132–144.

Ismard, Paulin. 2017. *Democracy's Slaves: A Political History of Ancient Greece*. Translated by Jane Todd. Cambridge, MA.

Ito, Gikyo. 1976. "Gathica XIV–XV: Syenian *Frataraka* and Persid *Fratarak*: New Iranian Elements in Ancient Aramaic." *Orient* 12:47–66.

Jacobson, David, 2015. *Antioch and Jerusalem: The Seleucids and Maccabees in Coins*. London.

Jameson, Fredric. 2003. "The End of Temporality." *Critical Inquiry* 29:695–718.

Janzen, Gerald. 2012. *When Prayer Takes Place: Forays into a Biblical World*. Eugene, OR.

Jasnow, Richard. 1997. "The Greek Alexander Romance and Demotic Egyptian Literature." *JNES* 56:95–103.

Jassen, Alex. 2011. "Prophecy after 'The Prophets': The Dead Sea Scrolls and the History of Prophecy in Judaism." In *The Dead Sea Scrolls in Context: Integrating the Dead Sea Scrolls in the Study of Ancient Texts, Languages, and Cultures*, edited by Armine Lange, Emanuel Tov, Matthias Weigold, and Beenie Reynolds III, 2:577–593. Leiden.

Jenni, Hanna. 1986. *Das Dekorationsprogramm des Sarkophages Nektanebos' II*. Geneva.

Joannès, Francis. 1979–1980. "Les successeurs d'Alexandre le Grand en Babylonie. Essai de détermination chronologique d'après les documents cunéiformes." *Anatolica* 7:99–115.

———. 2006. "La Babylonie méridionale: Continuité, déclin ou rupture?" In *La transition entre l'empire achéménide et les royaumes hellénistiques*, edited by Pierre Briant and Francis Joannès, 101–135. Paris.

Johnson, Janet. 1984. "Is the Demotic Chronicle an Anti-Greek Tract?" In *Grammata Demotika: Festschrift für Erich Lüddeckens zum 15. Juni 1983*, edited by Heinz-J. Thissen and Karl-Th. Zauzich, 107–124. Würzburg.

Johnstone, Steven. 2011. *A History of Trust in Ancient Greece*. Chicago.

Jones, Alexander. 2006. "IG XII,1 913: An Astronomical Inscription from Hellenistic Rhodes." *ZPE* 158:104–110.

———. 2016. Introduction to *Time and Cosmos in Greco-Roman Antiquity*, edited by Alexander Jones, 19–43. New York.

Jones, Barry. 1995. *The Formation of the Book of the Twelve: A Study in Text and Canon*. Atlanta.

Jones, Christopher. 2009. "The Inscription from Tel Maresha for Olympiodoros." *ZPE* 171:100–104.

Jones, Prudence. 2005. *Reading Rivers in Roman Literature and Culture*. Lanham, MD.

Joosten, Jan. 2007. "The Original Language and Historical Milieu of the Book of Judith." In "A Festschrift for Devorah Dimant," edited by Moshe Bar-Asher and Emanuel Tov, special issue, *Meghillot* 5–6: 159–176.

Jursa, Michael. 2002. "*Florilegium babyloniacum*: Neue Texte aus hellenistischer und spätachämenidischer Zeit." In *Mining the Archives: Festschrift for Christopher Walker on the Occasion of His 60th Birthday, 4th October 2002*, edited by Cornelia Wunsch, 107–130. Dresden.

Kahn, Victoria. 2009. "Political Theology and Fiction in *The King's Two Bodies*." *Representations* 106:77–101.

Kallet-Marx, Robert. 1995. "Quintus Fabius Maximus and the Dyme Affair (*Syll.*³ 684)." *CQ* 45:129–153.

Kämmerer, Thomas, and Kai Metzler. 2012. *Das babylonische Weltschöpfungsepos Enūma elîš*. Münster.

Kantorowicz, Ernst. 1957. *The King's Two Bodies: A Study in Mediaeval Political Theology*. Princeton.

Kawkabani, Ibrahim. 2003. "Les anses timbrées de Jal el-Bahr." *Archaeology and History in Lebanon* 17:95–99.

Kellens, Jean. 2015. "The *Gāthās*, Said to Be of Zarathustra." In *The Wiley-Blackwell Companion to Zoroastrianism*, edited by Michael Stausberg and Yuhan Vevaina, 44–50. Oxford.

Kellermann, Ulrich. 1967. *Nehemia. Quellen, Überlieferung und Geschichte*. Berlin.

Kermode, Frank. 2000. *The Sense of an Ending: Studies in the Theory of Fiction*. Oxford.

Kessler, Karlheinz. 2002. "Ḥarinê—Zu einer problematischen Passage der Nabonid-Chronik." In *"Sprich doch mit deinen Knechten aramäisch, wir verstehen es!" 60 Beiträge zur Semitistik. Festschrift für Otto Jastrow zum 60. Geburtstag*, edited by Werner Arnold and Hartmut Bobzin, 389–393. Wiesbaden.

Khatchadourian, Lori. 2007. "Unforgettable Landscapes: Attachments to the Past in Hellenistic Armenia." In *Negotiating the Past in the Past: Identity, Memory, and Landscape in Archaeological Research,* edited by Norman Yoffee, 43-75. Tucson.

———. 2016. *Imperial Matter: Ancient Persia and the Archaeology of Empires.* Berkeley.

Khachatrian, Žores. 1998. "Artaxata, capitale dell'Armenia antica (II sec. a.C.-IV d.C.)." In *Ai piedi dell'Ararat: Artaxata e l'Armenia ellenistico-romana,* edited by Antonio Invernizzi, 97-115. Florence.

Kim, Jin. 2011. "Doxology as Rhetoric of Resistance in the Aramaic Tales of Daniel 2-6." PhD diss., Lutheran School of Theology at Chicago.

Kindler, A. 1968. "Addendum to the Dated Coins of Alexander Janneus." *IEJ* 18:188-191.

King, Preston. 2000. *Thinking Past a Problem: Essays on the History of Ideas.* London.

Kingsley, Peter. 1995. "Meetings with Magi: Iranian Themes among the Greeks, from Xanthus of Lydia to Plato's Academy." *Journal of the Royal Asiatic Society* 5:173-209.

Kippenberg, Hans. 1978. "Die Geschichte der mittelpersischen apokalyptischen Traditionen." *Studia Iraniaca* 7:49-80.

Kleiner, Fred. 1972. "The Dated Cistophoroi of Ephesus." *American Numismatic Society: Museum Notes* 18:17-32.

Klose, Dietrich. 2005. "Statthalter, Könige, Rebellen—Die Münzen der Persis vom 3. vorchristlichen Jahrhundert bis zum Beginn des Sasanidenreichs." *Numismatisches Nachrichtenblatt* 54:93-106.

Klose, Dietrich, and Wilhelm Müseler. 2008. *Statthalter, Rebellen, Könige: Die Münzen aus Persepolis von Alexander dem Großen zu den Sasaniden.* Munich.

Klotchkoff, Igor. 1982. "The Late Babylonian List of Scholars." In *Schriften zur Geschichte und Kultur des alten Orients,* edited by Horst Klengel, 143-148. Berlin.

Knausgaard, Karl Ove. 2014. *A Time for Everything.* Translated by James Anderson. Brooklyn.

Knibb, Michael. 1976. "The Exile in the Literature of the Intertestamental Period." *Heythrop Journal* 17:253-272.

———. 1999. "Eschatology and Messianism in the Dead Sea Scrolls." In *The Dead Sea Scrolls after Fifty Years: A Comprehensive Assessment,* edited by Peter Flint and James VanderKam, 2:379-402. Leiden.

Knobloch, Johann. 1985. "Eine etymologische Fabel im Sintflutbericht bei Berossos." *Glotta* 63:1.

Koch, Klaus. 1961. "Spätisraelitisches Geschichtsdenken am Beispiel des Buches Daniel." *Historische Zeitschrift* 193:1-32.

———. 1983. "Sabbatstruktur der Geschischte: Die sogennante Zehn-Wochen-Apokalypse (1 Hen 93 1-10, 91 11-17) und das Ringen um die alttestamentlichen Chronologien im späten Israelitentum." *ZAW* 95:403-430.

Kockelman, Paul, and Anya Bernstein. 2012. "Semiotic Technologies, Temporal Reckoning, and the Portability of Meaning. Or: Modern modes of temporality—Just How Abstract Are They?" *Anthropological Theory* 12:320-348.

Koenen, Ludwig. 1994. "Greece, the Near East, and Egypt: Cyclic Destruction in Hesiod and the *Catalogue of Women*." *TAPA* 124:1-34.

———. 2002. "Die Apologie des Töpfers an König Amenophis oder das Töpferorakel." In *Apokalyptik und Ägypten. Eine kritische Analyse der relevanten Texte aus dem griechisch-römsichen Ägypten,* edited by A. Blasius and B. Schipper, 139-187. Leuven.

Koenig, Jean. 1962. "L'activité herméneutique des scribes dans la transmission du texte de l'Ancient Testament." *Revue de l'histoire des religions* 161:141-174.

Korzakova, Hava. 2010. "Lead Weights." In Amos Kloner, Esther Eshel, Hava Korzakova, and Gerald Finkielsztejn, *Maresha Excavations Final Report III. Epigraphic Finds from the 1989-2000 Seasons,* 159-173. Jerusalem.

Kose, Arno. 2000. "Das 'Palais' auf Tell A von Girsu—Wohnstätte eines hellenistisch-parthischen Sammlers von Gudeastatuen?" *Baghdader Mitteilungen* 31:377-446.

———. 2013a. "Das seleukidische Resch-Heiligtum." In *Uruk: 5000 Jahre Megacity,* 333-339. Petersburg.

———. 2013b. "Himmelsgott, König, Tempelgemeinde: Die Nachblüte der altorientalischen Stadt Uruk in der Seleukidenzeit." In *Uruk: 5000 Jahre Megacity,* 320-329. Petersburg.

Koselleck, Reinhart. 2002. *The Practice of Conceptual History: Timing History, Spacing Concepts.* Translated by Todd Presner et al. Stanford.

———. 2004. *Futures Past: On the Semantics of Historical Time.* Translated by Keith Tribe. New York.

Kosmetatou, Elisabeth. 1995. "The Legend of the Hero Pergamus." *Ancient Society* 26:133-144.

———. 2000. "Lycophron's *Alexandra* Reconsidered: The Attalid Connection." *Hermes* 128:32-53.

———. 2004. "Vision and Visibility: Art Historical Theory Paints a Portrait of New Leadership in Posidippus' *Andriantopoiika*." In *Labored in Papyrus Leaves: Perspectives on an Epigram Collection Attributed to Posidippus (P.Mil.Vog. VIII 309),* edited by Benjamin Acosta-Hughes, Elizabeth Kosmetatou, and Manuel Baumbach, 187-211. Cambridge, MA.

Kosmin, Paul. 2013a. "Apologetic Ethnography: Megasthenes' *Indica* and the Seleucid Elephant." In *Ancient Ethnography: New Approaches,* edited by Eran Almagor and Joseph Skinner, 97-115. London.

———. 2013b. "Rethinking the Hellenistic Gulf: The New Greek Inscription from Bahrain." *JHS* 133:61-79.

———. 2013c. "Seleucid Ethnography and Indigenous Kingship: The Babylonian Education of Antiochus I." In *The World of Berossos*, edited by Johannes Haubold, Giovanni Lanfranchi, Robert Rollinger, and John Steele, 193–212. Wiesbaden.

———. 2014a. *The Land of the Elephant Kings: Space, Territory, and Ideology in the Seleucid Empire.* Cambridge, MA.

———. 2014b. "Seeing Double in Seleucid Babylonia: Rereading the Borsippa Cylinder of Antiochus I." In *Patterns of the Past*, edited by Alfonso Moreno and Rosalind Thomas, 173–198. Oxford.

———. 2015. "A Phenomenology of Democracy: Ostracism as Political Ritual." *CA* 34:121–161.

———. 2016. "Indigenous Revolts in *2 Maccabees:* The Persian Version." *CP* 111:32–53.

———. 2017a. "An Encomium for Albert." In *Albert's Anthology*, edited by Kathleen Coleman, 99–100. Cambridge, MA.

———. 2017b. "The Politics of Science: Eratosthenes' Geography and Ptolemaic Imperialism." *Orbis Terrarum* 15:99–100.

Kosmin, Paul, and Ian Moyer. Forthcoming. "Imperial and Indigenous Temporalities in the Ptolemaic and Seleucid Dynasties: A Comparison of Times." In *Comparing the Ptolemaic and Seleucid Empires: Centers of Power, Local Elites and Populations*, edited by Christelle Fischer-Bovet and Sitta von Reden. Cambridge.

Kovacs, Frank. 2016. *Armenian Coinage in the Classical Period.* Lancaster.

Krahmalkov, C. 1976. "Notes on the Rule of the *šōfṭīm* in Carthage." *Rivista di Studi Fenici* 4:153–157.

Kratz, Reinhard. 1991. *Tranlatio imperii. Untersuchungen zu den aramäischen Danielerzählungen und ihrem theologiegeschichtlichen Umfeld.* Düsseldorf.

Kraus, Hans-Joachim. 1989. *Psalms 60–150. A Commentary.* Translated by H. C. Oswald. Minneapolis.

Kraybill, Donald. 1994. "The Amish Encounter with Modernity." In *The Amish Struggle with Modernity*, edited by Donald Kraybill and Marc Olshan, 21–33. Hanover, NH.

Krüger, Thomas. 2000. *Kohelet (Prediger).* Neukirchen-Vluyn.

Krul, Julia. 2014. "'The Beautiful Image Has Come Out': The Nocturnal Fire Ceremony and the Revival of the Anu Cult at Late Babylonian Uruk." PhD diss., Westfälischen Wilhelms-Univeristät.

Kugel, James. 1995. "The Ladder of Jacob." *Harvard Theological Review* 88:209–227.

Kugler, Franz. 1907–1924. *Sternkunde und Sterndienst in Babel: Assyriologische, astronomische und astralmythologische Untersuchungen.* 3 vols. Münster.

Kuhrt, Amélie. 1987. "Berossus' *Babyloniaka* and Seleucid Rule in Babylonia." In *Hellenism in the East: The Interaction of Greek and Non-Greek civilizations from Syria to Central Asia after Alexander*, edited by Amélie Kuhrt and Susan Sherwin-White, 32–56. London.

Kushnir-Stein, Alla. 1997. "On the Chronology of Some Inscribed Lead Weights from Palestine." *ZDPV* 113:88–91.

———. 2001. "Dates on Ancient Palestinian Coins." In *Measuring and Weighing in Ancient Times*, edited by Yisrael Ronen, 47–50. Haifa.

———. 2002. "New Hellenistic Lead Weights from Palestine and Phoenicia." *IEJ* 52:225–230.

———. 2005a. "City Eras on Palestinian Coinage." In *Coinage and Identity in the Roman Provinces*, edited by Christopher Howgego, Volker Heuchert, and Andrew Burnett, 157–161. Oxford.

———. 2005b. "Two Hellenistic Weights from Phoenicia in the Hecht Museum Collection." *Michmanim* 19:15–20.

Kvanvig, Helge. 1988. *Roots of Apocalyptic: The Mesopotamian Background of the Enoch Figure and of the Son of Man*. Neukirchen-Vluyn.

———. 2007. "Cosmic Laws and Cosmic Imbalance: Wisdom, Myth and Apocalyptic in the Early Enochic Writings." In *The Early Enochic Literature*, edited by Gabriele Boccaccini and John Collins, 139–158. Leiden.

Lambdin, Thomas. 2006. *Introduction to Classical Ethiopic (Ge'ez)*. Winona Lake.

Lambert, Wilfred. 1957. "Ancestors, Authors, and Canonicity." *JCS* 11:1–14.

———. 1962. "A Catalogue of Texts and Authors." *JCS* 16:59–77.

———. 1978. *The Background of Jewish Apocalyptic*. London.

———. 1997. "The Assyrian Recension of Enūma Eliš." In *Assyrien im Wandel der Zeiten: XXXIXᵉ Rencontre Assyriologique Internationale, Heidelberg, 6.-10. Juli 1992*, edited by Hartmut Waetzoldt and Harald Hauptmann, 77–79. Heidelberg.

———. 2005. "No.44: Historical Literature: Letter of Sîn-šarru-iškun to Nabopolassar." In *Cuneiform Texts in the Metropolitan Museum of Art II Literary and Scholastic Texts of the First Millennium B.C.*, edited by Ira Spar and Wilfred Lambert, 203–210. New York.

Lambert, Wilfred, and A. Millard. 1969. *Atrahasis: The Babylonian Story of the Flood*. Oxford.

Lambert, Wilfred, and P. Walcot. 1965. "A New Babylonian Theology and Hesiod." *Kadmos* 4:64–72.

Lämmer, M. 1967. "Der Diskos des Asklepiades aus Olympia und das Marmor Parium." *ZPE* 1:107–109.

Landau, Y. 1966. "A Greek Inscription Found near Hefzibah." *IEJ* 16:54–70.

Landauer, Carl. 1994. "Ernst Kantorowicz and the Sacralization of the Past." *Central European History* 27:1–25.

Lane Fox, Robin. 2011. "'Glorious Servitude . . .': The Reigns of Antigonos Gonatas and Demetrios II." In *Brill's Companion to Ancient Macedon: Studies in the Archaeology and History of Macedon, 650 BC—300 AD*, edited by Robin Lane Fox, 495–519. Leiden.

Lang, Mabel, and Margaret Crosby. 1964. *Weights, Measures and Tokens. The Athenian Agora X*. Princeton.

Lang, Martin. 2013. "Book Two: Mesopotamian Early History and the Flood Story." In *The World of Berossos*, edited by Johannes Haubold, Giovanni Lanfranchi, Robert Rollinger, and John Steele, 47–60. Wiesbaden.

Lange, Armin. 2004. "From Literature to Scripture: The Unity and Plurality of the Hebrew Scriptures in the Light of the Qumran Library." In *One Scripture or Many? Canon from Biblical, Theological and Philosophical Perspectives*, edited by Christine Helmer and Christof Landmesser, 51–107. Oxford.

———. 2008. "'The Law, the Prophets, and the Other Books of the Fathers' (Sir, Prologue): Canonical Lists in Ben Sira and Elsewhere?" In *Studies in the Book of Ben Sira: Papers of the Third International Conference on the Deuterocanonical Books, Shime'on Centre, Pápa, Hungary, 18–20 May, 2006*, edited by Géza Xeravits and József Zsengellér, 54–80. Leiden.

Le Goff, Jacques. 1960. "Au Moyen Âge: Temps de l'Église et temps du marchand." *Annales* 15:417–433.

Le Rider, Georges. 1965. *Suse sous les Séleucides et les Parthes*. Paris.

Lebram, J. 1975. "König Antiochus im Buch Daniel." *VT* 25:737–772.

Lee, Thomas. 1986. *Studies in the Form of Sirach 44–50*. Atlanta.

Lenzi, Alan. 2008. "The Uruk List of Kings and Sages and Late Mesopotamian Scholarship." *Journal of Ancient Near Eastern Religions* 8:137–169.

Leriche, Pierre. 1996. "Le *chreophylakeion* de Doura-Europos et la mise en place du plan hippodamien de la ville." In *Archives et sceaux du monde hellénistique*, edited by Marie-Françoise Boussac and Antonio Invernizzi, 157–169. Athens.

Lerner, Jeffrey. 1999. *The Impact of Seleucid Decline on the Eastern Iranian Plateau: The Foundations of Arsacid Parthia and Graeco-Bactria*. Stuttgart.

Leschhorn, W. 1993. *Antike Ären: Zeitrechnung, Politik und Geschichte im Schwarzmeerraum und in Kleinasien nördlich des Tauros*. Stuttgart.

Leslau, Wolf. 2010. *Concise Dictionary of Ge'ez (Classical Ethiopic)*. Wiesbaden.

Levenson, David, and Thomas Martin. 2009. "Akairos or Eukairos? The Nickname of the Seleucid King Demetrius III in the Transmission of the Texts of Josephus' *War* and *Antiquities*." *JSJ* 40:307–341.

Lewy, Hildegard. 1944. "The Genesis of the Faulty Persian Chronology." *JAOS* 64:197–214.

Liampi, Katerini. 1990. "Der makedonische Schild als propagandistisches Mittel in der hellenistischen Zeit." ΠΟΙΚΙΛΑ (Μελετήματα 10): 157–172.

Licht, Jacob. 1965. "Time and Eschatology in Apocalyptic Literature and Qumran." *JJS* 16:177–182.

Lichtenstein, Hans. 1931–1932. "Die Fastenrolle. Eine Untersuchung zur jüdisch-hellenistischen Geschichte." *Hebrew Union College Annual* 8–9:257–351.

Lincoln, Bruce. 1977. "Death and Resurrection in Indo-European Thought." *Journal of Indo-European Studies* 5:247–264.

———. 2012. *"Happiness for Mankind": Achaemenian Religion and the Imperial Project.* Leuven.

Lindström, Gunvor. 2003. *Uruk. Siegelabdrücke auf hellenistischen Tonbullen und Tontafeln.* Mainz.

Linssen, Marc. 2004. *The Cults of Uruk and Babylon: The Temple Ritual Texts as Evidence for Hellenistic Cult Practices.* Leiden.

Lipiński, Edward. 2004. *Itineraria Phoenicia.* Leuven.

Liverani, Mario. 1990. *Prestige and Interest: International Relations in the Near East, ca. 1600–1100 B.C.* Padua.

———. 2016. *Imagining Babylon: The Modern Story of an Ancient City.* Translated by Ailsa Campbell. Berlin.

Longman, Tremper, III. 1991. *Fictional Akkadian Autobiography: A Generic and Comparative Study.* Winona Lake.

Lorber, Catharine. 2007. "The Ptolemaic Era Coinage Revisited." *NC* 167:105–117.

———. 2015. "Royal Coinage in Hellenistic Phoenicia: Expressions of Continuity, Agents of Change." In *La Phénicie hellénistique. Actes du colloque international de Toulouse (18–20 février 2013)*, edited by Julien Aliquot and Corinne Bonnet, 55–88. Lyon.

Löwith, Karl. 1949. *Meaning in History: The Theological Implications of the Philosophy of History.* Chicago.

Löwy, Michael. 2005. *Fire Alarm: Reading Walter Benjamin's* On the Concept of History. Translated by Chris Turner. London.

Luther, Andreas. 1999. "Die ersten Könige von Osrhoene." *Klio* 81:437–454.

———. 2015. "Das Königreich Adiabene zwischen Parthern und Römern." In *Amici—socii—clientes? Abhängige Herrschaft im Imperium Romanum*, edited by Ernst Baltrusch and Julia Wilker, 275–300. Berlin.

Luz, Christine. 2010. *Technopaignia: Formspiele in der griechischen Dichtung.* Leiden.

Ma, John. 2002. *Antiochos III and the Cities of Western Asia Minor.* Oxford.

———. 2003. "Kings and Kingship." In *A Companion to the Hellenistic World*, edited by Andrew Erskine, 177–196. Oxford.

———. 2008. "Mysians on the Çan Sarcophagus? Ethnicity and Domination in Achaemenid Military Art." *Historia* 57:243–254.

———. 2010. "Autour des balles de fronde 'camiréennes.'" *Chiron* 40:155–173.

———. 2012. "Relire les *Institutions des Séleucides* de Bikerman." In *Rome, a City and Its Empire in Perspective: The Impact of the Roman World through Fergus Millar's Research*, edited by Stéphane Benois, 59–84. Leiden.

———. 2013a. "Re-Examining Hanukkah." *The Marginalia Review of Books*, July 9.

———. 2013b. *Statues and Cities: Honorific Portaits and Civic Identity in the Hellenistic World*. Oxford.

Machinist, Peter. 1983. "Assyria and Its Image in the First Isaiah." *JAOS* 103: 719–737.

Mack, Burton. 1985. *Wisdom and the Hebrew Epic: Ben Sira's Hymn in Praise of the Fathers*. Chicago.

MacRae, Duncan. 2016. *Legible Religion: Books, Gods, and Rituals in Roman Culture*. Cambridge, MA.

Maddoli, G. 1963–1964. "Le cretule del Nomophylakion di Cirene." *Annuario della Scuola Archeologica di Atene e delle Missioni Italiane in Oriente*, 39–145.

Mahé, Jean-Pierre. 1994. "Moïse de Khorène et les inscriptions grecques d'Armawir." *Topoi* 4:567–586.

Maier, Charles. 1999. "I paradossi del 'prima' e del 'poi.' Periodizzazioni e rotture nella storia." *Contemporanea* 2:715–722.

Mairs, Rachel. 2014. *The Hellenistic Far East: Archaeology, Language, and Identity in Greek Central Asia*. Berkeley.

Malitz, Jürgen. 1983. *Die Historien des Poseidonios*. Munich.

Manni, E. 1949. "Tre note di cronologia ellenistica." *Rendiconti della Classe di Scienze morali, storiche e filologiche dell'Accademia dei Lincei* 8:53–85.

Manning, Joseph. 2010. *The Last Pharaohs: Egypt under the Ptolemies, 305–30 BC*. Princeton.

Mansfield, J., and D. Runia. 2009. *Aëtiana: Method and Intellectual Context of a Doxographer II. The Compendium*. Leiden.

Marasco, G. 1985. "La 'Profezia Dinastica' e la resistenza babilonese alla reconquista di Alessandro." *Annali della Scuola Normale Superiore di Pisa. Classe di Lettere e Filosofia* 15:529–537.

Marciak, Michał. 2012. "The Historical Geography of Sophene." *Acta Antiqua Academiae Scientiarum Hungaricae* 52:295–338.

Marciak, Michał, and Robert Wójcikowski. 2016. "Images of Kings of Adiabene: Numismatic and Sculptural Evidence." *Iraq* 78:79–101.

Marcus, Ralph. 1943. *Josephus VII (LCL 365): Josephus, Jewish Antiquities, Books XII–XIII*. Cambridge, MA.

Martinez-Sève, Laurianne. 2002. "La ville de Suse à l'époque hellénistique." *Revue archéologique* 33:31–53.

———. 2014. "Remarques sur la transmission aux Parthes des pratiques de gouvernement séleucides: Modalités et chronologie." *Ktema* 39:123–142.

———. 2015. "SUSA iv. The Hellenistic and Parthian Periods." *Encyclopaedia Iranica*. http://iranicaonline.org/articles/susa-iv-hellenistic-parthian-periods.

Mastin, B. 1973. "Daniel 2 46 and the Hellenistic World." *ZAW* 85:80–93.

Mathys, Hans-Peter. 2000. *Vom Anfang und vom Ende: Fünf alttestamentliche Studien.* Frankfurt.

Mayer, Werner. 1988. "Ein neues Königsritual gegen feindliche Bedrohung." *Orientalia* 57:145–164.

Mbembe, Achille. 2001. *On the Postcolony.* Berkeley.

McCarthy, Cormac. 2005. *No Country for Old Men.* New York.

McDonald, Lee. 2006. "Canon." In *The Oxford Handbook of Biblical Studies,* edited by J. Rogerson and Judith Lieu, 777–809. Oxford.

McDowell, Robert. 1935. *Stamped and Inscribed Objects from Seleucia on the Tigris.* Ann Arbor.

McEwan, Gilbert. 1981a. *Priest and Temple in Hellenistic Babylonia.* Wiesbaden.

———. 1981b. "Arsacid Temple Records." *Iraq* 43:131–143.

———. 1982. "An Official Seleucid Seal Reconsidered." *JNES* 41:51–53.

———. 1985. "The First Seleucid Document from Babylonia." *Journal of Semitic Studies* 30:169–180.

McIntyre, Andrew. 2007. "The Eras of the Alexanders of Aspendos and Perge." *NC* 167:93–98.

McKenzie, Leah. 1994. "Patterns in Seleucid Administration: Macedonian or Near Eastern." *Mediterranean Archaeology* 7:61–68.

Meade, David. 1986. *Pseudonymity and Canon: An Investigation into the Relationship of Authorship and Authority in Jewish and Early Christian Tradition.* Tübingen.

Meadows, Andrew. 2009. "The Eras of Pamphylia and the Seleucid Invasions of Asia Minor." *American Journal of Numismatics* 21:51–88.

Meeus, Alexander. 2012. "Diodorus and the Chronology of the Third Diadoch War." *Phoenix* 66:74–96.

Mehl, Andreas. 1986. *Seleukos Nikator und sein Reich I. Seleukos' Leben und die Entwicklung seiner Machtposition.* Leuven.

Meißner, B. 1992. *Historiker zwischen Polis und Königshof: Studien zur Stellung der Geschichtsschreiber in der griechischen Gesellschaft in spätklassischer und frühhellenistischer Zeit.* Göttingen.

Melvin, David. 2013. *The Interpreting Angel Motif in Prophetic and Apocalyptic Literature.* Minneapolis.

Menninger, Karl. 1969. *Number Words and Number Symbols: A Cultural History of Numbers.* Translated by Paul Broneer. Cambridge, MA.

Meshel, Ze'ev. 2000. "The Nabataean 'Rock' and the Judaean Desert Fortresses." *IEJ* 50:109–115.

Meshorer, Ya'akov. 1982. *Ancient Jewish Coinage.* New York.

Meshorer, Ya'akov, and Shraga Qedar. 1991. *The Coinage of Samaria in the Fourth Century BCE.* Jerusalem.

Messina, Vito. 2006. *Seleucia al Tigri: L'edificio degli archivi.* Florence.

———. 2010. *Seleucia al Tigri: Il monumento di Tell 'Umar: Lo scavo e le fasi architettoniche.* Florence.

Messina, Vito, and Paolo Mollo. 2004. *Selucia al Tigri: Le impronte di sigillo dagli archivi 1. Sigilli ufficiali, ritratti.* Alessandria.

Metzger, Bruce. 1972. "Literary Forgeries and Canonical Pseudepigrapha." *JBL* 91:3–24.

Meyer, Eduard. 1921–1923. *Ursprung und Anfänge des Christentums.* 3 vols. Stuttgart.

Miano, David. 2010. *Shadow on the Steps: Time Measurement in Ancient Israel.* Atlanta.

Mildenberg, L. 1984. *The Coinage of the Bar Kokhba War.* Aarau.

Milik, Joseph. 2003. "Une bilingue araméo-grecque de 105/104 avant J.-C." In *Hauran II. Les installations de Si' 8. Du sanctuaire à l'établissement viticole,* edited by Jacqueline Dentzer-Feydy, Jean-Marie Dentzer, and Pierre-Marie Blanc, 269–275. Beirut.

Millar, Fergus. 1983. "The Phoenician Cities: A Case-Study of Hellenisation." *Proceedings of the Cambridge Philological Society* 29:55–71.

Miller, Kenneth. 2011. "A First Jewish Revolt *Prutah* Overstrike." *Israel Numismatic Research* 6:127–132.

Mitchell, Stephen. 2017. "The Greek Impact in Asia Minor 400–250 BCE." In *Hellenism and the Local Communities of the Eastern Mediterranean 400 BCE–250 CE,* edited by Boris Chrubasik and Daniel King, 13–28. Oxford.

Mitsuma, Yasayuki. 2008. "69. Ištar of Babylon in 'Day-One Temple.'" *NABU* §69.

———. 2013. "The Offering for the Ritual of King Seleucus III and His Offspring." In *Time and History in the Ancient Near East: Proceedings of the 56th Rencontre Assyriologique Internationale at Barcelona, 26–30 July 2010,* edited by L. Feliu, J. Llop, A. Millet Albà, and J. Sanmartín, 739–744. Winona Lake.

Möller, Astrid. 2005. "Epoch-Making Eratosthenes." *Greek, Roman and Byzantine Studies* 45:245–260.

Mollo, Paolo. 1997. "Sigilli e timbri ufficiali nella Mesopotamia seleucide." In *Sceaux d'orient et leur emploi,* edited by Rika Gyselen, 89–107. Bures-sur-Yvette.

Momigliano, Arnaldo. 1966. "Time in Ancient Historiography." *History and Theory* 6:1–23.

———. 1970. "J. G. Droysen between Greeks and Jews." *History and Theory* 9:139–153.

———. 1976. "The Date of the First Book of Maccabees." In *L'Italie préromaine et la Rome républicaine. I. Mélanges offerts à Jacques Heurgon,* 657–661. Rome.

———. 1980. "Daniele e la teoria greca della successione degli imperi." *Atti della Accademia Nazionale dei Lincei, Classe di scienze morali, storiche e filologiche. Rendiconti* 35:157–162.

———. 1982. "Biblical Studies and Classical Studies: Simple Reflections about Historical Method." *Biblical Archaeologist* 45:224-228.

———. 1987. "Some Preliminary Remarks on the 'Religious Opposition' to the Roman Empire." In *Opposition et résistances à l'empire d'Auguste à Trajan*, edited by Adalberto Giovannini, 103-133. Geneva.

Monachov, Sergej. 2005. "Rhodian Amphoras: Developments in Form and Measurements." In *Chronologies of the Black Sea Area in the Period c. 400-100 BC*, edited by V. Stolba, 69-95. Aarhus.

Monerie, Julien. 2014. *D'Alexandre à Zoilos. Dictionnaire prosopographique des porteurs de nom grec dans les sources cunéiformes*. Stuttgart.

Monson, Andrew. 2016. "The Jewish High Priesthood for Sale: Farming Out Temples in the Hellenistic Near East." *JJS* 67:15-35.

Montgomery, James. 1927. *A Critical and Exegetical Commentary on the Book of Daniel*. New York.

Montiglio, Silvia. 2005. *Wandering in Ancient Greek Culture*. Chicago.

Moore, Stewart. 2015. *Jewish Ethnic Identity and Relations in Hellenistic Egypt: Within Walls of Iron?* Leiden.

Mørkholm, Otto. 1966. *Antiochus IV of Syria*. Gyldendal.

———. 1991. *Early Hellenistic Coinage from the Accession of Alexander to the Peace of Apamea (336—188 B.C.)*. Cambridge.

Mosès, Stéphane. 1994. *Der Engel der Geschichte: Franz Rosenzweig, Walter Benjamin, Gershom Scholem*. Frankfurt.

Mousavi, Ali. 2012. *Persepolis: Discovery and Afterlife of a World Wonder*. Berlin.

Moyer, Ian. 2002. "Herodotus and an Egyptian Mirage: The Genealogies of the Theban Priests." *JHS* 122:70-90.

———. 2011. *Egypt and the Limits of Hellenism*. Cambridge.

Moyn, Samuel, and Andrew Sartori. 2013. "Approaches to Global Intellectual History." In *Global Intellectual History*, edited by Samuel Moyn and Andrew Sartori, 3-30. New York.

Muccioli, Fredericomaria. 2004. "'Il re dell' Asia': Ideologia e propaganda da Alessandro Magno a Mitridate VI." *Simblos* 4:105-158.

Mueller, Simone. 2016. "Time Is Money—Everywhere? Analysing Time Metaphors across Varieties of English." In *Metaphor and Communication*, edited by Elisabetta Gola and Francesca Ervas, 79-104. Amsterdam.

Mulder, Otto. 2003. *Simon the High Priest in Sirach 50: An Exegetical Study of the Significance of Simon the High Priest as Climax to the Praise of the Fathers in Ben Sira's Concept of the History of Israel*. Leiden.

Müller, Helmut. 2000. "Der hellenistische Archiereus." *Chiron* 30:519-542.

Müller, Sabine. 2009. "Inventing Traditions: Genealogie und Legitimation in den hellenistischen Reichen." In *Genealogisches Bewusstsein als Legitimation: Inter-*

und intragenerationelle Auseinandersetzungen sowie die Bedeutung von Verwandtschaft bei Amtswechseln, edited by Hartwin Brandt, Katrin Köhler, and Ulrike Siewert, 61–82. Bamberg.

Murray, Oswyn. 1969. "Cultural Imperialism." Review of *Kulturgeschichte des Hellenismus,* by Carl Schneider. *Classical Review* 19:69–72.

Murthy, Viren. 2015. "Looking for Resistance in All the Wrong Places? Chibber, Chakrabarty, and a Tale of Two Histories." *Critical Historical Studies* 2:113–153.

Müseler, Wilhelm. 2005-2006. "Die sogenannten dunklen Jahrhunderte der Persis—Anmerkungen zu einem lange vernachlässigten Thema." *Jahrbuch für Numismatik und Geldgeschichte* 55-56:75–103.

———. 2018. "The Dating and Sequence of the Persid *Frataraka* Revisited." *Koinon* 1.

Myers-Scotton, Carol. 1993. "Common and Uncommon Ground: Social and Structural Factors in Code-Switching." *Language in Society* 22:475–503.

Na'aman, Nadav. 1984. "Statements of Time-Spans by Babylonian and Assyrian Kings and Mesopotamian Chronology." *Iraq* 46:115–123.

Nagy, Gregory. 1990. *Pindar's Homer: The Lyric Possession of an Epic Past.* Baltimore.

Najman, Hindy. 2003. *Seconding Sinai: The Development of Mosaic Discourse in Second Temple Judaism.* Leiden.

Naveh, Joseph. 1968. "Dated Coins of Alexander Janneus." *IEJ* 18:20–26.

———. 1971. "The Aramaic Inscriptions on Boundary Stones in Armenia." *Die Welt des Orients* 6:42–46.

Neujahr, Matthew. 2012. *Predicting the Past in the Ancient Near East: Mantic Historiography in Ancient Mesopotamia, Judah, and the Mediterranean World.* Providence.

Neusner, Jacob. 1988. "Beyond Myth, after Apocalypse: The Mishnaic Conception of History." In *The Social World of Formative Christianity and Judaism: Essays in Tribute to Howard Clark Kee,* edited by Jacob Neusner, Peder Borgen, Ernest Frerichs and Richard Horsley, 17–35. Philadelphia.

———. 1990. "The Mishnah's Generative Mode of Thought: *Listenwissenschaft* and Analogical-Contrastive Reasoning." *JAOS* 110:317–321.

———. 2001. *A Theological Commentary to the Midrash IV: Leviticus Rabbah.* Lanham, MD.

Newsom, Carol. 2014. *Daniel: A Commentary.* Louisville.

Nickelsburg, George. 2001. *1 Enoch I. A Commentary on the Book of 1 Enoch, Chapters 1–36; 81–108.* Minneapolis.

———. 2007. "Enochic Wisdom and Its Relationship to the Mosaic Torah." In *The Early Enochic Literature,* edited by Gabriele Boccaccini and John Collins, 81–94. Leiden.

Nicolaou, Ino. 1979. "Inscribed Clay Sealings from the Archeion of Paphos." In *Actes du VIIᵉ congrès international d'épigraphie grecque et latine, Constantza, 9–15 septembre 1977,* edited by D. Pippidi, 413–416. Paris.

———. 1997. "The Contribution of the Numismatic Evidence to the Dating of the Seal Impressions from the 'Archives' of the City of Ancient Paphos." In *Numismatic Archaeology, Archaeological Numismatics: Proceedings of an International Conference Held to Honour Dr. Mando Oeconomides in Athens 1995*, edited by K. Sheedy and Ch. Papageorgiadou-Banis, 47–53. Oxford.

Nielsen, John. 2015. "'I Overwhelmed the King of Elam': Remembering Nebuchadnezzar I in Persian Babylonia." In *Political Memory in and after the Persian Empire*, edited by Jason Silverman and Caroline Waerzeggers, 53–73. Atlanta.

Nihan, Christophe. 2013. "The 'Prophets' as Scriptural Collection and Scriptural Prophecy during the Second Temple Period." In *Writing the Bible: Scribes, Scribalism and Script*, edited by Philip Davies, 67–85. London.

Nissinen, Martti. 2002. "A Prophetic Riot in Seleucid Babylonia." In *"Wer darf hinaufsteigen zum Berg JHWHs?" Beiträge zu Prophetie und Poesie des Alten Testaments. Festschrift für Sigurður Örn Steingrímsson*, edited by Hubert Irsigler, 63–74. St. Ottilien.

Nixon, Lucia. 2004. "Chronologies of Desire and the Uses of Monuments: Eflatunpinar to Çatalhöyük and Beyond." In *Archaeology, Anthropology and Heritage in the Balkans and Anatolia: The Life and Times of F. W. Hasluck, 1878–1920*, edited by David Shankland, 429–452. Istanbul.

Noam, Vered. 2006. "Megillat Taanit: The Scroll of Fasting." In *The Literature of the Sages II: Midrash and Targum, Liturgy, Poetry, Mysticism, Contracts, Inscriptions, Ancient Science, and the Languages of Rabbinic Literature*, edited by Shmuel Safrai, 339–362. Assen.

Nongbri, Brent. 2005. "The Motivations of the Maccabees and Judean Rhetoric of Ancestral Tradition." In *Ancient Judaism in Its Hellenistic Context*, edited by Carol Bakhos, 85–111. Leiden.

Nozick, Robert. 1974. *Anarchy, State, and Utopia*. Oxford.

Nylander, Carl. 1983. "The Standard of the Great King—A Problem in the Alexander Mosaic." *Opuscula Romana* 14:19–37.

O'Brien, Steven. 1976. "Indo-European Eschatology: A Model." *Journal of Indo-European Studies* 4:295–320.

Oelsner, J. 1974. "Keilschriftliche Beiträge zur politischen Geschichte Babyloniens in den ersten Jahrzehnten der griechischen Herrschaft (331–305 v.u.Z.)." *Altorientalische Forschungen* 1:129–151.

Ogden, Daniel. 2017. *The Legend of Seleucus: Kingship, Narrative, and Mythmaking in the Ancient World*. Cambridge.

Ohana, David. 2012. *Modernism and Zionism*. London.

O'Kennedy, Daniel. 2014. "Haggai 2:20–23: Call to Rebellion or Eschatological Expectation?" *Old Testament Essays* 27:520–540.

Olbrycht, Marek. 2013. "The Titulature of Arsaces I, King of Parthia." *Parthica* 15:63–74.

Olshan, Marc. 1994. "Conclusion: What Good Are the Amish?" In *The Amish Struggle with Modernity*, edited by Donald Kraybill and Marc Olshan, 231–242. Hanover, NH.

Olshausen, Eckart. 1974. *Prosopographie der hellenistischen Königsgesandten I: Von Triparadeisos bis Pydna*. Leuven.

Olson, Daniel. 2013. *A New Reading of the* Animal Apocalypse *of 1 Enoch: "All Nations Shall be Blessed."* Leiden.

Overlaet, Bruno, Michael Macdonald, and Peter Stein. 2016. "An Aramaic-Hasaitic Bilingual Inscription from a Monumental Tomb at Mleiha, Sharjah, UAE." *Arabian Archaeology and Epigraphy* 27:127–142.

Overtoom, Nikolaus. 2016. "The Power-Transition Crisis of the 240s BCE and the Creation of the Parthian State." *International History Review* 38:984–1013.

Pamir, Hatice, and Nilüfer Sezgin. 2016. "The Sundial and Convivium Scene on the Mosaic from the Rescue Excavation in a Late Antique House of Antioch." *Adalya* 19:251–280.

Panaino, Antonio. 2003. "The *bayān* of the Fratarakas: Gods or Divine Kings?" In *Religious Themes and Texts of Pre-Islamic Iran and Central Asia: Studies in Honour of Professor Gherardo Gnoli on the Occasion of his 65th Birthday on 6th December 2002*, edited by Carlo Cereti, Mauro Maggi, and Elio Provasi, 265–288. Wiesbaden.

———. 2015. "Cosmologies and Astrology." In *The Wiley-Blackwell Companion to Zoroastrianism*, edited by Michael Stausberg and Yuhan Vevaina, 235–257. Oxford.

Pannenberg, Wolfhart. 1961. *Offenbarung als Geschichte*. Göttingen.

Parrot, André. 1948. *Tello. Vingt campagnes de fouilles (1877–1933)*. Paris.

Patterson, Dilys. 2008. "Re-Membering the Past: The Purpose of Historical Discourse in the Book of Judith." In *The Function of Ancient Historiography in Biblical and Cognate Studies*, edited by Patricia Kirkpatrick and Timothy Goltz, 111–123. New York.

Patterson, Lee. 2001. "Rome's Relationship with Artaxias I of Armenia." *Ancient History Bulletin* 15:154–162.

Pearce, Laurie, and L. Timothy Doty. 2000. "The Activities of Anu-belšunu, Seleucid Scribe." In *Assyriologica et Semitica: Festschrift für Joachim Oelsner*, edited by Joachim Marzahn and Hans Neumann, 331–341. Münster.

Pedersén, Olof. 1998. *Archives and Libraries in the Ancient Near East, 1500–300 B.C.* Bethesda, MD.

Périkhanian, Anahit. 1971. "Les inscriptions araméennes du roi Artachès." *Revue des études arméniennes* 8:169–174.

Peters, Janelle. 2009. "Hellenistic Imagery and Iconography in Daniel 12.5–13." *Journal for the Study of the Pseudepigrapha* 19:127–145.

Peters, John, and Hermann Thiersch. 1905. *Painted Tombs in the Necropolis of Marissa (Marêshah).* London.

Petersen, Jens. 1995. "Which Books *Did* the First Emperor of Ch'in Burn? The Meaning of *Pai Chia* in Early Chinese Sources." *Monumenta Serica* 43:1–52.

Pfeiffer, Stefan. 2004. *Das Dekret von Kanopos (238 v. Chr.). Kommentar und historische Auswertung eines dreisprachigen Synodaldekretes der ägyptischen Priester zu Ehren Ptolemaios' III. und seiner Familie.* Leipzig.

Pfrommer, Michael. 1993. *Metalwork from the Hellenized East.* Malibu.

Philonenko, Marc. 1992. "Jusqu'à ce que se lève un prophète digne de confiance (1. *Machabées* 14,41)." In *Messiah and Christos: Studies in the Jewish Origins of Christianity Presented to David Flusser,* edited by Ithamar Gruenwald, Shaul Shaked, and Gedaliahu Strouma, 95–98. Tübingen.

Piejko, Francis. 1991. "Antiochus III and Ptolemy son of Thraseas: The Inscription of Hefzibah Reconsidered." *L'antiquité classique* 60:245–259.

Pinès, S. 1963. "Un fragment de Séleucus de Séleucie conservé en version arabe." *Revue d'histoire des sciences et de leurs applications* 16:193–209.

Piras, Daniela. 2010. "Who Were the Karians in Hellenistic Times? The Evidence from Epichoric Language and Personal Names." In *Hellenistic Karia,* edited by Riet van Bremen and Jan-Mathieu Carbon, 217–233. Bourdeaux.

Pirngruber, Reinhard. 2013. "The Historical Sections of the Astronomical Diaries in Context: Developments in a Late Babylonian Scientific Text Corpus." *Iraq* 75:197–210.

———. 2017. *The Economy of Late Achaemenid and Seleucid Babylonia.* Cambridge.

Platt, Verity. 2006. "Making an Impression: Replication and the Ontology of the Graeco-Roman Seal Stone." *Art History* 29:233–257.

Plischke, Sonja. 2014. *Die Seleukiden und Iran: Die seleukidische Herrschaftspolitik in den östlichen Satrapien.* Wiesbaden.

Plöger, Otto. 1968. *Theocracy and Eschatology.* Translated by S. Rudman. Oxford.

Pomian, Krzysztof. 1984. *L'ordre du temps.* Paris.

———. 1992. "Les Archives, du Trésor des chartes au CARAN." In *Les lieux de mémoire,* edited by Pierre Nora, 3.3:162–233. Paris.

Pongratz-Leisten, Beate. 1994. *Ina šulmi īrub. Die kulttopographische und ideologische Programmatik der akītu-Prozession in Babylonien und Assyrien im I. Jahrtausend v. Chr.* Mainz.

Pope, Marvin. 1955. *El in the Ugaritic Texts.* Leiden.

Portier-Young, Anathea. 2010. "Languages of Identity and Obligation: Daniel as Bilingual Book." *VT* 60:98–115.

——. 2011. *Apocalypse against Empire: Theologies of Resistance in Early Judaism.* Grand Rapids.

——. 2014. "Symbolic Resistance in the *Book of Watchers.*" In *The Watchers in Jewish and Christian Traditions,* edited by Angela Kim Harkins, Kelley Coblentz Bautch, and John Endres, 39–49. Minneapolis.

Postgate, J. 1974. "The *bīt akīti* in Assyrian Nabu Temples." *Sumer* 30:51–74.

Potts, Daniel. 1999. "Elamite Ulā, Akkadian Ulaya, and Greek Choaspes: A Solution to the Eulaios Problem." *Bulletin of the Asia Institute* 13:27–44.

——. 2007. "Foundation Houses, Fire Altars and the *Frataraka:* Interpreting the Iconography of Some Post-Achaemenid Persian Coins." *Iranica Antiqua* 42:271–300.

Prasad, Ritika. 2013. "'Time-Sense': Railways and Temporality in Colonial India." *Modern Asian Studies* 47:1252–1282.

Priebatsch, Hans. 1974. "Das Buch Judith und seine hellenistischen Quellen." *ZDPV* 90:50–60.

Primo, Andrea. 2009. *La Storiografia sui Seleucidi da Megastene a Eusebio di Cesarea.* Pisa.

Pritchett, W. Kendrick. 1991. *The Greek State at War. Part V.* Berkeley.

Puerto, Mercedes. 2006. "Reinterpreting the Past: Judith 5." *Deuterocanonical and Cognate Literature Yearbook,* 115–140.

Qedar, Shraga. 1983. *Münz Zentrum Auktion XLIX. Gewichte aus drei Jahrtausenden JV.* Cologne.

Radner, Karen. 2005. *Die Macht des Namens: Altorientalische Strategien zur Selbsterhaltung.* Wiesbaden.

——. 2008. "The Delegation of Power: Neo-Assyrian Bureau Seals." In *L'archive des Fortifications de Persépolis. État des questions et perspectives de recherches,* edited by Pierre Briant, Wouter Henkelman and Matthew Stolper, 481–515. Paris.

Rajak, Tessa. 1996. "Hasmonean Kingship and the Invention of Tradition." In *Aspects of Hellenistic Kingship,* edited by Per Bilde, Troels Engberg-Pedersen, Lise Hannestad, and Jan Zahle, 99–115. Aarhus.

Rapin, Claude. 2010. "L'ère Yavana d'après les parchemins gréco-bactriens d'Asangorna et d'Amphipolis." In *The Traditions of East and West in the Antique Cultures of Central Asia,* edited by Kazim Abdullaev, 234–252. Tashkent.

Rapp, Stephen, Jr. 2009. "The Iranian Heritage of Georgia: Breathing New Life into the Pre-Bagratid Historiographical Tradition." *Iranica Antiqua* 44:645–692.

Rappaport, Roy. 1992. "Ritual, Time, and Eternity." *Zygon* 27:5–30.

Raulff, Ulrich. 1999. *Der unsichtbare Augenblick. Zeitkonzepte in der Geschichte.* Göttingen.

Raup Johnson, Sara. 2004. *Historical Fictions and Hellenistic Jewish Identity: Third Maccabees in Its Cultural Context.* Berkeley.

Rea, John R. Senior, and A. Hollis. 1994. "A Tax Receipt from Hellenistic Bactria." *ZPE* 104:261–280.

Redditt, Paul. 1998. "Daniel 11 and the Sociohistorical Setting of the Book of Daniel." *Catholic Biblical Quarterly* 60:463–474.

Redgate, A. 1998. *The Armenians.* Oxford.

Reed, Annette Yoshiko. 2003. "The Textual Identity, Literary History, and Social Setting of 1 Enoch: Reflections on George Nickelsburg's Commentary on 1 Enoch 1–36; 81–108." *Archiv für Religionsgeschichte* 5:279–296.

———. 2005a. *Fallen Angels and the History of Judaism and Christianity: The Reception of Enochic Literature.* Cambridge.

———. 2005b. "Interrogating 'Enochic Judaism': 1 Enoch as Evidence for Intellectual History, Social Realities, and Literary Tradition." In *Enoch and Qumran Origins: New Light on a Forgotten Connection,* edited by Gabriele Boccaccini, 336–344. Grand Rapids.

———. 2005c. "'Revealed Literature' in the Second Century BCE: Jubilees, 1 Enoch, Qumran, and the Prehistory of the Biblical Canon." In *Enoch and Qumran Origins: New Light on a Forgotten Connection,* edited by Gabriele Boccaccini, 94–98. Grand Rapids.

Reese, Günter. 1999. *Die Geschichte Israels in der Auffassung des frühen Judentums. Eine Untersuchung der Tiervision und der Zehnwochenapokalypse des äthiopischen Henochbuches, der Geschichtsdarstellung der Assumptio Mosis und der des 4Esrabuches.* Berlin.

Regev, Eyal. 2008. "Ḥanukkah and the Temple of the Maccabees: Ritual and Ideology from Judas Maccabeus to Simon." *Jewish Studies Quarterly* 15:87–114.

———. 2013. *The Hasmoneans: Ideology, Archaeology, Identity.* Göttingen.

Reiner, Erica. 1961. "The Etiological Myth of the 'Seven Sages.'" *Orientalia* 30:1–11.

Rey, Sébastien. 2016. *For the Gods of Girsu: City-State Formation in Ancient Sumer.* Oxford.

Reynolds, Bennie, III. 2011. *Between Symbolism and Realism: The Use of Symbolic and Non-Symbolic Language in Ancient Jewish Apocalypses 333–63 B.C.E.* Göttingen.

Rezakhani, Khodadad. 2013. "Arsacid, Elymaean, and Persid Coinage." In *The Oxford Handbook of Ancient Iran,* edited by Daniel Potts, 766–777. Oxford.

Ricoeur, Paul. 2004. *Memory, History, Forgetting.* Translated by Kathleen Blamey and David Pellauer. Chicago.

Rieder, John. 2011. "Race and Revenge Fantasies in *Avatar, District 9* and *Inglourious Basterds.*" *Science Fiction Film and Television* 4:41–56.

Ristvet, Lauren. 2015. *Ritual, Performance, and Politics in the Ancient Near East.* New York.

Robert, Jeanne, and Louis Robert. 1952. "Bulletin épigraphique #176 (Arménie)." *REG* 65:181–185.

Robert, Louis. 1962. *Villes d'Asie Mineure: Études de géographie antique.* Paris.

———. 1973. "Sur les inscriptions de Délos." *Études déliennes*, supplement 1, *BCH*: 435–489.

Robinson, David. 1941. *Excavations at Olynthus. Part X: Metal and Minor Miscellaneous Finds.* Baltimore.

Robson, Eleanor. 2011. "The Production and Dissemination of Scholarly Knowledge." In *The Oxford Handbook of Cuneiform Culture*, edited by Karen Radner and Eleanor Robson, 557–576. Oxford.

Rochberg-Halton, F. 1991. "Between Observation and Theory in Babylonian Astronomical Texts." *JNES* 50:107–120.

Roitman, Adolfo. 1992. "Achior in the Book of Judith: His Role and Significance." In *"No One Spoke Ill of Her": Essays on Judith*, edited by James VanderKam, 31–45. Atlanta.

Röllig, Wolfgang. 1991. "Hellenistic Babylonia: The Evidence from Uruk." In *Ο Ελληνισμος στην Ανατολη. Πρακτικά Α´ διεθνούς αραχαιολογιχού συνεδρίου: Δελφοί 6–9 Νοεμβρίου 1986*, 121–129. Athens.

Rooke, Deborah. 2000. *Zadok's Heirs: The Role and Development of the High Priesthood in Ancient Israel.* Oxford.

Root, Margaret. 1979. *The King and Kingship in Achaemenid Art: Essays on the Creation of an Iconography of Empire.* Leiden.

Rosa, Hartmut. 2003. "Social Acceleration: Ethical and Political Consequences of a Desynchronized High-Speed Society." *Constellations* 10:3–33.

Rosenberger, Veit. 2008. "Panhellenic, Athenian, and Local Identities in the Marmor Parium?" In *Religion and Society: Rituals, Resources and Identity in the Ancient Graeco-Roman World*, edited by Anders Holm Rasmussen and Susanne William Rasmussen, 225–234. Rome.

Rosenzweig, Franz. 1921. *Der Stern der Erlösung.* Frankfurt.

Roshwald, Aviel. 2006. *The Endurance of Nationalism: Ancient Roots and Modern Dilemmas.* Cambridge.

Rostovtzeff, M. 1932. "Seleucid Babylonia: Bullae and Seals of Clay with Greek Inscriptions." *Yale Classical Studies* 3:1–114.

———. 1935. "ΠΡΟΓΟΝΟΙ." *JHS* 55:56–66.

Rostovtzeff, M., A. Bellinger, F. Brown, and C. Welles, eds. 1944. *The Excavations at Dura-Europos. Preliminary Report of the Ninth Season of Work 1935–1936. Part I, The Agora and Bazaar.* London.

Rotstein, Andrea. 2016. *Literary History in the Parian Marble.* Cambridge, MA.

Rousso, Henry. 2001. *The Haunting Past: History, Memory, and Justice in Contemporary France.* Translated by Ralph Schoolcraft. Philadelphia.

Ruiz, Teofilo. 2011. *The Terror of History: On the Uncertainties of Life in Western Civilization*. Princeton.

Rung, Eduard. 2015. "Some Notes on *Karanos* in the Achaemenid Empire." *Iranica Antiqua* 50:333–356.

Russell, Brian. 2007. *The Song of the Sea: The Date of Composition and Influence of Exodus 15:1–21*. New York.

Russell, D. 1964. *The Method & Message of Jewish Apocalyptic 200 BC–AD 100*. Philadelphia.

Russell, James. 1986–1987. "Some Iranian Images of Kingship in the Armenian Artaxiad Epic." *Revue des études arméniennes* 20:253–270.

———. 1987. *Zoroastrianism in Armenia*. Cambridge, MA.

Ruzicka, Stephen. 2012. *Trouble in the West: Egypt and the Persian Empire, 525–332 BCE*. Oxford.

Sacchi, Paolo. 2007. "Measuring Time among the Jews: The Zadokite Priesthood, Enochism, and the Lay Tendencies of the Maccabean Period." In *The Early Enochic Literature*, edited by Gabriele Boccaccini and John Collins, 95–118. Leiden.

Sachs, A. 1952. "Babylonian Horoscopes." *JCS* 6:49–75.

Sachs, A., and D. Wiseman. 1954. "A Babylonian King List of the Hellenistic Period." *Iraq* 16:202–212.

Sahlins, Marshall. 1981. *Historical Metaphors and Mythical Realities: Structure in the Early History of the Sandwich Islands Kingdom*. Ann Arbor.

———. 1985. *Islands of History*. Chicago.

Sancisi-Weerdenburg, H. 1993. "Alexander and Persepolis." In *Alexander the Great: Reality and Myth*, edited by J. Carlsen, 177–188. Rome.

Sanders, Donald, ed. 1996. *Nemrud Daği. The Hierothesion of Antiochus I of Commagene. Volume 1: Text*. Winona Lake.

Sartre, Maurice. 2001. *D'Alexandre à Zénobie: Histoire du Levant antique, IVe siècle av. J.-C.—IIIe siècle ap. J.-C*. Paris.

Sauter, Michael. 2007. "Clockwatchers and Stargazers: Time Discipline in Early Modern Berlin." *American Historical Review* 112:685–709.

Savalli-Lestrade, Ivana. 2010. "Intitulés royaux et intitulés civiques dans les inscriptions de cités sujettes de Carie et de Lycie (Amyzon, Eurômos, Xanthos). Histoire politique et mutations institutionnelles." *Studi ellenistici* 24:127–148.

Schalles, Hans-Joachim. 1985. *Untersuchungen zur Kulturpolitik der pergamenischen Herrscher im dritten Jahrhundert vor Christus*. Tübingen.

Schaudig, Hanspeter. 2001. *Die Inschriften Nabonids von Babylon und Kyros' des Großen samt den in ihrem Umfeld entstandenen Tendenzschriften—Textausgabe und Grammatik*. Münster.

———. 2003: "Nabonid, der 'Archäologe auf dem Königsthron.' Zum Geschichts-bild des ausgehenden neubabylionischen Reiches." In *Festschrift für Burkhart Kienast (Alter Orient und Altes Tesament 274)*, edited by Gebhard Selz, 447-497. Münster.

———. 2010. "The Restoration of Temples in the Neo- and Late Babylonian Pe-riods: A Royal Prerogative as the Setting for the Polticial Argument." In *From the Foundations to the Crenellations: Essays on Temple Building in the Ancient Near East and Hebrew Bible*, edited by Mark Boda and Jamie Novotny, 141-164. Münster.

Schiffman, Zachary. 2011. *The Birth of the Past*. Baltimore.

Schmidt, Erich. 1941. "Report an Sitzung am 22. April 1941." *Archäologischer Anzeiger* 3-4:786-843.

———. 1953. *Persepolis I: Structures, Reliefs, Inscriptions*. Chicago.

Schnabel, P. 1923. *Berossos und die babylonisch-hellenistische Literatur*. Leipzig.

Schniedewind, William. 2004. *How the Bible Became a Book: The Textualization of An-cient Israel*. Cambridge.

———. 2013. *A Social History of Hebrew: Its Origins through the Rabbinic Period*. New Haven.

Scholem, Gershom. 1971. *The Messianic Idea in Judaism and Other Essays on Jewish Spir-ituality*. New York.

Schuol, Monika. 2000. *Die Charakene: Ein mesopotamisches Königreich in hellenistisch-parthischer Zeit*. Stuttgart.

Schwartz, Daniel. 1998. "The Other in 1 and 2 *Maccabees*." In *Tolerance and Intoler-ance in Early Judaism and Christianity*, edited by Graham Stanton and Guy Stroumsa, 30-37. Cambridge.

———. 2008. *2 Maccabees*. Berlin.

Schwartz, J. 1982. "Numismatique et renouveaux nationalistes dans l'empire séleucide au IIe s. a.C." In *Actes du 9ème congrès international de numismatique, Berne, septembre 1979 I*, edited by Tony Hackens and Raymond Weiller, 243-249. Louvain-la-Neuve.

Schwartz, Seth. 1991. "Israel and the Nations Roundabout: 1 Maccabees and the Hasmonean Expansion." *JJS* 42:16-38.

———. 1995. "Language, Power and Identity in Ancient Palestine." *Past and Present* 148:3-47.

———. 2001. *Imperialism and Jewish Society, 200 BCE to 640 CE*. Princeton.

Schwentzel, Christian-Georges. 2005. "Les thèmes du monnayage royal nabatéen et le modèle monarchique hellénistique." *Syria* 82:149-166.

Scolnic, Benjamin. 2014a. "Antiochus IV as the Man Who Will Overflow the Flood and Break Its Arms (Daniel 11.22)." *Bible Translator* 65:24-33.

———. 2014b. "Is Daniel 11:1-19 Based on a Ptolemaic Narrative?" *JSJ* 45:157-184.

Scott, James. 1990. *Domination and the Arts of Resistance: Hidden Transcripts.* New Haven.

Scurlock, JoAnn. 2006. "Whose Truth and Whose Justice? The Uruk and Other Late Akkadian Prophecies Re-Visited." In *Orientalism, Assyriology and the Bible,* edited by Steven Holloway, 447–465. Sheffield.

Segal, Michael. 2009. "From Joseph to Daniel: The Literary Development of the Narrative in Daniel 2." *VT* 59:123–149.

Sens, Alexander. 2005. "The Art of Poetry and the Poetry of Art: The Unity and Poetics of Posidippus' Statue-Poems." In *The New Posidippus: A Hellenistic Poetry Book,* edited by Kathryn Gutzwiller, 206–225. Oxford.

Sergueenkova, Valeria, and Felipe Rojas. 2017. "Persia on Their Minds: Achaemenid Memory Horizons in Roman Anatolia." In *Persianism in Antiquity* (*Oriens et Occidens* 25), edited by Rolf Strootman and Miguel Versluys, 269–288. Stuttgart.

Seri, Andrea. 2012. "The Role of Creation in Enūma eliš." *Journal of Ancient Near Eastern Religions* 12:4–29.

Sève, Michel. 1993. "Bulletin épigraphique #117." *REG* 106:476.

Seyrig, Henri. 1950. "Antiquités syriennes 42. Sur les ères de quelques villes de Syrie: Antioche, Apamée, Aréthuse, Balanée, Épiphanie, Laodicée, Rhosos, Damas, Béryte, Tripolis, l'ère de Cléopâtre, Chalcis du Liban, Doliché." *Syria* 27:5–56.

Shachar, Ilan. 2004. "The Historical and Numismatic Significance of Alexander Jannaeus's Later Coinage as Found in Archaeological Excavations." *Palestine Exploration Quarterly* 136:5–33.

Shahbazi, A. 2003. "Iranians and Alexander." *American Journal of Ancient History* 2:5–38.

Shaked, Shaul. 2004. *Le satrape de Bactriane et son gouverneur: Documents araméens du IVᵉ s. avant notre ère provenant de Bactriane.* Paris.

Shaw, Matthew. 2011. *Time and the French Revolution: The Republican Calendar, 1789–Year XIV.* Woodbridge.

Shayegan, M. 2011. *Arsacids and Sasanians: Political Ideology in Post-Hellenistic and Late Antique Persia.* Cambridge.

Shear, T., Jr. 1973. "The Athenian Agora Excavations of 1971." *Hesperia* 42:121–179.

Sherwin-White, Susan. 1983. "Aristeas Ardibelteios: Some Aspects of the Use of Double Names in Seleucid Babylonia." *ZPE* 50:209–221.

———. 1987. "Seleucid Babylonia: A Case-Study for the Installation and Development of Greek Rule." In *Hellenism in the East: The Interaction of Greek and Non-Greek Civilizations from Syria to Central Asia after Alexander,* edited by Amélie Kuhrt and Susan Sherwin-White, 1–31. London.

Sherwin-White, Susan, and Amélie Kuhrt. 1993. *From Samarkhand to Sardis: A New Approach to the Seleucid Empire.* London.

Siebeneck, Robert. 1959. "May Their Bones Return to Life!—Sirach's Praise of the Fathers." *Catholic Biblical Quarterly* 21:411–428.

Siegel, Jonathan. 1971. "The Employment of Palaeo-Hebrew Characters for the Divine Names at Qumran in the Light of Tannaitic Sources." *Hebrew Union College Annual* 42:159–172.

Siegert, Bernhard. 2015. *Cultural Techniques: Grids, Filters, Doors, and Other Articulations of the Real.* Translated by Geoffrey Winthrop-Young. New York.

Sievers, Joseph. 1990. *The Hasmoneans and Their Supporters: From Mattathias to the Death of John Hyrcanus I.* Atlanta.

Silverman, Jason. 2012. *Persepolis and Jerusalem: Iranian Influence on the Apocalyptic Hermeneutic.* New York.

———. 2015. "From Remembering to Expecting the 'Messiah': Achaemenid Kingship as (Re)formulating Apocalyptic Expectations of David." In *Political Memory in and after the Persian Empire,* edited by Jason Silverman and Caroline Waerzeggers, 419–446. Atlanta.

Simmel, Georg. 1977. *The Problems of the Philosophy of History: An Epistemological Essay.* Translated by Guy Oakes. New York.

Sims-Williams, Nicholas. 2004. "The Bactrian Inscription of Rabatak: A New Reading." *Bulletin of the Asia Institute* 18:53–68.

Sims-Williams, Nicholas, and Joe Cribb. 1995–1996. "A New Bactrian Inscription of Kanishka the Great." *Silk Road Art and Archaeology* 4:75–142.

Sinos, Rebecca. 1998. "Divine Selection: Epiphany and Politics in Archaic Greece." In *Cultural Poetics in Archaic Greece: Cult, Performance, Politics,* edited by Carol Dougherty and Leslie Kurke, 73–91. Oxford.

Skjærvø, Prods. 1997. "The Joy of the Cup: A Pre-Sasanian Middle Persian Inscription on a Silver Bowl." *Bulletin of the Asia Institute* 11:93–104.

———. 1999. "Avestan Quotations in Old Persian? Literary Sources of the Old Persian Inscriptions." *Irano-Judaica* 4:1–64.

Smail, Dan. 2005. "In the Grip of Sacred History." *American Historical Review* 110:1337–1361.

Smend, Rudolf. 1906. *Die Weisheit des Jesus Sirach.* Berlin.

Smith, Adam. 2003. *The Political Landscape: Constellations of Authority in Early Complex Polities.* Berkeley.

Smith, Jonathan. 1975. "Wisdom and Apocalyptic." In *Religious Syncretism in Antiquity: Essays in Conversation with Geo Widengren,* edited by Birger Pearson, 131–156. Missoula.

———. 1976. "A Pearl of Great Price and a Cargo of Yams: A Study in Situational Incongruity." *History of Religions* 16:1–19.

———. 1978. *Map Is Not Territory: Studies in the History of Religions.* Leiden.

———. 1982. *Imagining Religion: From Babylon to Jonestown.* Chicago.

Sommer, Benjamin. 1996. "Did Prophecy Cease? Evaluating a Reevaluation." *JBL* 115:31–47.

———. 2000. "The Babylonian Akitu Festival: Rectifying the King or Renewing the Cosmos?" *JANES* 27:81–95.

Spengler, Oswald. 1923. *Der Untergang des Abendlandes: Umrisse einer Morphologie der Weltgeschichte.* Munich.

Sperber, Daniel. 1977. "On the Office of the Agoranomos in Roman Palestine." *Zeitschrift der deutschen Morgenländischen Gesellschaft* 127:227–243.

Staub, U. 1978. "Das Tier mit den Hörnen. Ein Beitrag zu Dan 7, 7f." *Freiburger Zeitschrift für Philosophie und Theologie* 25:351–397.

Stausberg, Michael. 2002. *Die Religion Zarathushtras: Geschichte—Gegenwart— Rituale 1.* Stuttgart.

Steele, John. 2013. "The 'Astronomical Fragments' of Berossos in Context." In *The World of Berossos,* edited by Johannes Haubold, Giovanni Lanfranchi, Robert Rollinger, and John Steele, 99–113. Wiesbaden.

Steinkeller, Piotr. 1988. "The Date of Gudea and His Dynasty." *JCS* 40:47–53.

Stern, Sacha. 2012. *Calendars in Antiquity: Empires, States, and Societies.* Oxford.

Stevens, Kathryn. 2013. "Secrets in the Library: Protected Knowledge and Professional Identity in Late Babylonian Uruk." *Iraq* 75:211–253.

———. 2016. "Empire Begins at Home: Local Elites and Imperial Ideologies in Hellenistic Greece and Babylonia." In *Cosmopolitanism and Empire: Universal Rulers, Local Elites, and Cultural Integration in the Ancient Near East and Mediterranean,* edited by Myles Lavan, Richard Payne, and John Weisweiler, 65–88. New York.

Stewart, Andrew. 1993. *Faces of Power: Alexander's Image and Hellenistic Politics.* Berkeley.

———. 2004. *Attalos, Athens, and the Akropolis: The Pergamene 'Little Barbarians' and Their Roman and Renaissance Legacy.* Cambridge.

———. 2005. "Posidippus and the Truth in Sculpture." In *The New Posidippus: A Hellenistic Poetry Book,* edited by Kathryn Gutzwiller, 183–205. Oxford.

Stewart, Susan. 1984. *On Longing: Narratives of the Miniature, the Gigantic, the Souvenir, the Collection.* Durham, NC.

Stoler, Ann. 2002. "Colonial Archives and the Arts of Governance." *Archival Science* 2:87–109.

Stolper, Matthew. 1990. "In the Chronicle of the Diadochi r. 3 f." *NABU* §7.

Stone, Michael. 2013. *4 Ezra and 2 Baruch: Translations, Introductions, and Notes.* Minneapolis.

Stowasser, Barbara. 2014. *The Day Begins at Sunset: Perceptions of Time in the Islamic World.* London.

Strasburger, Hermann. 1965. "Poseidonios on Problems of the Roman Empire." *JRS* 55:40–53.

Streck, Maximilian. 1916. *Assurbanipal und die letzten assyrischen Könige bis zum Untergange Nineveh's II.* Leipzig.

Strootman, Rolf. 2017. "Imperial Persianism: Seleukids, Arsakids and *Fratarakā*." In *Persianism in Antiquity*, edited by Rolf Strootman and Miguel Versluys, 177–200. Stuttgart.

Stuckenbruck, Loren. 2007. *1 Enoch 91–108*. Berlin.

Stylianou, P. 1991. "The Date of the Abolition of the Kingdoms of Lapethos and Keryneia in Cyprus." *Ἐπετηρὶς τοῦ Κέντρου Ἐπιστημονικῶν Ἐρευνῶν* 18:81–83.

Suleiman, Susan. 2014. "The Stakes in Holocaust Representation: On Tarantino's *Inglourious Basterds*." *Romanic Review* 105:69–86.

Summers, Carol. 2006. "Radical Rudeness: Ugandan Social Critiques in the 1940s." *Journal of Social History* 39:741–770.

Sundermann, Werner. 2011. "Bahman Yašt." *Encyclopaedia Iranica*. http://www.iranicaonline.org/articles/bahman-yast-middle-persian-apocalyptical-text.

———. 2012. "Hystaspes, Oracles of." *Encyclopaedia Iranica*. http://www.iranicaonline.org/articles/hystaspes-oracles-of.

Suter, Claudia. 2000. *Gudea's Temple Building: The Representation of an Early Mesopotamian Ruler in Text and Image*. Groningen.

———. 2012. "Gudea of Lagash: Iconoclasm or Tooth of Time?" In *Iconoclasm and Text: Destruction in the Ancient Near East and Beyond*, edited by Natalie May, 57–87. Chicago.

Swain, Joseph. 1940. "The Theory of the Four Monarchies: Opposition History under the Roman Empire." *CP* 35:1–21.

Sznycer, Maurice. 1975. "'L'assemblée du peuple' dans les cités puniques d'après les témoignages épigraphiques." *Semitica* 25:47–68.

———. 1981. "Le problème de la royauté dans le monde punique." *Bulletin archéologique du Comité des Travaux Historiques et Scientifiques* 17:291–301.

Tadmor, Hayim. 1981. "Addendum." *AJP* 102:338–339.

———. 1999. "World Dominion: The Expanding Horizon of the Assyrian Empire." In *Landscapes: Territories, Frontiers, and Horizons in the Ancient Near East. Part 1: Invited Lectures*, edited by L. Milano, S. de Martino, F. Fales, and G. Lanfranchi, 55–62. Padua.

Talmon, Shemaryahu. 1966. "The 'Desert Motif' in the Bible and Qumran Literature." In *Biblical Motifs: Origins and Transformations*, edited by Alexander Altmann, 31–63. Cambridge, MA.

Tarn, William. 1951. *The Greeks in Bactria and India*. Cambridge.

Taubes, Jacob. 2009. *Occidental Eschatology*. Translated by David Ratmoko. Stanford.

Taylor, Jonathan. 2011. "Tablets as Artefacts, Scribes as Artisans." In *The Oxford Handbook of Cuneiform Culture*, edited by Karen Radner and Eleanor Robson, 5–31. Oxford.

Tcherikover, Victor. 1961. *Hellenistic Civilization and the Jews*. Translated by S. Applebaum. Philadelphia.

Thissen, Heinz-Josef. 2002. "Das Lamm des Bokchoris." In *Apokalyptik und Ägypten. Eine kritische Analyse der relevanten Texte aus dem griechisch-römsichen Ägypten*, edited by A. Blasius and B. Schipper, 113–138. Leuven.

Thompson, E. P. 1967. "Time, Work-Discipline, and Industrial Capitalism." *Past and Present* 38:56–97.

Thomson, Robert. 1978. *Moses Khorenats'i: History of the Armenians*. Cambridge, MA.

Thonemann, Peter. 2005. "The Tragic King: Demetrios Poliorketes and the City of Athens." In *Imaginary Kings: Royal Images in the Ancient Near East, Greece and Rome*, edited by Olivier Hekster and Richard Fowler, 63–86. Munich.

———. 2013. "The Attalid State, 188–133 BC." In *Attalid Asia Minor: Money, International Relations, and the State*, edited by Peter Thonemann, 1–47. Oxford.

———. 2015. *The Hellenistic World: Using Coins as Sources*. Cambridge.

Tilia, Ann Britt. 1969. "Reconstruction of the Parapet on the Terrace Wall at Persepolis, South and West of Palace H." *East and West* 9:9–43.

———. 1972. *Studies and Restorations at Persepolis and Other Sites of Fārs*. Rome.

———. 1977. "Recent Discoveries at Persepolis." *American Journal of Archaeology* 81:67–77.

Tiller, Patrick. 1993. *A Commentary on the Animal Apocalypse of 1 Enoch*. Atlanta.

———. 2007. "The Sociological Settings of the Components of 1 Enoch." In *The Early Enoch Literature*, edited by Gabriele Boccaccini and John Collins, 237–255. Leiden.

Tilly, Michael. 2015. *1 Makkabäer*. Berlin.

Tirats'yan, Gevork. 2003. *From Urartu to Armenia: Florilegium Gevork A. Tirats'yan*. Edited by Rouben Vardanyan. Neuchâtel.

Tober, Daniel. 2017. "Greek Local Historiography and Its Audiences." *CQ* 67:460–484.

———. Forthcoming. "Ἡρακλῆς κάρρων, Σέλευκε: Resistance and History in Pontic Herakleia." In *The Maccabean Moment*, edited by Paul Kosmin and Ian Moyer.

Tod, Marcus. 1918–1919. "The Macedonian Era." *BSA* 23:206–217.

———. 1919/1920–1920/1921. "The Macedonian Era II." *BSA* 24:54–67.

———. 1950. "The Alphabetic Numeral System in Attica." *BSA* 45:126–139.

Tonikian, Armen. 1992. "The Layout of Artashat and Its Historical Development." *Mesopotamia* 27:161–187.

Tov, Emanuel. 1988. "Hebrew Biblical Manuscripts from the Judaean Desert: Their Contribution to Textual Criticism." *JJS* 39:5–37.

Traina, Giusto. 2007. "Artachat (Artaxata)." *Les dossiers d'archéologie* 321:79.

Trigger, David, and Cameo Dalley. 2010. "Negotiating Indigeneity: Culture, Identity, and Politics." *Reviews in Anthropology* 39:46–65.

Trombley, Frank. 2000. "Byzantine and Islamic Era Systems in the Seventh Century." *Storia della storiografia* 37:41–53.

Tubach, Jürgen. 2009. "Die Anfänge des Königsreichs von Edessa. Vom Zelt- zum Palastbewohner, oder: Erfolgreiche Migration in hellenistischer Zeit." In *Edessa in hellenistisch-römischer Zeit: Religion, Kultur und Politik zwischen Ost und West. Beiträge des internationalen Edessa-Symposiums in Halle an der Saale, 14.-17. Juli 2005*, edited by Lutz Greisiger, Claudia Rammelt, and Jürgen Tubach, 279–311. Beirut.

Tuplin, Christopher. 2009. "The Seleucids and Their Achaemenid Predecessors. A Persian Inheritance?" In *Ancient Greece and Ancient Iran: Cross-Cultural Encounters*, edited by Seyed Mohammad Reza Darbandi and Antigoni Zournatzi, 109–136. Athens.

———. 2013. "Berossos and Greek Historiography." In *The World of Berossos*, edited by Johannes Haubold, Giovanni Lanfranchi, Robert Rollinger, and John Steele, 177–197. Wiesbaden.

Tzoref, Shani. 2011. "*Pesher* and Periodization." *Dead Sea Discoveries* 18:129–154.

Van de Mieroop, Marc. 1993. "The Reign of Rim-Sin." *Revue d'assyriologie et d'archéologie orientale* 87:47–69.

———. 2016. *Philosophy before the Greeks: The Pursuit of Truth in Ancient Babylonia.* Princeton.

Van der Horst, Pieter. 2002. "Antediluvian Knowledge: Jewish Speculations about Wisdom from before the Flood in Their Ancient Context." In *Jüdische Schriften in ihren antik-jüdischen und urchristlichen Kontext*, edited by Hermann Lichtenberger and Gerbern Oegema, 163–181. Gütersloh.

Van der Kooij, A. 1998. "The Canonization of Ancient Books Kept in the Temple of Jerusalem." In *Canonization and Decanonization: Papers Presented to the International Conference of the Leiden Institute for the Study of Religions (Lisor), Held at Leiden 9-10 January 1997*, edited by A. van der Kooij and K. van der Toorn, 17–40. Leiden.

Van der Spek, R. J. 1985. "The Babylonian Temple during the Macedonian and Parthian Domination." *Bibliotheca Orientalis* 42:541–562.

———. 1993. "New Evidence on Seleucid Land Policy." In *De Agricultura. In Memoriam Pieter Willem de Neeve*, edited by H. Sancisi-Weerdenburg, R. van der Spek, H. Teitler, and H. Wallinga, 61–79. Amsterdam.

———. 1995. "Land Ownership in Babylonian Cuneiform Documents." In *Legal Documents of the Hellenistic World*, edited by Markham Geller and Herwig Maehler, 173-245. London.

———. 1998. "Cuneiform Documents on Parthian History: The Rahimesu Archive. Materials for the Study of the Standard of Living." In *Das Partherreich und seine Zeugnisse*, edited by Josef Wiesehöfer, 205-258. Stuttgart.

———. 2000a. "The Effect of War on the Prices of Barley and Agricultural Land in Hellenistic Babylonia." In *Économie antique: La guerre dans les économies antiques*, edited by J. Andreau, P. Briant, and R. Descat, 293-313. Toulouse.

———. 2000b. "The *Šatammus* of Esagila in the Seleucid and Arsacid Periods." In *Assyriologica et Semitica: Festschrift für Joachim Oelsner*, edited by Joachim Marzahn and Hans Neumann, 437-446. Münster.

———. 2001. "The Theatre of Babylon in Cuneiform." In *Veenhof Anniversary Volume: Studies Presented to Klaas R. Veenhof on the Occasion of His Sixty-Fifth Birthday*, edited by W. H. van Soldt, J. G. Dercksen, N. Kouwenberg, and T. Krispijn, 445-456. Leiden.

———. 2003. "Darius III, Alexander the Great and Babylonian Scholarship." In *Achaemenid History XIII. A Persian Perspective. Essays in Memory of Heleen Sancisi-Weerdenburg*, edited by Wouter Henkelman and Amélie Kuhrt, 289-346. Leiden.

———. 2005. "Ethnic Segregation in Hellenistic Babylon." In *Ethnicity in Ancient Mesopotamia: Papers Read at the 48th rencontre assyriologique internationale, Leiden, 1-4 July, 2002*, edited by W. van Soldt, R. Kalvelagen, and D. Katz, 393-408. Leiden.

———. 2009. "Multi-Ethnicity and Ethnic Segragation in Hellenistic Babylon." In *Ethnic Constructs in Antiquity: The Role of Power and Tradition*, edited by Tom Derks and Nico Roymans, 101-115. Aarhus.

———. 2014a. *"Ik bin een boodschapper van Nanaia!" Een Babylonische profeet als teken des tijds (133 voor Christus)*. Amsterdam.

———. 2014b. "Seleukos, Self-Appointed General *(Strategos)* of Asia (311-305 B.C.), and the Satrapy of Babylonia." In *The Age of the Successors and the Creation of the Hellenistic Kingdoms (323-276 B.C.)*, edited by Hans Hauben and Alexander Meeus, 323-342. Leuven.

———. 2015. "Coming to Terms with the Persian Empire: Some Concluding Remarks and Responses." In *Political Memory in and after the Persian Empire*, edited by Jason Silverman and Caroline Waerzeggers, 447-477. Atlanta.

———. 2016. "The Cult for Seleucus II and His Sons in Babylon." *NABU* §27.

Van Deventer, H. 2013. "Another Look at the Redaction History of the Book of Daniel, or, Reading Daniel from Left to Right." *Journal for the Study of the Old Testament* 38:239-260.

Van Dijk, Jan. 1962. "Die Inschriftenfunde." In *XVIII. Vorläufiger Bericht über die von dem Deutschen Archäologischen Institut und der Deutschen Orient-Gesellschaft aus Mitteln der Deutschen Forschungsgemeinschaft unternommenen Ausgrabungen in Uruk-Warka*, 39–62. Berlin.

———. 1963. "Die Tontafelfunde der Kampagne 1959/60." *Archiv für Orientforschung* 20:217–218.

van Dijk, Jan, and Werner Mayer. 1980. *Texte aus dem Rēš-Heiligtum in Uruk-Warka*. Berlin.

Van Nuffelen, Peter. 2001. "Un culte royal municipal de Séleucie du Tigre à l'époque séleucide." *Epigraphica Anatolica* 33:85–87.

———. 2004. "Le culte royal de l'empire des Séleucides: Une réinterprétation." *Historia* 53:278–301.

VanderKam, J. 1979. "The Origin, Character, and Early History of the 364-Day Calendar: A Reassessment of Jaubert's Hypotheses." *Catholic Biblical Quarterly* 41:390–411.

———. 1981. "2 Maccabees 6, 7a and Calendrical Change in Judaism." *JSJ* 12:52–74.

———. 1984. *Enoch and the Growth of an Apocalyptic Tradition*. Washington, DC.

———. 2004. "Open and Closed Eyes in the Animal Apocalypse (1 Enoch 85–90)." In *The Idea of Biblical Interpretation: Essays in Honor of James L. Kugel*, edited by Hindy Najman and Judith Newman, 279–292. Leiden.

———. 2007. "Mapping Second Temple Judaism." In *The Early Enoch Literature*, edited by Gabriele Boccaccini and John Collins, 1–20. Leiden.

Vanstiphout, H. 1992. "Enuma Elish as a Systematic Creed: An Essay." *Orientalia Lovanensia Periodica* 23:37–61.

Vardanyan, Karen, and Ruben Vardanyan. 2008. "Newly-Found Groups of Artaxiad Copper Coins." *Armenian Numismatic Society* 4:77–95.

Venter, P. 1997. "Daniel and Enoch: Two Different Reactions." *Hervormde Teologiese Studies* 53:68–91.

Vevaina, Yuhan. 2009. "Resurrecting the Resurrection: Eschatology and Exegesis in Late Antique Zoroastrianism." *Bulletin of the Asia Institute* 19:219–227.

———. 2011. "Miscegenation, 'Mixture,' and 'Mixed Iron': The Hermeneutics, Historiography, and Cultural Poesis of the 'Four Ages' in Zoroastrianism." In *Revelation, Literature, and Community in Late Antiquity*, edited by Philippa Townsend and Moulie Vidas, 237–269. Tübingen.

Vlassopoulos, Kostas. 2013. *Greeks and Barbarians*. Cambridge.

Virgilio, Biagio. 1993. *Gli Attalidi di Pergamo. Fama, eredità, memoria*. Pisa.

———. 2011. *Le roi écrit: Le correspondance du souverain hellénistique, suivie de deux lettres d'Antiochos III à partir de Louis Robert et d'Adolf Wilhelm*. Pisa.

Virno, Paolo. 2015. *Déjà vu and the End of History*. Translated by David Broder. London.

Visscher, Marijn. 2016. "Royal Presence in the Astronomical Diaries." Paper presented at the Keeping Watch in Babylon: From Evidence to Text in the Astronomical Diaries conference, Durham, June 24.

Vittmann, Günter. 1998. *Der demotische Papyrus Rylands 9.* Wiesbaden.

Wacholder, Ben Zion. 1968. "Biblical Chronology in the Hellenistic World Chronicles." *Harvard Theological Review* 61:451–481.

———. 1975. "Chronomessianism: The Timing of Messianic Movements and the Calendar of Sabbatical Cycles." *Hebrew Union College Annual* 46:201–218.

Waerzeggers, Caroline. 2003-2004. "The Babylonian Revolts Against Xerxes and the 'End of Archives.'" *Archiv für Orientforschung* 50:150–173.

———. 2010. *The Ezida Temple of Borsippa: Priesthood, Cult, Archives.* Leiden.

———. 2015. "Facts, Propaganda, or History? Shaping Political Memory in the Nabonidus Chronicle." In *Political Memory in and after the Persian Empire*, edited by Jason Silverman and Caroline Waerzeggers, 95–124. Atlanta.

Walbank, Frank. 1967. *Philip V of Macedon.* Cambridge.

Wallenfels, Ronald. 1996. "Private Seals and Sealing Practices at Hellenistic Uruk." In *Archives et sceaux du monde hellénistique,* edited by Marie-Françoise Boussac and Antonio Invernizzi, 113–129. Athens.

Walthall, D. Alex. 2011. "Magistrate Stamps on Grain Measures in Early Hellenistic Sicily." *ZPE* 179:159–169.

Wardle, David. 2006. *Cicero on Divination.* De Divinatione *Book 1 Translated with Introduction and Historical Commentary.* Oxford.

Watts, James. 2003. "The Rhetoric of Ritual Instruction in Leviticus 1–7." In *The Book of Leviticus: Composition and Reception,* edited by Rolf Rendtorff and Robert Kugler, 79–100. Leiden.

Weinberg, Joel. 2007. "Gedaliah, the Son of Ahikam, in Mizpah: His Status and Role, Supporters and Opponents." *ZAW* 119:356–368.

Weiss, Peter, and Kay Ehling. 2006. "Marktgewichte im Namen seleukidischer Könige." *Chiron* 36:369–378.

Weissert, Elnathan. 1997. "Creating a Political Climate: Literary Allusions to *Enūma Eliš* in Sennacherib's Account of the Battle of Halule." In *Assyrien im Wandel der Zeiten: XXXIXᵉ Rencontre assyriologique internationale, Heidelberg, 6.-10. Juli 1992,* edited by Hartmut Waetzoldt and Harald Hauptmann, 191–202. Heidelberg.

Weitzman, Steven. 2004. "Plotting Antiochus's Persecution." *JBL* 123:219–234.

Werman, Cana. 2006. "Epochs and End-Time: The 490-Year Scheme in Second Temple Literature." *Dead Sea Discoveries* 13:229–255.

West, Martin. 1997. *The East Face of Helicon: West Asiatic Elements in Greek Poetry and Myth.* Oxford.

Westermann, W., C. Keyes, and H. Liebesny. 1940. *Zenon Papyri: Business Papers of the Third Century* B.C. *Dealing with Palestine and Egypt.* Vol. 2. New York.

Westh-Hansen, Sidsel Maria. 2017. "Hellenistic Uruk Revisited: Sacred Architecture, Seleucid Policy, and Cross-Cultural Interaction." In *Contextualizing the Sacred in the Hellenistic and Roman Near East: Religious Identities in Local, Regional, and Imperial Settings*, edited by Rubina Raja, 155–168. Turnhout.

Wheatley, Pat. 1995. "Ptolemy Soter's Annexation of Syria 320 BC." *CQ* 45:433–440.

———. 2002. "Antigonus Monophthalmus in Babylonia, 310–308 BC." *JNES* 61:39–47.

———. 2003. "The Year 22 Tetradrachms of Sidon and the Date of the Battle of Gaza." *ZPE* 144:268–276.

White, Hayden. 1987. *The Content of the Form: Narrative Discourse and Historical Representation*. Baltimore.

Widengren, Geo. 1983. "Leitende Ideen und Quellen der iranischen Apokalyptik." In *Apocalypticism in the Mediterranean World and the Near East: Proceedings of the International Colloquium on Apocalypticism, Uppsala, August 12–17, 1979*, edited by David Hellholm, 77–162. Tübingen.

Wiesehöfer, Josef. 1994. *Die "dunklen Jahrhunderte" der Persis*. Munich.

———. 1996. *Ancient Persia from 550 BC to 650 AD*. Translated by Azizeh Azodi. London.

———. 2003. "The Medes and the Idea of the Succession of Empires in Antiquity." In *Continuity of Empire: Assyria, Media, Persia*, edited by Giovanni Lanfranchi, Michael Roaf, and Robert Rollinger, 391–396. Padua.

———. 2011. "Frataraka Rule in Seleucid Persis: A New Appraisal." In *Creating a Hellenistic World*, edited by Andrew Erskine and Lloyd Llewellyn-Jones, 107–121. Swansea.

———. 2013a. "Fratarakā and Seleucids." In *The Oxford Handbook of Ancient Iran*, edited by Daniel Potts, 718–727. Oxford.

———. 2013b. "Law and Religion in Achaemenid Iran." In *Law and Religion in the Eastern Mediterranean from Antiquity to Early Islam*, edited by Anselm Hagedorn and Reinhard Kratz, 41–57. Oxford.

Wilburn, Andrew. 2012. *Materia Magica: The Archaeology of Magic in Roman Egypt, Cyprus, and Spain*. Ann Arbor.

Will, Édouard. 1966–1967. *Histoire politique du monde hellénistique (323–30 av. J.-C.)*. 2 vols. Nancy.

Willi-Plein, I. 1977. "Das Geheimnis der Apokalyptik." *VT* 27:62–81.

Williamson, Christina. 2013. "Labraunda as Memory Theatre for Hellenistic Mylasa: Code-Switching between Past and Present?" *HEROM* 2:143–167.

Wills, Lawrence. 1990. *The Jew in the Court of the Foreign King: Ancient Jewish Court Legends*. Minneapolis.

———. 2011. "Jewish Novellas in a Greek and Roman Age: Fiction and Identity." *JSJ* 42:141–165.

Wilson, Elliot. 2015. "Constructing Antigonid Kingship: Monarchy, Memory, and Empire in Hellenistic Macedonia." Senior thesis, Harvard University.

Winckler, Hugo. 1898. *Altorientalische Forschungen* 2.1. Leipzig.

Windisch, Hans. 1929. *Die Orakel des Hystaspes.* Amsterdam.

Wittstruck, Thorne. 1978. "Influence of Treaty Curse Imagery on the Beast Imagery of Daniel 7." *JBL* 97:100-102.

Wolski, Józef. 1956-1957. "The Decay of the Iranian Empire of the Seleucids and the Chronology of the Parthian Beginnings." *Berytus* 12:35-52.

———. 1959. "L'Historicité d'Arsace Ier." *Historia* 8:222-238.

Wolters, Al. 1991a. "The Riddle of the Scales in Daniel 5." *Hebrew Union College Annual* 62:155-177.

———. 1991b. "Untying the King's Knots: Physiology and Wordplay in Daniel 5." *JBL* 110:117-122.

Wood, Rachel. Forthcoming. *Art in the Hellenistic East.*

Woods, Christopher. 2012. "Sons of the Sun: The Mythological Foundations of the First Dynasty of Uruk." *Journal of Ancient Near Eastern Religions* 12:78-96.

Wooten, Cecil. 1974. "The Speeches in Polybius: An Insight into the Nature of Hellenistic Oratory." *AJP* 95:235-251.

Wörrle, Michael. 2000. "Eine hellenistische Inschrift aus Gadara." *Archäologischer Anzeiger*, 267-271.

Wright, Benjamin, III. 1997. "'Fear the Lord and Honor the Priest': Ben Sira as Defender of the Jerusalem Priesthood." In *The Book of Ben Sira in Modern Research: Proceedings of the First International Ben Sira Conference 28-31 July 1996 Soesterberg, Netherlands*, edited by Pancratius Benntjes, 189-222. Berlin.

Wyatt, N. 1987. "Sea and Desert: Symbolic Geography in West Semitic Religious Thought." *Ugarit-Forschungen* 19:375-389.

———. 1996. *Myths of Power: A Study of Royal Myth and Ideology in Ugaritic and Biblical Tradition.* Münster.

Yadin, Yigal. 1962. *The Scroll of the War of the Sons of Light against the Sons of Darkness.* Translated by Batya Rabin and Chaim Rabin. Oxford.

Yarbro Collins, Adela. 1996. *Cosmology and Eschatology in Jewish and Christian Apocalypticism.* Leiden.

Yerushalmi, Yosef. 1982. *Zakhor: Jewish History and Jewish Memory.* Seattle.

Yon, Jean-Baptiste. 2015. "De Marisa à Byblos avec le courrier de Séleucos IV: quelques données sur Byblos hellénistique." In *La Phénicie hellénistique. Actes du colloque international de Toulouse (18-20 février 2013)*, edited by Julien Aliquot and Corinne Bonnet, 89-105. Lyon.

Zeitlin, Solomon. 1918. "Megillat Taanit as a Source for Jewish Chronology and History in the Hellenistic and Roman Periods." *Jewish Quarterly Review* 9:71-102.

———. 1974. "An Historical Study of the Canonization of the Hebrew Scriptures." In *The Canon and Masorah of the Hebrew Bible,* edited by Sid Leiman, 164-201. New York.

Zerubavel, Eviatar. 1977. "The French Republican Calendar: A Case Study in the Sociology of Time," *American Sociological Review* 42: 868-877.

Zerubavel, Yael. 1995. *Recovered Roots: Collective Memory and the Making of Israeli National Tradition.* Chicago.

Zgoll, Annette. 2006. "Königslauf und Götterrat. Struktur und Deutung des babylonischen Neujahrfestes." In *Festtraditionen in Israel und im Alten Orient,* edited by Erhard Blum and Rüdiger Lux, 11-80. Munich.

Ziegler, Ruprecht. 1993. "Ären kilikischer Städte und Politik des Pompeius in Südostkleinasien." *Tyche* 8:203-219.

Zimansky, Paul. 1995. "Xenophon and the Urartian Legacy." In *Dans les pas des dix-mille: Peuples et pays du Proche-Orient vus par un Grec,* edited by Pierre Briant, 255-268. Toulouse.

Zimmern, H. 1922. "Zu einigen neueren assyriologischen Fragen." ZA 34:189-205.

Zissu, Boaz, and Omri Abadi. 2014. "Paleo-Hebrew Script in Jerusalem and Judea from the Second Century BCE through the Second Century CE: A Reconsideration." *Journal for Semitics* 23:653-664.

MAPS, ILLUSTRATIONS, AND TABLES

Maps

Illustrations

Tables

INDEX

Page references to figures, tables, and maps are in italics.